ELEMENTARY STRUCTURAL ANALYSIS

ELEMENTARY STRUCTURAL ANALYSIS

Robert G. Boggs, Ph.D.
PROFESSOR OF CIVIL ENGINEERING
UNITED STATES COAST GUARD ACADEMY

HOLT RINEHART AND WINSTON
New York Chicago San Francisco Philadelphia
Montreal Toronto London Sydney Tokyo
Mexico City Rio de Janeiro Madrid

Copyright © 1984 by CBS College Publishing
All rights reserved
Address correspondence to:
383 Madison Avenue, New York, N.Y. 10017

Library of Congress Cataloging in Publication Data

Boggs, Robert G.
    Elementary structural analysis.

    Includes index.
    1. Structures, Theory of. I. Title.
TA645.B625 1984   624.1'7   83-18477
ISBN 0-03-063933-6

Printed in the United States of America

Published simultaneously in Canada
4 5 6 7   038   9 8 7 6 5 4 3 2 1

CBS COLLEGE PUBLISHING
Holt, Rinehart and Winston
The Dryden Press
Saunders College Publishing

Cover photo by Jet Lowe for the Historic American Engineering Record

Contents

Chapter 1 INTRODUCTION

1.0 Structures, Structural Design, and Structural Engineering   2
1.1 Scope of Coverage   2
1.2 Assumptions   3

PART I: BASIC PRINCIPLES AND TECHNIQUES

Chapter 2 BASIC PRINCIPLES

2.0 Introduction   8
2.1 Free-Body Diagrams   8
2.2 Equilibrium Equations   10
2.3 Internal Forces and Moments in Beams   10
2.4 Virtual Work   15

Chapter 3 STABILITY, DETERMINANCY AND CLASSIFICATION OF STRUCTURES

3.0 Introduction   28
3.1 Stability of a Single Rigid Body   28
3.2 Trusses   30
    3.2.1 Truss Stability and Determinacy   30
    3.2.2 Classification of Trusses   33
3.3 Beams   34
    3.3.1 Beam Classification   35
    3.3.2 Beam Stability and Determinacy   35

v

3.4 Rigid Frames   38
   3.4.1 Stability and Determinacy   39

Chapter 4   STRESS AND FORCE ANALYSIS OF STATICALLY DETERMINATE STRUCTURES

4.0 Introduction   48
4.1 Trusses   48
   4.1.1 Method of Joints   48
   4.1.2 Method of Sections   52
   4.1.3 Compound Truss Analysis   54
4.2 Axial Force, Shear Force and Bending Moment in Statically Determinate Rigid Frames   56

Chapter 5   STRUCTURAL LOADS

5.0 Introduction   70
5.1 Load Classification and Determination   70
   5.1.1 Dead Load   71
   5.1.2 Building Use and Occupancy Live Loads   71
   5.1.3 Bridge Live Loads   72
   5.1.4 Snow, Ice and Rain Live Loads   74
   5.1.5 Wind Loads   76
   5.1.6 Earthquake Loads   78
   5.1.7 Earth Pressure Loads   80
   5.1.8 Shrinkage, Settlement, and Misalignment Loads   80
   5.1.9 Load Combinations   80
5.2 Influence Lines   81
   5.2.1 Influence Lines by Virtual Work: The Müller–Breslau Principle   86
   5.2.2 Influence Lines for Girders and Members of Trusses   91
   5.2.3 Use of Influence Lines   98

Chapter 6   ELASTIC DEFLECTIONS OF PLANE STATICALLY DETERMINATE STRUCTURES

6.0 Introduction   106
6.1 Beam Deflection   106
   6.1.1 Conjugate Beams   106
   6.1.2 Moment-Area Methods   114
6.2 The Method of Virtual Work (Unit Load Method)   118
6.3 Castigliano's Theorem   132
6.4 Flexibility Coefficients   137

PART II: STATICALLY INDETERMINATE STRUCTURAL ANALYSIS: FORCE/FLEXIBILITY METHODS

Chapter 7 STATICALLY INDETERMINATE ANALYSIS: METHOD OF CONSISTENT DEFORMATIONS

7.0 Introduction 148
7.1 Beam and Frame Analysis by Consistent Deformations 149
7.2 Selection of Primary Structures 156
7.3 Truss Analysis by Consistent Deformations 160

Chapter 8 THE METHOD OF LEAST WORK

8.0 Introduction 170
8.1 Statically Indeterminate Beams by the Method of Least Work 171
8.2 Statically Indeterminate Rigid Frames by the Method of Least Work 173
8.3 Least Work Analysis of Statically Indeterminate Composite Structures 176
8.4 The Method of Least Work vis-a-vis the Method of Consistent Deformations 178

Chapter 9 STATICALLY INDETERMINATE STRUCTURES: CONJUGATE BEAM ANALYSIS

9.0 Introduction 184
9.1 Statically Indeterminate Beam Analysis by Conjugate Beams 184
9.2 Influence Lines for Statically Indeterminate Structures: The Müller–Breslau Principle 186

PART III: STATICALLY INDETERMINATE STRUCTURAL ANALYSIS: DISPLACEMENT/STIFFNESS METHODS

Chapter 10 STATICALLY INDETERMINATE STRUCTURES: THE SLOPE DEFLECTION METHOD

10.0 Introduction 202
10.1 Statically Indeterminate Beam Analysis Using Slope Deflection 208
10.2 Statically Indeterminate Frames Using Slope Deflection 214
10.3 Symmetry and Antisymmetry 220
10.4 Multiple Degrees of Sidesway 225

Chapter 11 MOMENT DISTRIBUTION: STRUCTURES WITHOUT SIDESWAY

11.0 Introduction 240
11.1 Stiffness, Distribution Factors, and Carry-Overs 240
11.2 Fixed End Moments 244
11.3 Moment Distribution Procedure 244
11.4 Modified Stiffness 251

CHAPTER 12 MOMENT DISTRIBUTION: STRUCTURES WITH JOINT TRANSLATION

12.0 Introduction   262
12.1 Sidesway Induced Fixed End Moments and Their Distribution   262
12.2 Superposition of Sidesway Prevented and Sidesway Induced   266
12.3 Sidesway Induced Fixed End Moments; Members with Far End Pinned   270
12.4 Sidesway Involving Symmetry or Antisymmetry   270
12.5 Multiple Degrees of Sidesway   282

Chapter 13 BEAMS AND RIGID FRAMES WITH NONPRISMATIC MEMBERS

13.0 Introduction   296
13.1 Published Properties for Nonprismatic Members   296
13.2 Integral Formulas for Carry-Over Factors, Rotational Stiffnesses, and Fixed End Moments   296
13.3 Numerical Procedures   301
13.4 Moment Distribution for Structures with Nonprismatic Members   306
13.5 Relationship between Rotational Stiffnesses and Carry-Over Factors   307
13.6 Fixed End Moments from Joint Translation   309
13.7 Modified Stiffnesses   310
13.8 Moment Distribution for Structures with Nonprismatic Members, with Sidesway, and Using Modified Stiffnesses   312
13.9 Slope Deflection for Structures with Nonprismatic Members   313

Chapter 14 MATRIX FORMULATION FOR STRUCTURAL ANALYSIS

14.0 Introduction   318
14.1 Systems of Axial Force Members; Local and Global Stiffnesses   318
14.2 Plane Truss Analysis   323
14.3 Computer Considerations   333
14.4 Space Truss Analysis   335
14.5 Rigid Plane Frame Analysis   343
14.6 Loads between Joints   355
14.7 Support Displacements   358
14.8 Closure   363

SEQUENCED PROBLEMS   369

Appendix A: MATRIX ALGEBRA

A.1 Introduction   396
A.2 Definitions   396
A.3 Algebraic Operations   398
A.4 Matrix Partitioning   399
A.5 Solution of Simultaneous Equations by Gauss Elimination   400

ANSWERS   403

INDEX   417

Preface

The objective of a first course in structural analysis should be to develop in the student an understanding of basic principles so that they can be applied competently. This book is intended to serve such a course and is written for the student. Often the style used is conversational; I hope you will like that. This book parallels my teaching style: relaxed and friendly, with occasional attempts at humor. As I often begin my classes with a joke or an anecdote, so too I begin each chapter with a joke or cartoon.

An important feature of this text is its sequenced problems. The analysis of real structures is often arduous, involving calculations too extensive to permit their use as daily homework exercises. In other textbooks it is common practice to devise homework problems shortened to fit available time. As a result, students work through problems that illustrate various procedures, but never learn how the procedures are connected. Thus students may become proficient at performing each step, yet remain unable to sort out which approach to use or how to carry out several steps involving different procedures in a logical sequence.

The sequenced problems here break real structural problems into homework-sized parts. As the course progresses, the parts are assigned in connection with current topics. Frequently, one part will require the use of the solutions of earlier parts. By the end of the course the student will have solved the entire sequenced problem, and in doing so will have worked completely through a major structural analysis problem. In addition, the student will have seen how the procedural steps fit together. Faculty are urged to assign at east one truss and one rigid-frame sequenced problem with each course offering.

Not all assigned homework need come from the sequenced problems. A number of the more traditional homework problems are provided at the end of each chapter, their solutions aided by the unusually large number of carefully worked out examples that have been used throughout the book.

The book is divided into three parts. Part I provides a transition from the usual prerequisite courses in statics and strength of materials. Some review material is included along with several new methods of deflection analysis. Load analysis is developed with unusual depth and is connected to the use and interpretation of influence lines. Since Part I treats only statically determinate structures, several of its topics are only begun in Part I. They are recalled and extended to statically indeterminate structures in Parts II and III. This early introduction with later recall and extension is valuable in setting the concepts firmly in the mind of the student.

Part II develops methods for analyzing statically indeterminate structures using several of the classical force/flexibility methods: consistent deformations, least work, and conjugate beams.

Part III also treats statically indeterminate structures, but uses displacement/stiffness methods. It begins with the classical slope deflection and moment distribution methods and follows with an extensive development of the displacement/stiffness methods in matrix form.

The deferral of matrix methods to the end of Part III is intentional. Matrix methods are inconvenient for hand calculations. I believe their earlier introduction would only add complexity and could, therefore, undermine achieving our goal of developing a sound and usable knowledge of the basic principles. But matrix methods cannot be dismissed. They are very important since they are used almost exclusively in computer-based structural analysis. Today, with user-friendly software for structural analysis becoming increasingly available, the trend is more and more toward computer-based analysis of structures. With this software so available there seems little need for analysts to develop their own programs, leaving one to wonder why matrix methods are studied at all. Indeed, one might even ask why structural analysis itself is studied if it can all be done on a computer.

I believe both classical methods and matrix methods must be studied since they, together, provide the base for careful, correct use of the software. A working knowledge, an intuition, perhaps, is needed to describe the structure properly and to interpret and use the results properly when a computer analysis is done.

A number of persons have helped me with this book, some directly, others indirectly. Many of my students have used this book or portions of it in manuscript form during various developmental stages. The students always seemed pleased to let me know when they had found an error or when they felt that some section made little sense. I know their real purpose was to help me, and that help was sincerely received and appreciated.

Louise Pittaway typed the manuscript and made editorial suggestions while suffering through subscripts, exponents, and Greek letters, and sharing a multitude of rewrites with me. Her good spirits and professional skills were invaluable.

The indirect contributors are a few of my own teachers, who helped me set the highest of standards for myself by following their examples. I will mention Miss Effie May Olsen, my junior high school geometry teacher; Professors Domina Eberle Spencer, Ronald S. Brand, and E. Russell Johnstone, Jr., all of the University of Connecticut; and Professor J. L. Meriam of the University of California at Berkeley. All are teacher's teachers; all profoundly influenced my professional life. This book is dedicated to them.

Robert G. Boggs

Introduction

A painfully shy young man who wanted to date women of his own age but neither knew how to ask them nor had the courage to do so, decided to buy a book on the subject for guidance. Because he was too shy to describe to the store clerk what it was he wanted, he made his own selection: a large, beautifully bound volume titled How To Hug. It was, he felt certain, just what he needed. He bought it and took it home. But when he began to read it, he discovered that he had purchased volume nine of an encyclopedia.

1.0 STRUCTURES, STRUCTURAL DESIGN, AND STRUCTURAL ENGINEERING

This is a book on structural analysis. There should be no confusion from its title. Occasionally there is confusion between the terms structural design, structural analysis, structural engineering and structures. Let us try to eliminate that confusion from the start.

There are several definitions of the word structure. In this book we will limit ourselves to one: *a structure is something constructed of one or more parts to carry loads.* That definitions is valid in all engineering fields, although the kinds of structures vary from field to field. They go from bridges or buildings in civil engineering to gyroscope frames or penstock supports in mechanical engineering.

Structures may be as large as the Sears Tower in Chicago, or as small as a lens support and alignment frame in a portable laser device. They may be fixed in position, as in an electrical transmission line tower, or movable, as in a radar antenna.

Most of the structures used for problems and illustrations in this book have been taken from civil engineering since this is the curriculum for which the book is primarily intended. This should not imply, however, that the principles and methods presented are exclusive to civil engineering. Indeed, they are used regularly in all engineering fields.

Often structural engineering is but one of several engineering efforts performed in an engineering project. Almost always it is a significant effort. In building construction, for example, the structural engineer is one of the principal collaborators with the architect. He or she selects the structural system, determines suitable proportioning, and plans and oversees construction. The structural engineer provides information to engineers involved in other aspects of the project and responds, in turn, to information received from them by refining the structural requirements and design.

Structural design is the detailed design of the structure itself and its components. It includes material selection and member proportioning. Individual members must be selected or designed to carry the loads applied to them. Often individual member loading depends on properties of the members and their physical arrangement. Then various methods of structural analysis are used to determine the individual member loading.

Thus structural analysis consists of a variety of mathematical procedures for determining such quantities as the member forces and various structural displacements as a structure responds to its loads. Estimating realistic loads for the structure considering its use and location is often a part of structural analysis.

1.1 SCOPE OF COVERAGE

Only three general categories of structures will be considered: beams, trusses, and plane rigid frames, with statically determinate and indeterminate systems treated in each category. This coverage is typical of most first courses in structural analysis. It is a beginning only; such important topics as shell structures, space frames, cable structures, arches, and membrane structures have not been included. A secondary goal of this book is to excite the reader about structural analysis so that he or she will pursue these subjects in later studies.

No attempt has been made to address member selection, for that process is extremely dependent on the peculiarities of the material to be used. Member selection should be studied in courses related to design with a specific material. It is considered essential that studies be undertaken in steel design, reinforced concrete design, or both to complement the material of this book.

1.2 ASSUMPTIONS

Only two assumptions are made regarding the materials used in the structures of this book. First, the material has a linear stress–strain relationship. Second, there is no difference in the material behavior when stressed in tension vis-à-vis compression.

The frames and trusses studied are plane structural systems. It will be assumed that there is adequate bracing perpendicular to the plane so that no member will fail due to an elastic instability. The very important consideration regarding such instabilities will be left for the specific design courses.

All structures are assumed to undergo only small deformations as they are loaded. As a consequence we assume no change in the position or direction of a force as a result of structural deflections.

Finally, since linear elastic materials and small displacements are assumed, the principle of superposition will apply in all cases. Thus the displacements or internal forces that arise from two different force systems applied one at a time may be added algebraically to determine the structure's response when both systems are applied simultaneously.

Bon voyage.

Part I
BASIC PRINCIPLES AND TECHNIQUES

In Chapters 2 through 6 a variety of principles and techniques will be examined. Some will be familiar to you, others will start as familiar topics but will be expanded into new ones, while still others will be entirely new. Most could be considered under the title "statically determinate" structures. What is important about these first chapters is that they will form the foundation for the techniques that follow. A thorough understanding of them will be essential before moving into Part II.

Basic Principles

*Why does a chicken cross the road?
Why do firemen wear red suspenders?*

2.0 INTRODUCTION

Structural analysis is built on the basic principles that are the "old saws" you studied in courses in statics and strength of materials. They should be as familiar to you as those two ancient riddles. Some of these principles are presented in this chapter. Most of you will recognize them as old friends. They are presented to refresh your memory, to establish a few of the sign conventions that will be used in this book, and to reveal any of those topics that may be new to you. This chapter is intended as a review; so it is brief. If you find a topic that seems to be new territory, you are urged to study it in more detail in statics or strength of materials textbooks.

2.1 FREE-BODY DIAGRAMS

Solutions of all problems involving forces and moments applied to bodies should start with free-body diagrams. In statics, correctly drawn and carefully used free-body diagrams lead to correct equilibrium equations. Those diagrams ensure the inclusion of all forces and aid in establishing useful and correct moment equations by suggesting logical moment centers and correct moment arms.

Free-body diagrams are easily drawn by using these steps:

(1) Select the body or the assembly of bodies to be isolated and sketch its outline.
(2) Show on this sketch all the forces that act on the body, and all the forces and couples caused by objects that are in contact with the body but not part of it.
(3) Label all the forces and couples known in magnitude and direction by their actual values. Label all unknown forces and couples with letters.

When isolating bodies, the supports and the connections with other bodies must be replaced with all of the forces and couples that they could provide and none that they could not provide. The replacement forces for typical structural support devices are shown and described here.

Pin Connections and Knife Edges. These supports can apply a single force in an arbitrary direction or, in the more usual view, two mutually perpendicular and independent forces. Thus, both the pin

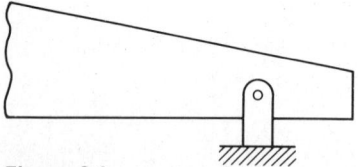

Figure 2.1

and the knife edge

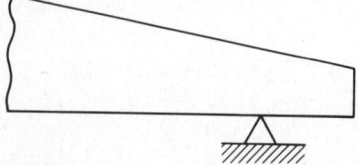

Figure 2.2

can be drawn as a single force F in the direction θ

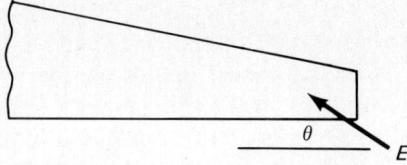

Figure 2.3

or as two independent force components F_x and F_y

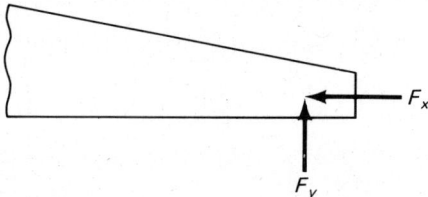

Figure 2.4

Roller Supports. Roller supports provide a force only in the direction normal to the surface in contact with the roller. Thus, the roller

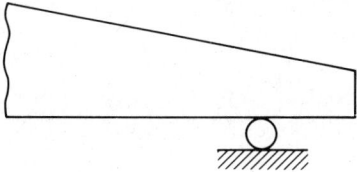

Figure 2.5

is replaced on the free-body diagram by the single force F

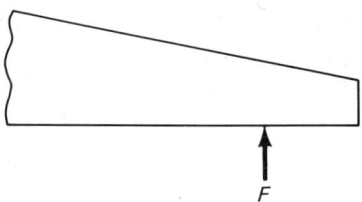

Figure 2.6

Clamped Supports. Clamped supports provide forces in two directions, as does the pin. But, in addition, they provide a couple. The clamped support

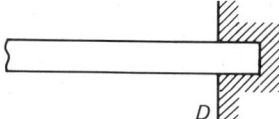

Figure 2.7

can be replaced with

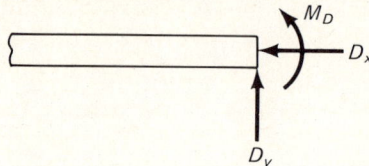

Figure 2.8

2.2 EQUILIBRIUM EQUATIONS

A structure is in equilibrium if its state of velocity is not changing. In basic structural situations this unchanging velocity is usually zero, that is, not moving.

In plane problems, free-body diagrams are used to write three simultaneous equilibrium equations. One set of these is

$$\sum F_x = 0 \quad \sum F_y = 0 \quad \sum M_A = 0 \qquad (2.1a)$$

where $\sum F_x$ and $\sum F_y$ are the summations of all the x components and y components of force, respectively, as found on the free-body diagram. $\sum M_A$ is the summation of the moments of all forces on the free-body diagram, using an arbitrary point A as a moment center.

While it will not be proven here, an equivalent set of equations is

$$\sum F_x = 0 \quad \sum M_A = 0 \quad \sum M_B = 0 \qquad (2.1b)$$

where A and B are two arbitrary points that do not lie on a line perpendicular to the x direction. The x direction can be taken as any convenient direction.

A third set of equilibrium equations is

$$\sum M_A = 0 \quad \sum M_B = 0 \quad \sum M_C = 0 \qquad (2.1c)$$

where A, B, and C are three arbitrary points that do not all lie on a single line.

Often there are occasions where more than three unknown forces and couples are found on a single free-body diagram. In these it is tempting to use both Equations (2.1a) and (2.1c) to get enough equations to solve for all of the unknowns. But when this is tried, it fails, usually giving

$$0 = 0$$

or its equivalent. This always occurs because Equations (2.1a) and (2.1c) are not five independent equations. Rather, three of them are independent, while the other two are dependent, that is, the two can be written by starting with the three independent equations and rearranging them algebraically.

2.3 INTERNAL FORCES AND MOMENTS IN BEAMS

Many structural engineers believe that beams are the most important of all the structural elements. Key items in their analysis are internal shear force and bending moment, which usually vary as functions of position along the beam. Their values may be expressed as sets of algebraic expressions or displayed as curves, called shear and moment diagrams, which plot the values of shear and moment versus position along the beam.

For portions of this book the sign conventions for transverse load, shear force, and bending moment are shown in the diagram below.[1]

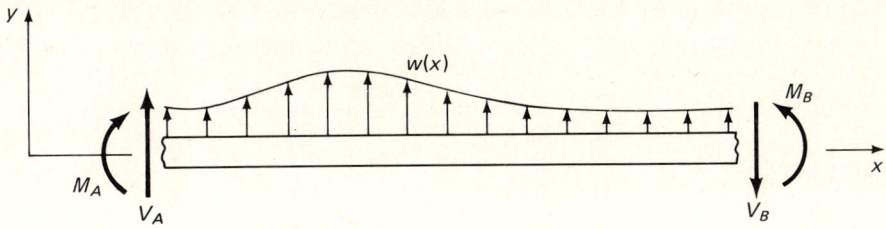

Figure 2.9

Transverse loading w is considered positive if it is directed upward; shear forces, such as V_A and V_B, are positive in the directions and locations shown, that is, upward on the left end, downward on the right end; and bending moments, such as M_A and M_B, are positive as shown, that is, clockwise on the left end and counterclockwise on the right.

Transverse loading, shear force, and bending moment are related. For the short element of the beam

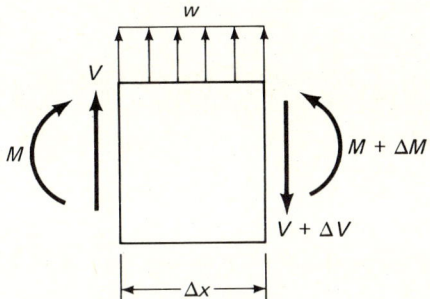

Figure 2.10

we have

$$+\uparrow \sum F_y = w\,\Delta x + V - (V + \Delta V) = 0$$

so that

$$\frac{\Delta V}{\Delta x} = w$$

which, in the limit, becomes

$$\frac{dV}{dx} = \lim_{\Delta x \to 0} \frac{\Delta V}{\Delta x} = w$$

---

[1] Different sign conventions will be used for the slope-deflection and moment-distribution methods.

Hence,

$$\frac{dV}{dx} = w \qquad (2.2)$$

which shows that the rate of change of shear with position (that is, the slope of the shear diagram) is equal to the distributed transverse force w.

Next, by taking moments about the right end of the element,

$$+\!\sum M_{\text{Rend}} = -M + (M - \Delta M) - w\,\Delta x\left(\frac{\Delta x}{2}\right) - V(\Delta x) = 0$$

which simplifies to

$$\frac{\Delta M}{\Delta x} = w\frac{\Delta x}{2} + V$$

and again taking limits,

$$\frac{dM}{dx} = \lim_{\Delta x \to 0} \frac{\Delta M}{\Delta x} = \lim_{\Delta x \to 0}\left(w\frac{\Delta x}{2} + V\right)$$

Hence,

$$\frac{dM}{dx} = V \qquad (2.3)$$

showing that the rate of change of moment with respect to position (the slope of the moment diagram) is equal to the shear force.

Equation (2.2) rearranged and integrated gives

$$\int_{V_1}^{V_2} dV = \int_{x_1}^{x_2} w\,dx$$

or

$$V_2 - V_1 = \int_{x_1}^{x_2} w\,dx \qquad (2.4)$$

This shows that the change in shear between x_1 and x_2 equals the force added externally between x_1 and x_2.

Integration of equation (2.3) gives

$$\int_{M_1}^{M_2} dM = \int_{x_1}^{x_2} V\,dx$$

or

$$M_2 - M_1 = \int_{x_1}^{x_2} V\,dx \qquad (2.5)$$

which shows that the change in moment between x_1 and x_2 equals the area under the shear diagram between x_1 and x_2.

Concentrated forces and concentrated couples cause abrupt changes in the shear and moment diagrams, respectively. Consider a short element of beam on which the concentrated force P_1 and the concentrated couple C_1 are applied:

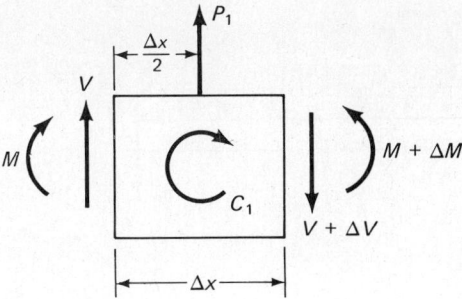

Figure 2.11

We satisfy the equilibrium equations first by summing forces vertically:
$$+\uparrow \sum F_y = V - (V + \Delta V) + P_1 = 0$$
$$\Delta V = P_1$$

This shows that ΔV is not affected by the length of the element, nor is it affected by C_1. It is exactly equal to P_1 as an abrupt change in shear.

Now satisfying moment equilibrium,
$$+\circlearrowleft \sum M_{\text{Rend}} = -M + M + \Delta M - P_1 \frac{\Delta x}{2} - C_1 = 0$$
$$\Delta M = P_1 \frac{\Delta x}{2} + C_1$$

If $\Delta x \to 0$, then $\Delta M = C_1$, which shows that the moment diagram shifts abruptly by the magnitude of a concentrated couple applied to the beam. It also shows that there is no abrupt shifting of the moment diagram locally brought on by the concentrated force.

EXAMPLE 2.3.1

This example illustrates all of the features discussed so far.

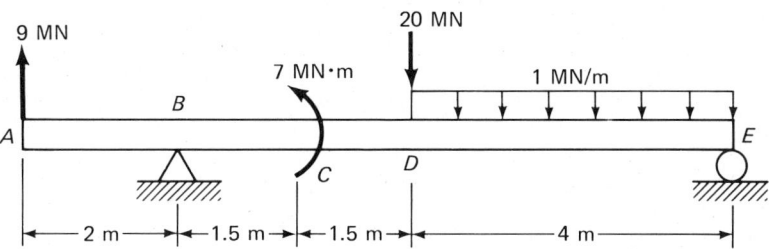

Figure 2.12

First the free-body diagram is drawn:

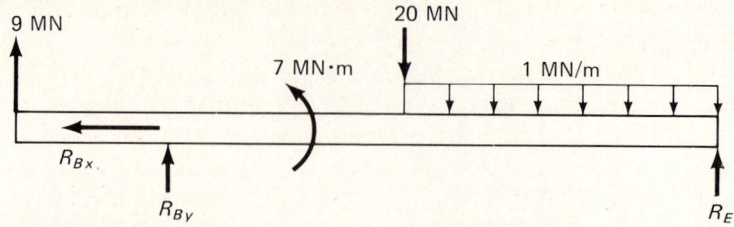

Figure 2.13

and then moments are taken about point E:

$$+\circlearrowleft \sum M_E = -9(9) - 7R_{By} + 7 + 20(4) + 4(2) = 0$$

The value 4(2) is the moment of the 1-MN/m distributed force, totaling 4 MN, considered to be acting at its centroid. This equation is solved, giving

$$R_{By} = 2 \text{ MN}$$

and then taking vertical forces,

$$+\uparrow \sum F_Y = 9 + 2 - 20 - 4 + R_E = 0$$

$$R_E = 13 \text{ MN}$$

The shear diagram and moment diagram are sketched next:

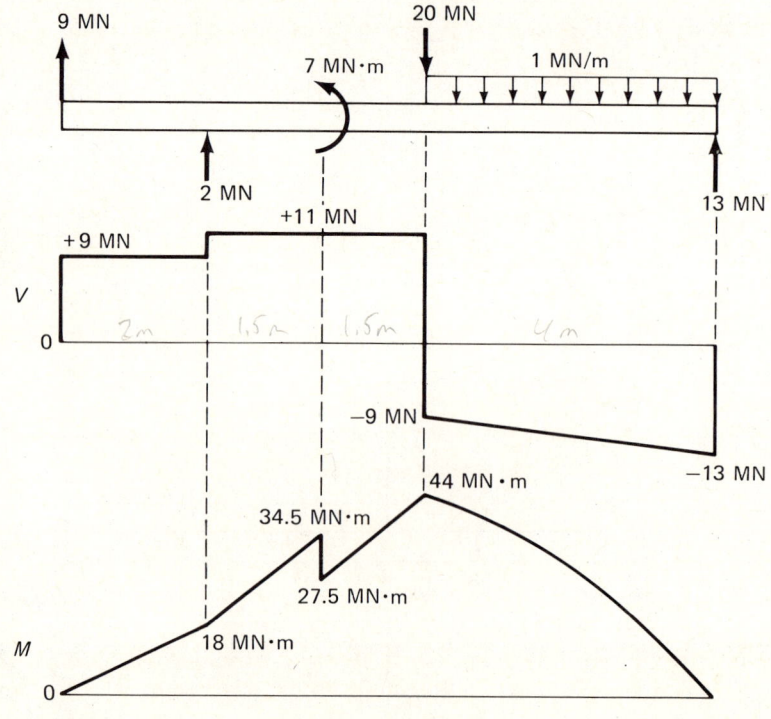

Figure 2.14

Construction of the shear diagram uses the following steps:

(1) At A the shear shifts upward 9 MN, in response to the 9-MN upward concentrated force there.
(2) Between A and B the shear remains constant because no distributed forces acts in the diagram, that is, $w=0=dV/dx$, so the slope is zero.
(3) At B an upward step of 2 MN comes because of the 2-MN upward support reaction there.
(4) The shear continues constant from B to D because no distributed force is present. Notice that it is undisturbed by the 7-MN·m couple acting at C.
(5) At D the shear shifts downward 20 MN in response to the 20-MN downward force there.
(6) From D to E the shear diagram has a downward slope of 1 MN/m, matching the downward distributed force in that region.
(7) It shifts upward 13 MN by the upward 13-MN at E.

Construction of the moment diagram was done using the following steps:

(1) No external couple or support moment is applied at A, so the moment there is zero and the diagram must begin at zero at A.
(2) From A to B it has a constant upward slope of 9 MN·m/m, equaling the value of the shear in that region ($dM/dx=V$).
(3) The area under the shear diagram from A to B is 18 MN·m, which is the change from the value $M=0$ at A to $M=18$ MN·m at B.
(4) From B to C it has a slope of 11 MN·m/m (equal to the shear), changing by $11 \times 1.5 = 16.5$ MN·m from B to a value of 34.5 MN·m.
(5) At C the counterclockwise couple of 7 MN·m causes an abrupt shift downward in the moment from 34.5 to 27.5 MN·m.
(6) Between C and D the 11-MN·m/m (value of V) upward slope continues, causing an additional $11 \times 1.5 = 16.5$ MN·m change, reaching $27.5 + 16.5 = 44$ MN·m at D.
(7) Between D and E the moment diagram slopes downward, starting with a slope of 9 MN·m/m near D and reaching a slope of 13 MN·m/m near E. The area under the shear in that region is -44 MN·m, which is found by the area of a -9 MN $\times$ 4 m rectangle and a -4 MN $\times$ 4 m triangle. Having started at $+44$ MN·m at D, the -44-MN·m change returned it to 0 at E.
(8) There is no source of moment at end E, so the zero moment found in step (7) agrees with what it must be, and a check is established.

2.4 VIRTUAL WORK

Virtual work is very important to structural analysis, but you may have overlooked it because its value first becomes evident in its simplifying of more complicated problems, ones you may not have had to face yet.

16 BASIC PRINCIPLES

The principle of virtual work is stated:

> If a system of rigid bodies in equilibrium, ideally supported and connected, is given a small virtual displacement consistent with its constraints, the total work done by the active forces and moments acting on it is zero.

imaginary ↗ (annotation pointing to "virtual")

Ideally supported and connected means that all the supports between the rigid bodies are smooth. If these conditions are not satisfied, energy that would not be included in the conservation of energy relation applied here could be stored or dissipated during the virtual displacement.

The virtual displacements imposed must be infinitesimally small so that the lines of action of the forces do not change.

Forces and moments can be considered in two classes: active and inactive. Active forces and moments do work during the virtual displacement, while inactive ones do not. Inactive forces and moments are those that (1) do not move, or (2) have no component of motion in the direction of the force, or (3) are internal to the system. In case (3) internal forces and moments are inactive because they always occur in equal and opposite pairs that must go through identical virtual displacements. Thus the positive work arising from one of the pairs is canceled by negative work from the other. Active forces are defined by inference from the preceding definition of inactive forces, that is, active forces can move and have components in the direction of that motion and are external forces for the system.

The validity of the statement of the principle of virtual work may be shown by examining a particle acted on by a set of n forces in equilibrium:

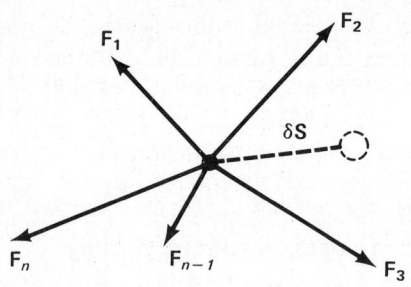

Figure 2.15

If the small virtual displacements $\delta \mathbf{S}$ is given, the virtual work done is

$$\delta W = \mathbf{F}_1 \cdot \delta \mathbf{S} + \mathbf{F}_2 \cdot \delta \mathbf{S} + \cdots + \mathbf{F}_n \cdot \delta \mathbf{S}$$
$$= (\mathbf{F}_1 + \mathbf{F}_2 + \cdots + \mathbf{F}_n) \cdot \delta \mathbf{S}$$
$$= (\sum \mathbf{F}) \cdot \delta \mathbf{S}$$
$$= 0$$

The substitution $\sum \mathbf{F} = 0$ follows the requirement that the particle be in equilibrium.

Next the idea is expanded to a rigid body. If a rigid body is in equilibrium, each particle of the many that make up the body must also be in equilibrium. If the body is given a small virtual displacement, consistent with its constraints, each particle undergoes a small virtual displacement consistent with that of the body. By adding for each particle the work done in this displacement, the total work is zero. It is important to note that most of the particles are subjected only to forces internal to the rigid body which, as noted before, occur in pairs so that the net work of each pair is zero. Thus it is necessary to consider only the active forces applied externally to the body; so their net work in the virtual displacement must be zero.

Finally, the concept is extended to a system of rigid bodies, which is done by analogy with the reasoning already stated for a single rigid body. If the net work for each rigid body is zero when displaced as part of the system's virtual displacement, then the system's work is also zero. Internal forces and moments at connections occur only as pairs of equal opposite forces or moments undergoing identical displacements whose net work cancels. Potential matching forces and moments at connections that could go through different displacements (such as in friction-induced couples in a hinge) would violate these assumptions. But the connections are assumed to be frictionless, eliminating these as existing forces or couples. Thus for an ideal system of rigid bodies only the active forces applied externally to the system need be considered.

EXAMPLE 2.4.1

The horizontal rod AB is supported in equilibrium at A by a smooth pin and at B by a force B_y. There is a 40-lb force applied downward at C.

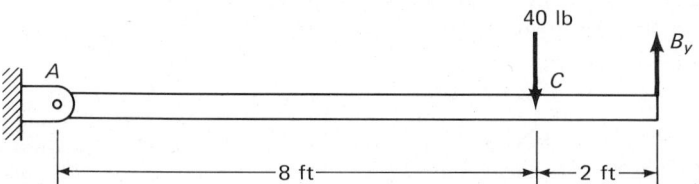

Figure 2.16

The figure shown for the problem is an *active force diagram*, that is, it shows the member and all forces that would do work if AB were given a small displacement consistent with its constraints

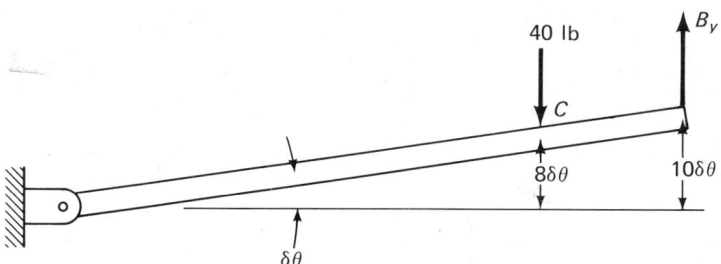

Figure 2.17

The only motion possible is a rigid body rotation about A. A small virtual rotation $\delta\theta$ is applied as shown, so that the 40-lb force moves upward a distance $8\delta\theta$, and B_y moves upward $10\delta\theta$. The virtual work is written and equated to zero:

$$\delta Wk = -40(8\delta\theta) + B_y(10\delta\theta) = 0$$

The minus sign in the first term is used because the 40-lb force and the $8\delta\theta$ displacement are in opposite directions. The other term has a positive sign since the force and the displacement are in the same direction. The expression is solved algebraically, giving $B_y = +32$ lb. The $+$ sign is interpreted to mean that B_y is upward—not because $+$ is defined as upward, but rather because B_y was assumed upward originally and the $+$ confirms that this assumption was correct.

EXAMPLE 2.4.2

This example should display some of the power of the method of virtual work.

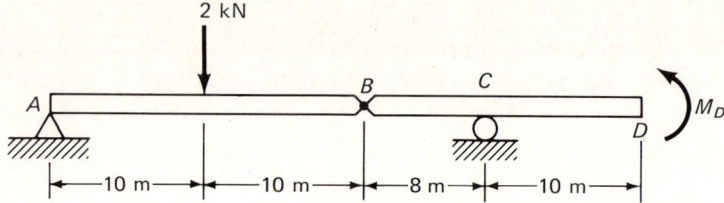

Figure 2.18

Beam $ABCD$ has a 2-kN force and an unknown couple M_D that maintains system equilibrium and is applied as shown. M_D is to be determined by the method of virtual work. The given figure is the active force diagram since no motion is possible for any of the support reaction forces at A or C. An upward virtual displacement δS is applied at B, and $ABCD$ deforms to the shape shown, accommodating δS at B, yet not displacing at A and C.

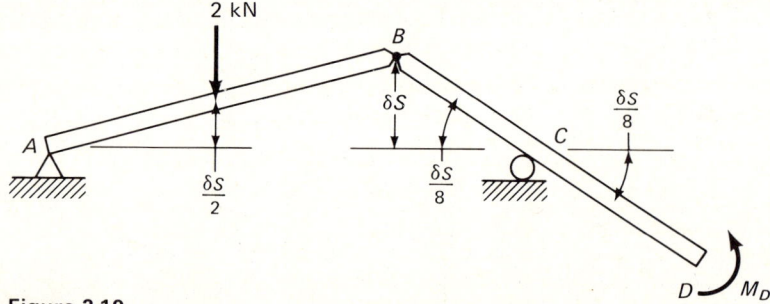

Figure 2.19

The virtual work expression is

$$\delta Wk = -2\left(\frac{\delta S}{2}\right) - M_D\left(\frac{\delta S}{8}\right) = 0$$

Both terms are negative since both virtual displacements are opposite in direction to those of the force and couple. The equation is solved algebraically for the unknown couple, giving

$M_D = -8$ kN·m

The negative sign shows that the actual direction of the couple M_D is opposite to the original direction assumed for M_D. Thus M_D really is 8 kN·m, clockwise.

You should have noticed that in the expression for virtual work, the parameters for the virtual displacement ($\delta\theta$ in Example 2.4.1 and δS in Example 2.4.2) appeared as factors in every term of the virtual work expressions. This is essential, for these parameters must cancel out of the virtual work expressions. If the system has more than one rigid body, the virtual displacements must be written in terms of a single parameter, and the displacements of each body must be prescribed so that they are consistent with the constraints.

If the unknown force is a reaction provided by a rigid support in a stable statically determinate structure, it is necessary to remove that support and replace it with the unknown force to be found. Once this has been done, the support force can be given a virtual displacement and its work included in the virtual work expression. Example 2.4.1 might have been posed originally as: find the reaction provided by support B for the beam

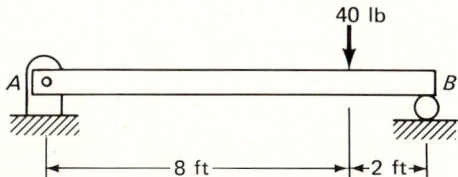

Figure 2.20

If support B is removed, permitting motion of end B, and the force B_y is used to replace the support, the problem becomes identical to Example 2.4.1.

Beam shear and moment can be evaluated conveniently by virtual work. Moments are found by adding a hinge where the moment is to be determined and, simultaneously, adding a pair of equal, opposite couples to replace the moment lost when adding the hinge. A virtual displacement consistent with the remaining supports is imposed, and the virtual work expression is written and solved for the moment.

EXAMPLE 2.4.3

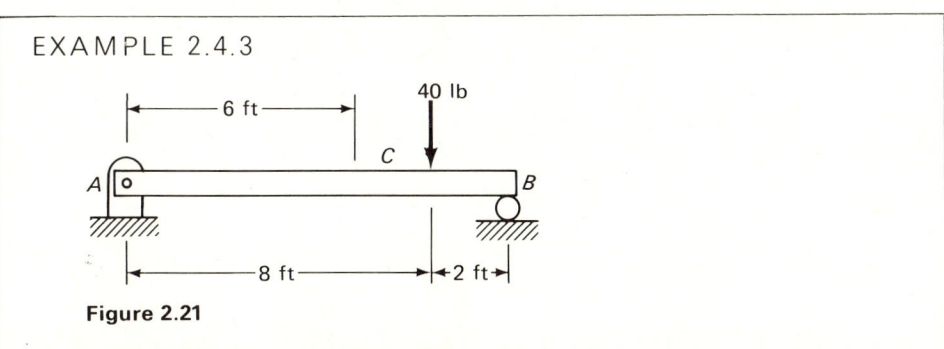

Figure 2.21

Use virtual work to calculate the bending moment at C.

20 BASIC PRINCIPLES

A hinge and equal, opposite moments M_C are added at C. Thus

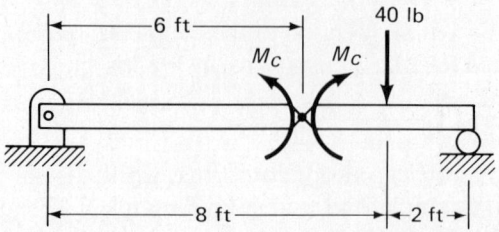

Figure 2.22

The structure is given a virtual displacement

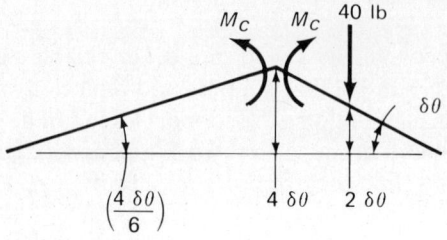

Figure 2.23

from which

$$\delta W k = M_C \left(\frac{4\delta\theta}{6} \right) + M_C \delta\theta - 40(2\delta\theta) = 0$$

$$M_C = 48 \text{ kip·ft}$$

The evaluation of internal beam shears has an added complication. To permit a virtual displacement of the shear forces, the beam must be severed at the point in question. In doing this, not only must equal, opposite shear forces be added, but equal, opposite bending moments must be added as well. Now when a virtual displacement is imposed, it must be such that a net virtual work is done only by the shear forces, that is, it must be such that the bending moments do no net work. This is satisfied if the severed ends are given a shear displacement and simultaneously maintained parallel. Then each of the equal, opposite moments will do equal virtual work, but of opposite sign, thereby canceling.

EXAMPLE 2.4.4

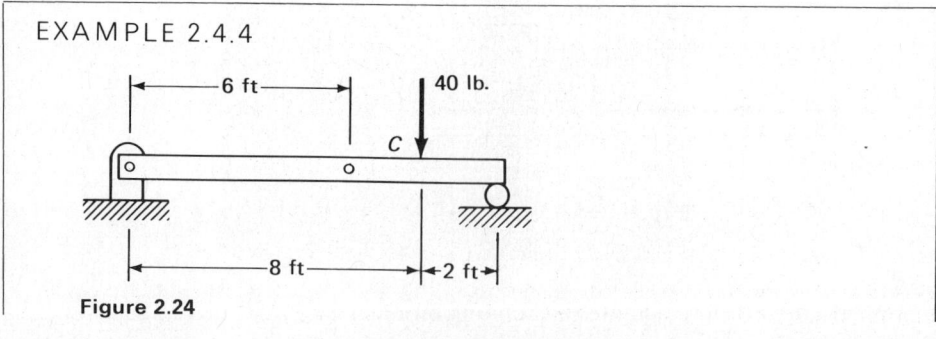

Figure 2.24

Use virtual work to determine the beam shear at C. The beam is severed at C; the shear force and the bending moments are added,

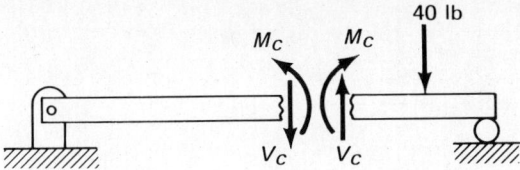

Figure 2.25

and a virtual displacement is imposed in such a way that AC and BC remain parallel:

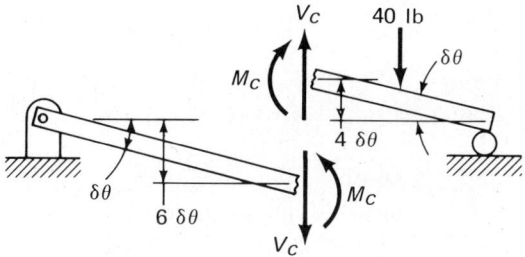

Figure 2.26

$$\delta Wk = V_C(6\delta\theta) + V_C(4\delta\theta) - 40(2\delta\theta) = 0$$
$$V_C = 8 \text{ lb}$$

Construction of the Dover-Newington Bridge near Portsmouth, NH. Note the haunched beams and the pin supports on top of the concrete pier. (Courtesy of Bethlehem Steel Corporation).

PROBLEMS

2.1

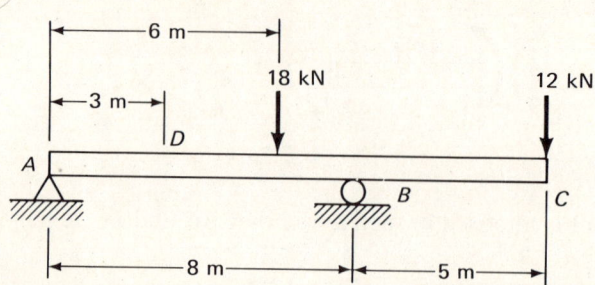

Figure P2.1

(a) Determine the reactions at A and B using only the method of virtual work.
(b) Determine the reactions at A and B using free-body diagrams and equilibrium equations.
(c) Compare results of parts (a) and (b). Resolve any differences.
2.2 Evaluate the following for the beam and loading of Problem 2.1.
(a) Shear force and bending moment at D using only the method of virtual work.
(b) Shear force and bending moment at D using an appropriate section, free-body diagram, and equilibrium equations.
(c) Compare the results of parts (a) and (b). Resolve any differences.
2.3

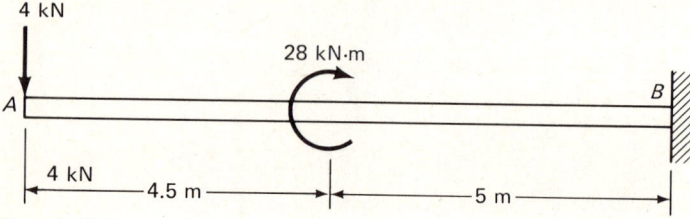

Figure P2.3

(a) Determine all of the reactions at B using the method of virtual work exclusively.
(b) Determine all of the reactions at B using free-body diagrams and equilibrium equations.
(c) Compare the results of parts (a) and (b). Resolve any differences.
2.4

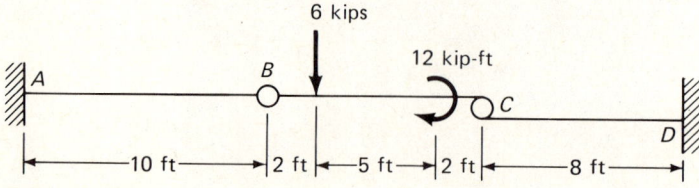

Figure P2.4

(a) Calculate all of the support reactions at A and D using the method of virtual work exclusively.
(b) Calculate all of the support reactions at A and D using free-body diagrams and equilibrium equations.
(c) Compare the results of parts (a) and (b). Resolve any differences.
2.5

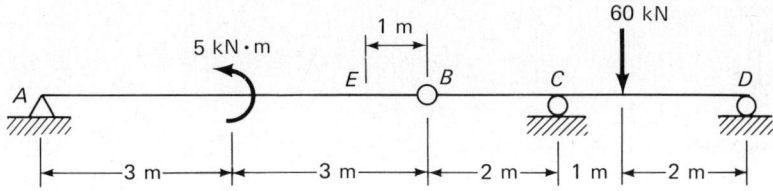

Figure P2.5

(a) Calculate the support reactions at A, C, and D using only the method of virtual work.
(b) Calculate the support reactions at A, C, and D using free-body diagrams and equilibrium equations.
(c) Compare the results of parts (a) and (b). Resolve any differences.
(d) Calculate the bending moment at C and the shear at E using only the method of virtual work.
(e) Calculate the bending moment at C and the shear at E by taking appropriate sections, free-body diagrams, and equilibrium equations.
2.6

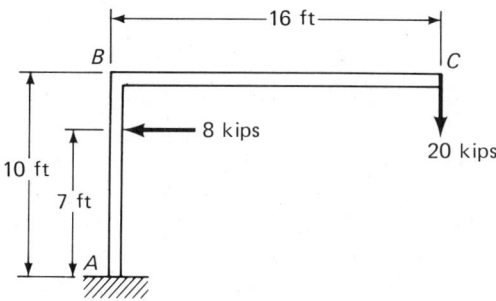

Figure P2.6

(a) Determine the support reactions at A using the method of virtual work exclusively.
(b) Determine the support reactions at A using free-body diagrams and equilibrium equations.
(c) Compare the results of parts (a) and (b). Resolve any differences.
2.7
Determine the support reactions.

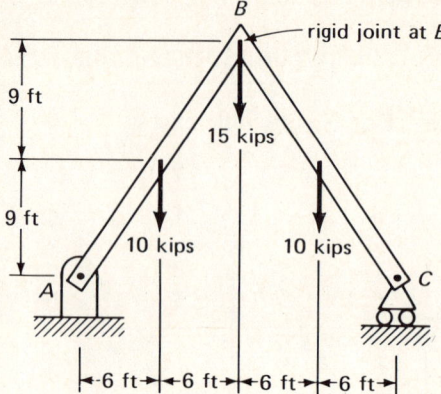

Figure P2.7

2.8

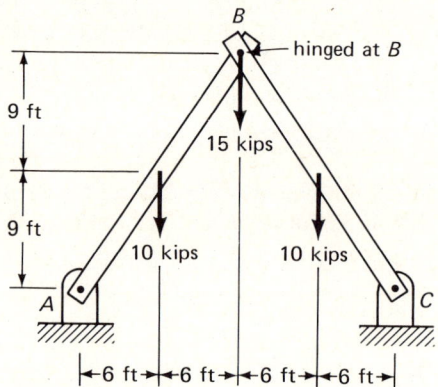

Figure P2.8

Determine the support reactions.

2.9 For the beams illustrated in Figures A–G
(1) Construct the shear and moment diagrams.
(2) Write algebraic expressions for the shear and moment as functions of position x along the beam. Where several different expressions are required, each for a different region of the beam, specify the applicable ranges of x.

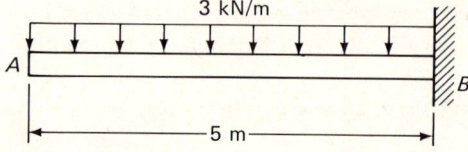

Figure 2.9A

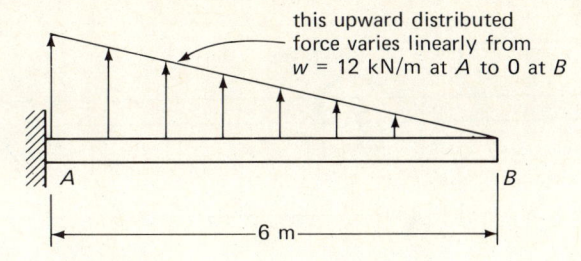

Figure P2.9B

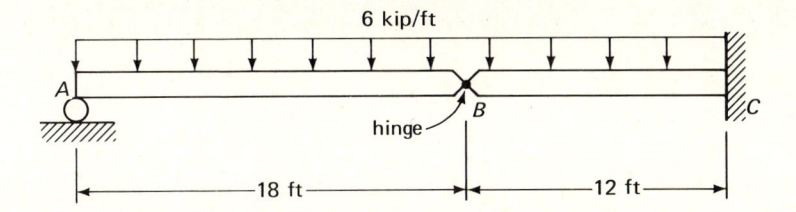

this upward distributed force varies linearly from $w = 12$ kN/m at A to 0 at B

6 m

Figure P2.9C

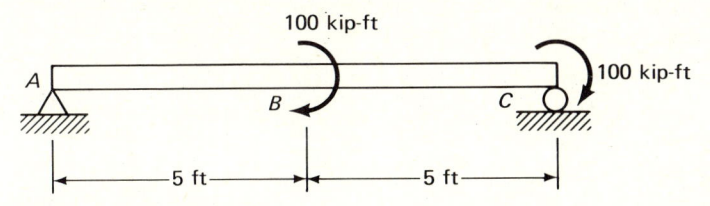

6 kip/ft

hinge

18 ft — 12 ft

Figure P2-9D

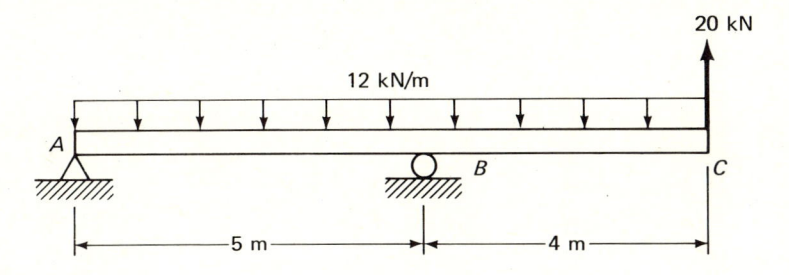

100 kip-ft

100 kip-ft

5 ft — 5 ft

Figure P2-9E

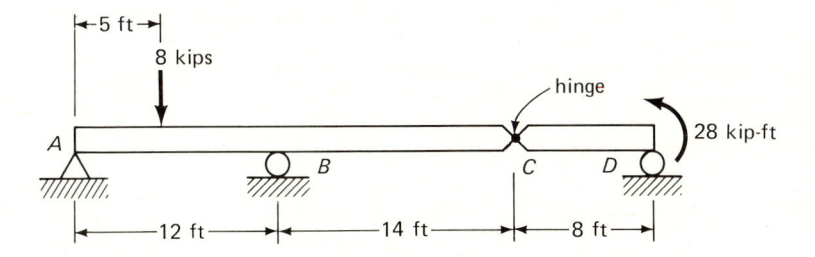

20 kN

12 kN/m

5 m — 4 m

Figure P2.9F

5 ft

8 kips

hinge

28 kip-ft

12 ft — 14 ft — 8 ft

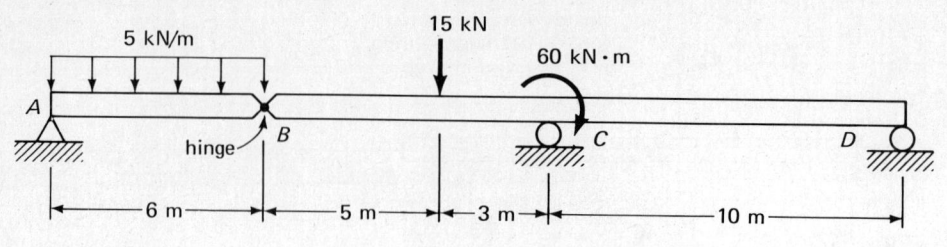

Figure P2.9G

Stability, Determinacy and Classification of Structures

... so far I had followed every one of his instructions, so that I had my left arm straight up, I was bent forward at the waist with my right arm between my legs and my right hand grasping the back of my belt. My left leg was raised backwards so that I was balanced precariously on my right foot. 'All right,' he said, 'lift your right foot so that ...'

28 STABILITY, DETERMINACY AND CLASSIFICATION OF STRUCTURES

3.0 INTRODUCTION

A first step in any structural analysis problem is to check the structure's stability and determinacy. Sometimes the structure's stability (or lack of it) is as obvious as that of the contortionist. At other times a very careful analysis is necessary. In every case it must be considered. To neglect it could be as disastrous as it would be for our tenuously balanced friend to lift his foot without thinking. To begin the structural analysis of an *unstable* structure is unproductive and pointless.

Static determinacy versus indeterminacy must also be evaluated. Your knowledge of this, including the degree of static indeterminacy, is very important in helping you decide which is the best method to be used for your analysis.

3.1 STABILITY OF A SINGLE RIGID BODY

A rigid body is stable when it is supported so that the reactions necessary to satisfy the equations of equilibrium for any arbitrary loading can be provided by its supports without changes in the geometry or orientation of the structure.

An instability exists if it can be shown that it is impossible to satisfy one or more of the equilibrium equations for any arbitrary loading of the body. Normally, analysis of stability takes the form of testing to see whether an equilibrium equation can be found that cannot be satisfied for the arbitrary loading. If one cannot be found, the body is presumed stable.

EXAMPLE 3.1.1

The circular cylinder is supported in a circular support on three rollers.

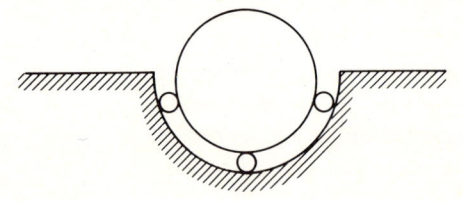

Figure 3.1

The free-body diagram shows that the three reaction forces are concurrent at 0. Any force or couple applied to cause a moment about 0 would make $\sum M_0 = 0$ impossible to satisfy, since none of the reaction forces can provide a moment about 0. Hence the body is unstable.

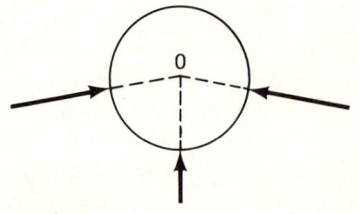

Figure 3.2

EXAMPLE 3.1.2

The beam rests on three rollers.

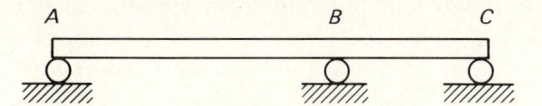

Figure 3.3

The free-body diagram shows that the three support reactions are parallel, so that an externally applied force having a component in the x direction would make $\sum F_x = 0$ impossible to satisfy. Hence the body is unstable.

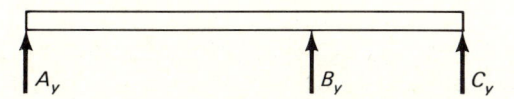

Figure 3.4

EXAMPLE 3.1.3

Beam BC is supported by a pin and a cable.

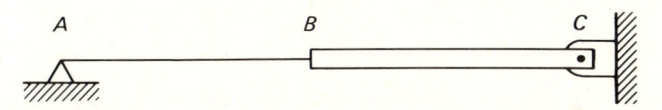

Figure 3.5

The free-body diagram of the beam shows that the reaction forces are all concurrent at C. A force causing a moment about C makes $\sum M_C = 0$ impossible to satisfy, showing it to be unstable.

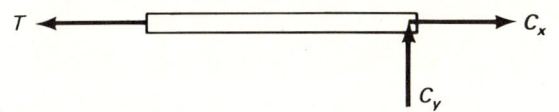

Figure 3.6

The instability in this example would take the form of a significant rotation about pin C. It may seem difficult for you to agree that this motion is possible, especially since rigid bodies have been assumed. It is necessary here to temper the rigid-body assumption to agree more closely with real bodies. All bodies deform when subjected to forces; the rigid-body assumption should really be interpreted to mean that these deformations are very small in comparison with the dimensions of the body itself. Under that more permissive rigid-body definition, a significant rotation of the beam can be accommodated by a very small stretch of the cable, an almost negligible stretch of the beam, or both.

3.2 TRUSSES

Trusses are structures composed of straight members connected at their ends by frictionless pins, and loaded by forces applied only at the joints. Usually their members are arranged to form a system of triangles, although there are exceptions.

3.2.1 Truss Stability and Determinacy

Each member of a truss is a two-force member, that is, forces are applied only at the two ends, couples are not applied anywhere, and the weight of the member is negligible compared to the truss' loads. Thus each member carries only an axial tension (or compression) force internally and represents one unknown element of internal force. In addition, the support reactions are unknowns, so that for a truss there are $m+r$ unknowns, where m is the number of members and r is the number of support reactions.

A complete static analysis of the truss will require $m+r$ equations to solve for the $m+r$ unknowns. Each joint may be considered a plane concurrent force system from which two equations of equilibrium may be written. If j is the number of joints, then $2j$ equations are available. Comparing unknowns to available equations, we establish criteria for the stability and determinacy of a truss based on the number of members, joints, and reactions:

$m+r<2j$ truss is unstable

$m+r=2j$ truss is stable and statically determinate

$m+r>2j$ truss is stable and statically indeterminate

In the statically indeterminate case, the degree of indeterminacy is $m+r-2j$, that is, how much $m+r$ exceeds $2j$.

The above criteria must be used with caution. Internal instabilities *may* exist where the geometry of the structure could change even though stability seems to be indicated.

EXAMPLE 3.2.1

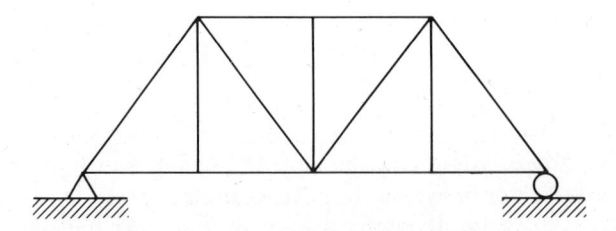

Figure 3.7

The truss has 13 members, 8 joints, and 3 reactions. Substitution in the criteria shows $13+3=2\times 8$, indicating that the truss is stable and statically determinate.

Now compare it to the truss below. Since it has the same number of members, joints, and reactions, our criteria suggest that it too should be stable and statically determinate.

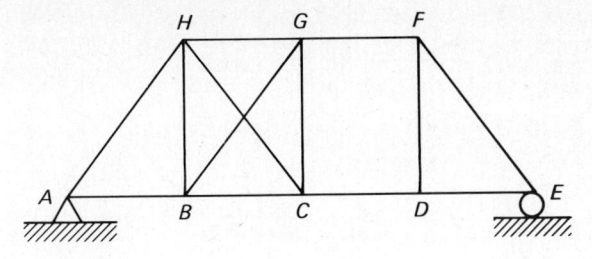

Figure 3.8

But it is not difficult to visualize it undergoing a sizable geometric change, such as shown.

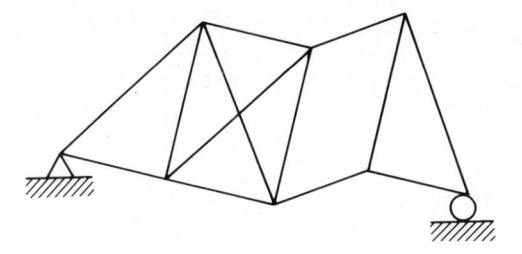

Figure 3.9

Recall our basic criterion for the stability of a rigid body given in Section 3.1.

> A rigid body is stable when it is supported so that the reactions necessary to satisfy the equations of equilibrium for any arbitrary loading can be provided by its supports without changes in the geometry or orientation of the structure.

Consider the free-body diagram of the rigid body that is a portion of the truss.

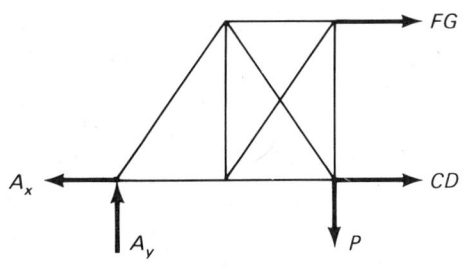

Figure 3.10

Force P is an arbitrarily applied force. Since all forces are concurrent through A, with the exception of P and FG, the moment equation taken about joint A shows that FG must have a nonzero value (unless $P=0$).

The free-body diagram for the other half of the structure is shown below.

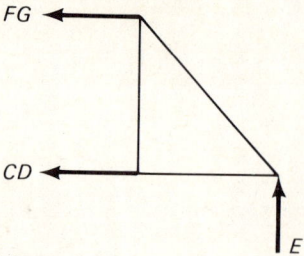

Figure 3.11

Moments taken about joint E show that FG must be zero. But clearly that is inconsistent with the previous finding that $FG \neq 0$. Our conclusion is that the structure is unstable, since for the supports and the geometry an equilibrium equation exists that cannot be satisfied for the arbitrary load P.

EXAMPLE 3.2.2

Don't be misled by the presence of a truss panel that is four (or more) sided, rather than three sided. While these always require investigation, support by other means may ensure stability.

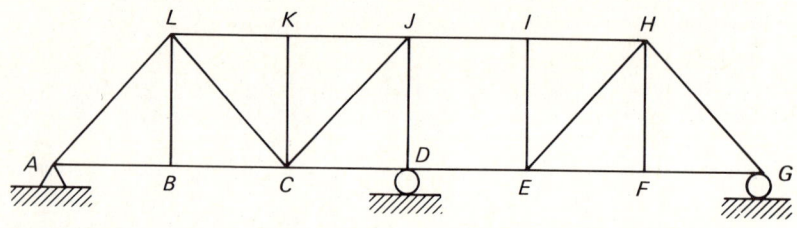

Figure 3.12

The truss has 20 members, 4 support reactions, and 12 joints. The stability criterion gives

$m + r \; ? \; 2j$

$20 + 4 = 2(12)$

$24 = 24$

Stability and static determinacy are both indicated, but panel $DEIJ$ gives us concern about a possible internal instability.

An arbitrary load P is applied at E, and the free-body diagrams are drawn.

Using the right-hand diagram and taking moments about G, we find $IJ \neq 0$. Then using the left-hand diagram and moments about A, we find that force D may adjust itself to satisfy any nonzero value of IJ. A similar analysis of DE using horizontal force summations discloses no impossible

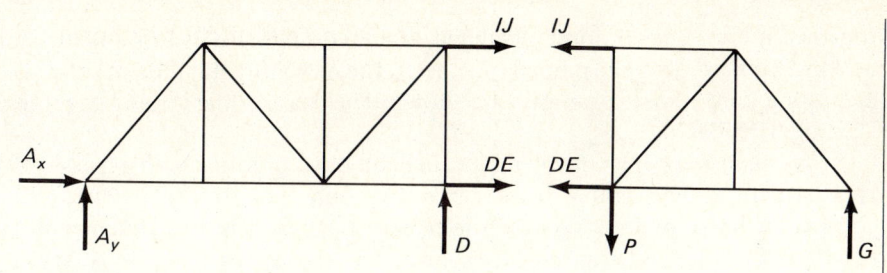

Figure 3.13

situations, so it is concluded that no geometric instabilities are present. The truss is stable and statically determinate.

EXAMPLE 3.2.3

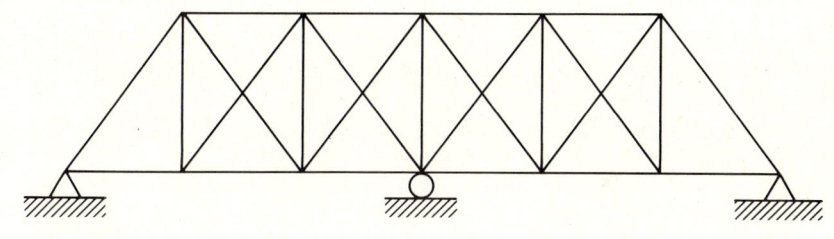

Figure 3.14

This truss has 33 members, 16 joints, and 5 support reactions:

$m + r \; ? \; 2j$

$33 + 5 \; ? \; 2(16)$

$38 > 32$

The truss is stable and statically indeterminate to the sixth degree.

3.2.2 Classification of Trusses

Trusses may be classified as simple, compound, or complex.

A *simple truss* is one that can be generated by starting with three members pinned

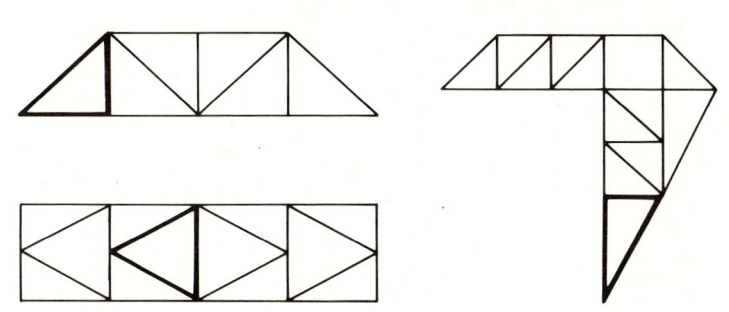

Figure 3.15

together at their ends to form a triangle, and then sequentially forming new joints by adding two new members pinned together at the new joint and their other ends pinned at existing joints. Some examples are shown. The heavier lined triangles represent the starting triangles.

Compound trusses are formed by connecting two or more simple trusses to form a rigid structure. The simple trusses may be connected by three nonparallel, non-concurrent links, or by a common joint and a single link not concurrent with the common joint. Several examples are shown, with the separate simple trusses indicated by different shading.

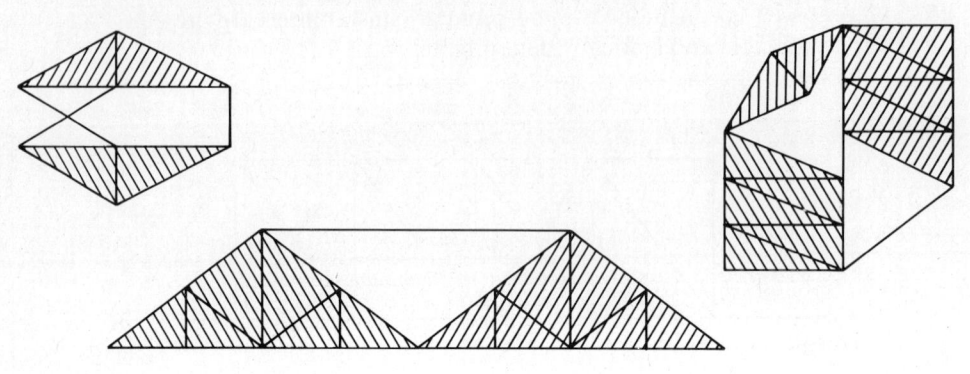

Figure 3.16

Complex trusses are those that do not fit the definitions of simple or compound trusses. Two examples are shown.

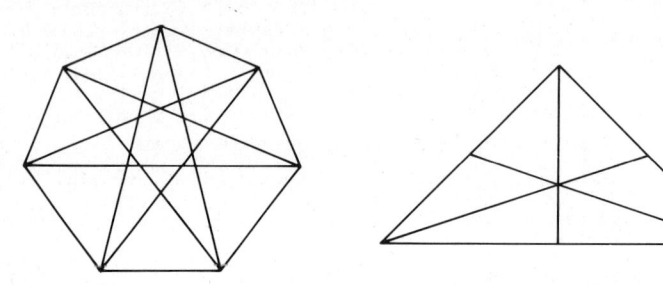

Figure 3.17

3.3 BEAMS

3.3.1 Beam Classification

Beams fall into the following classes:

(1) Simple beam—one supported at its ends by a pin and a roller.
(2) Cantilever beam—one supported at one end by a fixed (clamped) support, while the other end is free (not supported). An alternative form for the fixed end would be a pin support and a roller support placed quite close together at that end.

(3) *Overhanging beam*—similar to a simple beam, except that the pin support or roller support, or both, are not at the ends.

(4) *Compound beam*—a beam that may have a variety of supporting elements, such as pins, rollers, and clamps, and in addition consists of several structural pieces, connected by internal hinges, rollers, or links.

Except for compound beams, all of these are statically determinate and stable. Compound beams may be either statically determinate or indeterminate as well as stable or unstable. The following illustrates the classification.

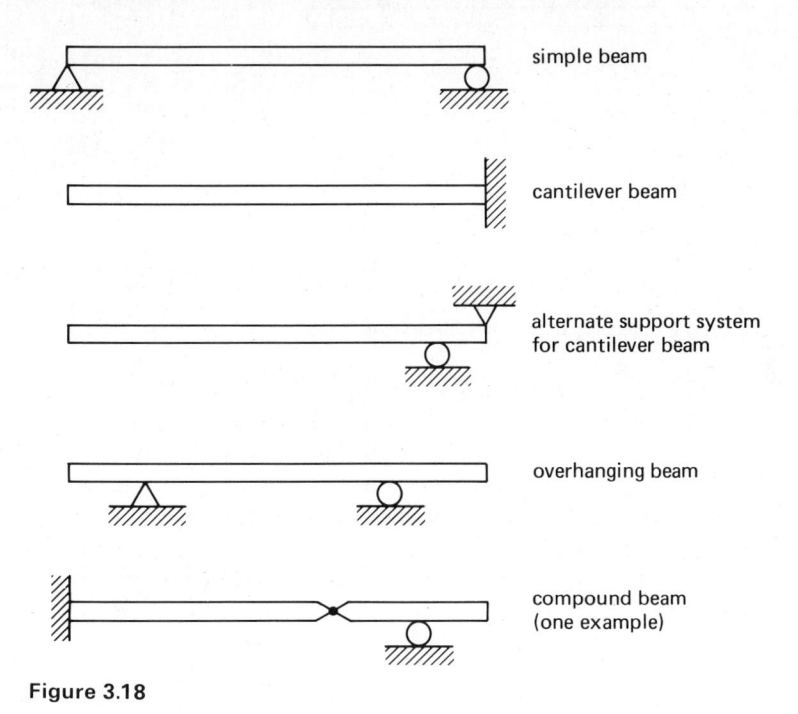

Figure 3.18

3.3.2 Beam Stability and Determinacy

As noted previously, simple, cantilever, and overhanging beams are stable, statically determinate beams. In addition, any beam made as a single structural piece may be analyzed for stability and determinacy using the techniques used for any single rigid body (see Section 3.1).

The stability and determinacy analysis for compound beams is more complicated. Made up of several structural pieces, they differ from other beams by having connections between the pieces that permit some relative motion between them. For example, an internal hinge allows rotation of the left piece relative to the right piece. An equation may be developed from each increased freedom of motion. They are called condition equations.

An internal hinge provides one condition equation, which is that $M=0$ at the hinge. Internal rollers and internal links, both shown, provide two condition equations, which are $M=0$ and $N=0$ at the roller (link), where M and N are the bending moment and the axial force, respectively.

36 STABILITY, DETERMINACY AND CLASSIFICATION OF STRUCTURES

internal roller

internal link

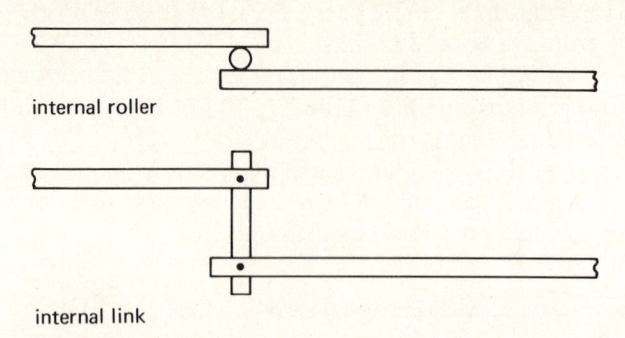

Figure 3.19

Any single-piece beam has the three equilibrium equations available, which can be supplemented by the available condition equations to solve for additional unknown reactions. Thus the criteria for stability and static determinacy are

$r < c + 3$ beam is unstable

$r = c + 3$ beam is stable and statically determinate

$r > c + 3$ beam is stable and statically indeterminate

where r is the number of support reactions, c is the number of condition equations, and the 3 represents the number of planar equilibrium equations.

Caution is necessary when these criteria are used, just as it was with trusses. There is danger that instabilities in the form of geometric changes can occur for some beam systems, even though the criterion indicates a stable system.

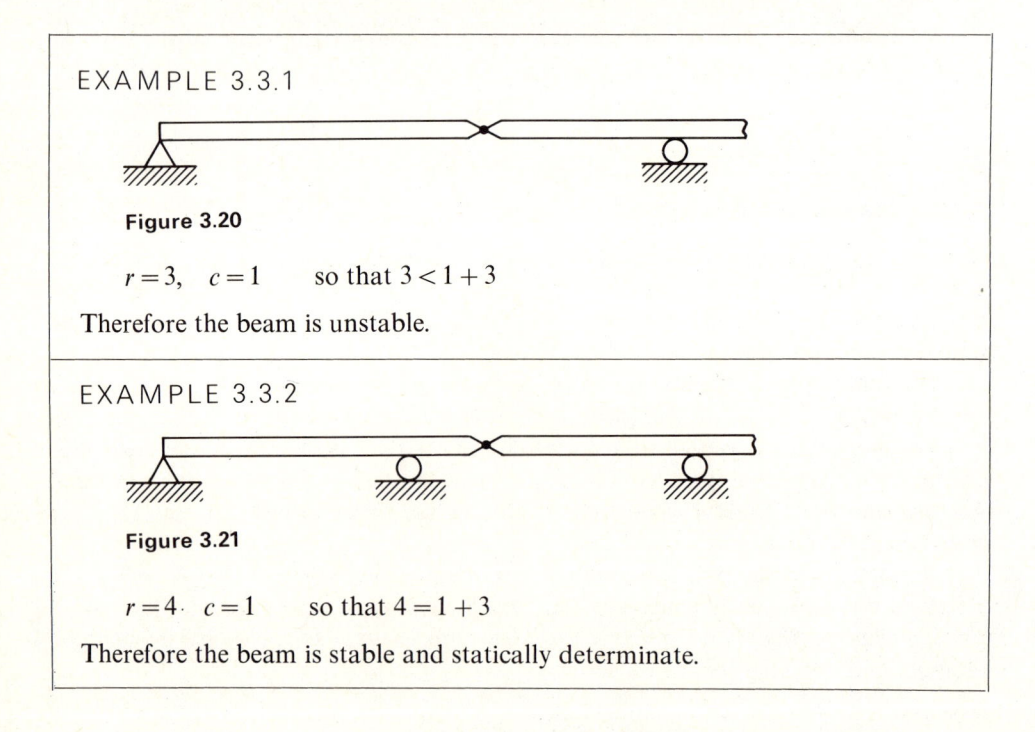

EXAMPLE 3.3.1

Figure 3.20

$r = 3, \quad c = 1 \quad$ so that $3 < 1 + 3$

Therefore the beam is unstable.

EXAMPLE 3.3.2

Figure 3.21

$r = 4, \quad c = 1 \quad$ so that $4 = 1 + 3$

Therefore the beam is stable and statically determinate.

EXAMPLE 3.3.3

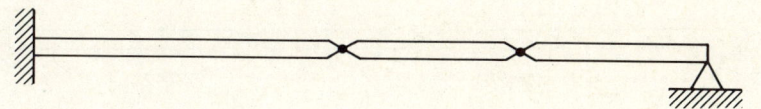

Figure 3.22

$r = 5$, $c = 2$ so that $5 = 2 + 3$

In this example the criteria suggest that the beam is stable and statically determinate. However, for a transverse loading of the center member, the free-body diagram of the center would be

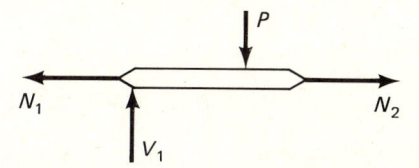

Figure 3.23

While N_1 and V_1 are both forces that could be provided by the cantilever section at the left, N_2 is the only possible force that can be provided by the right-hand piece since it is a two-force member. N_1, N_2, and V_1 are all concurrent forces, and the moment equilibrium equation taken about their point of concurrency cannot be satisfied until a geometric change in the structure occurs that will allow N_2 to have a new line of action which is not through the concurrency point. Thus the new shape would be

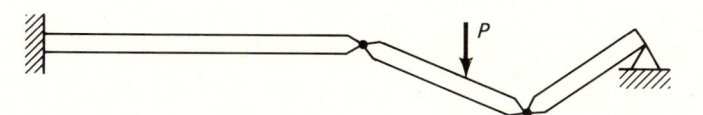

Figure 3.24

The free-body diagram of the center piece

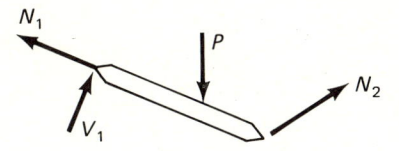

Figure 3.25

shows that the line of action of N_2 is not concurrent with N_1 and V_1.

EXAMPLE 3.3.4

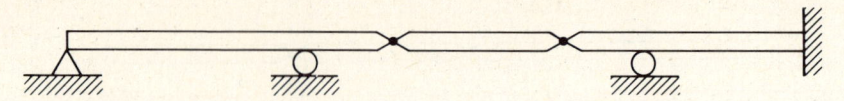

Figure 3.26

$r = 7$, $c = 2$ so that $7 > 2 + 3$

The beam is statically indeterminate to the second degree and stable. In this case a transverse load applied to the center piece would not require a physical reorientation of the structure, since neither the piece to its right nor that to its left is a two-force member. Both may apply forces in the transverse direction. This is shown in the free-body diagram of the center piece.

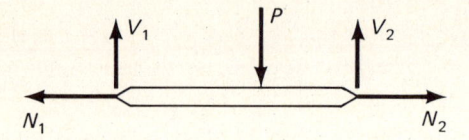

Figure 3.27

All forces are possible, and all equilibrium equations may be satisfied.

EXAMPLE 3.3.5

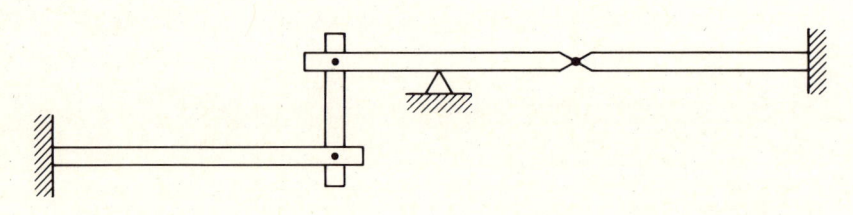

Figure 3.28

$r = 8$, $c = 3$ so that $8 > 3 + 3$

The beam is statically indeterminate to the second degree and stable.

3.4 RIGID FRAMES

A rigid frame is a structure composed of several members connected together at joints, some or all of which are rigid. Rigid joints are capable of transmitting both forces and moments from one member to another, as distinguished from pin joints, which can transmit forces but not moments. Furthermore, rigid joints not only require identical linear displacements of the ends of all the members connected at the joint, but they also *require identical rotations* of these ends. Unless stated to the contrary, it will be assumed that the locus of the centroids of each member lies along the line joining the centers of the joints at the ends of the members. This assumption is helpful since, through it, an axial force carried in a member will not contribute to the moments applied at the member's end.

3.4.1 Stability and Determinacy

The stability and determinacy for a rigid frame may be investigated using criteria analogous to those used for trusses and beams. Essentially the number of unknown support reactions, internal forces, and internal moments is compared to the number of equilibrium equations and supplementing condition equations available to solve them.

Three equilibrium equations may be written for each joint. These are two force equations and one moment equation (taken about the joint itself). The joint is assumed to be small, so that all forces can be assumed to pass through the joint center, and only the member end moments enter this equation.

Each member has three unknowns, since if the bending moment, shear force, and axial force are known anywhere in the member, they can be found anywhere else in the member by simple statics.

Condition equations are based on known internal moments or forces, such as a bending moment of zero at an internal hinge.

Based on the above and comparing available equations to the number of unknowns, the criteria for stability and determinacy are

$3m + r < 3j + c$ frame is unstable

$3m + r = 3j + c$ frame is stable and statically determinate

$3m + r > 3j + c$ frame is stable and statically indeterminate

In the latter case the difference between the left- and right-hand sides of the inequality gives the degree of indeterminacy.

As in the application of all the stability/determinacy criteria, care must be exercised. The potential for geometric instabilities exists in rigid frames, even if stability is indicated by this criterion. Thus each structure must be examined for geometric instabilities in addition to the use of the criteria.

Assigning a value to c in the stability determinacy criterion for rigid frames follows reasoning similar to that used when assigning value to c for compound beams. It is the number of independent equations that may be written that describe known internal conditions. A hinge in the middle of a member, for example, gives $c = 1$ since the single equation $M = 0$ at the hinge fully describes that condition. In cases where n members are pinned together at a common joint, use $c = n - 1$. This follows, since $M = 0$ in each member at the pin end, but there are only $n - 1$ independent equations. The nth question would come from $M = 0$ for the joint.

EXAMPLE 3.4.1

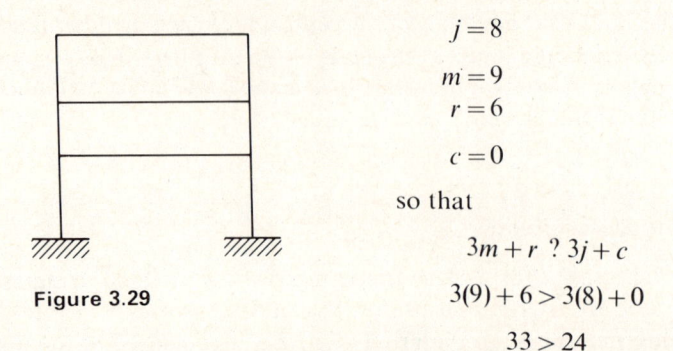

Figure 3.29

$j = 8$

$m = 9$

$r = 6$

$c = 0$

so that

$3m + r \; ? \; 3j + c$

$3(9) + 6 > 3(8) + 0$

$33 > 24$

The structure is stable and statically indeterminate to the ninth degree.

EXAMPLE 3.4.2

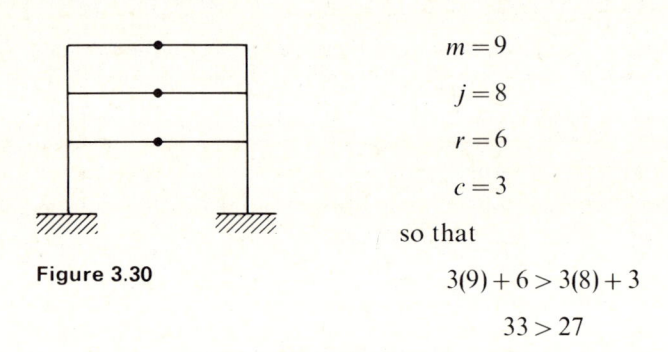

Figure 3.30

$m = 9$

$j = 8$

$r = 6$

$c = 3$

so that

$3(9) + 6 > 3(8) + 3$

$33 > 27$

The structure is stable and statically indeterminate to the sixth degree $(33 - 27 = 6)$.

An alternative viewpoint could count

$m = 12$

$j = 11$

$r = 6$

$c = 3$

so that

$3(12) + 6 > 3(11) + 3$

$42 > 36$

This gives the same result: **stable**, statically indeterminate to the sixth degree $(42 - 36 = 6)$.

EXAMPLE 3.4.3

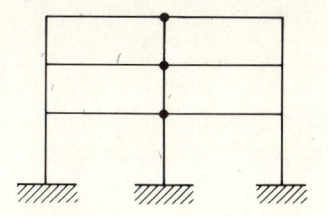

Figure 3.31

$m = 15$

$j = 12$

$r = 9$

$c = 8$

so that

$$3(15) + 9 > 3(12) + 8$$
$$54 > 44$$

The structure is stable and statically indeterminate to the tenth degree $(54 - 44 = 10)$.

The value $c = 8$ was found, having accounted for one hinged joint with three members and two hinged joints with four members pinned together. Thus $c = (3 - 1) + 2(4 - 1) = 8$.

EXAMPLE 3.4.4

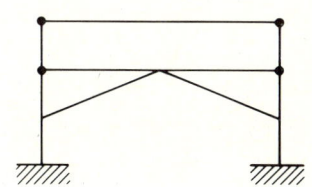

Figure 3.32

$m = 11$

$j = 9$

$r = 6$

$c = 6$

so that

$$3(11) + 6 > 3(9) + 6$$
$$39 > 33$$

The criteria indicate stability and static indeterminacy to the sixth degree. The structure is unstable, however, as examination of the three top members suggests. There is nothing to prevent a change in geometry such as shown.

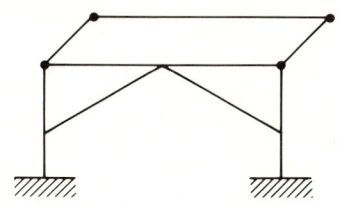

Figure 3.33

Examine the free-body diagram of the top member. Since the two members supporting it are two force members, it is

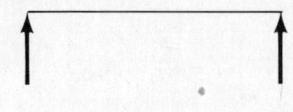

Figure 3.34

Any force with horizontal components applied to it would not be balanced, and $\sum F_{horiz} \neq 0$. Equilibrium would occur only after a large geometric change.

A useful alternate approach for analyzing determinacy, sometimes called the method of "trees," requires analytical cutting of members so that the original structure is separated into several stable and statically determinate parts. Wherever a member has been cut, it is presumed that three unknowns exist (the internal axial and shear forces and the internal bending moment). By counting the number of cuts and multiplying by 3, the degree of indeterminacy is established.

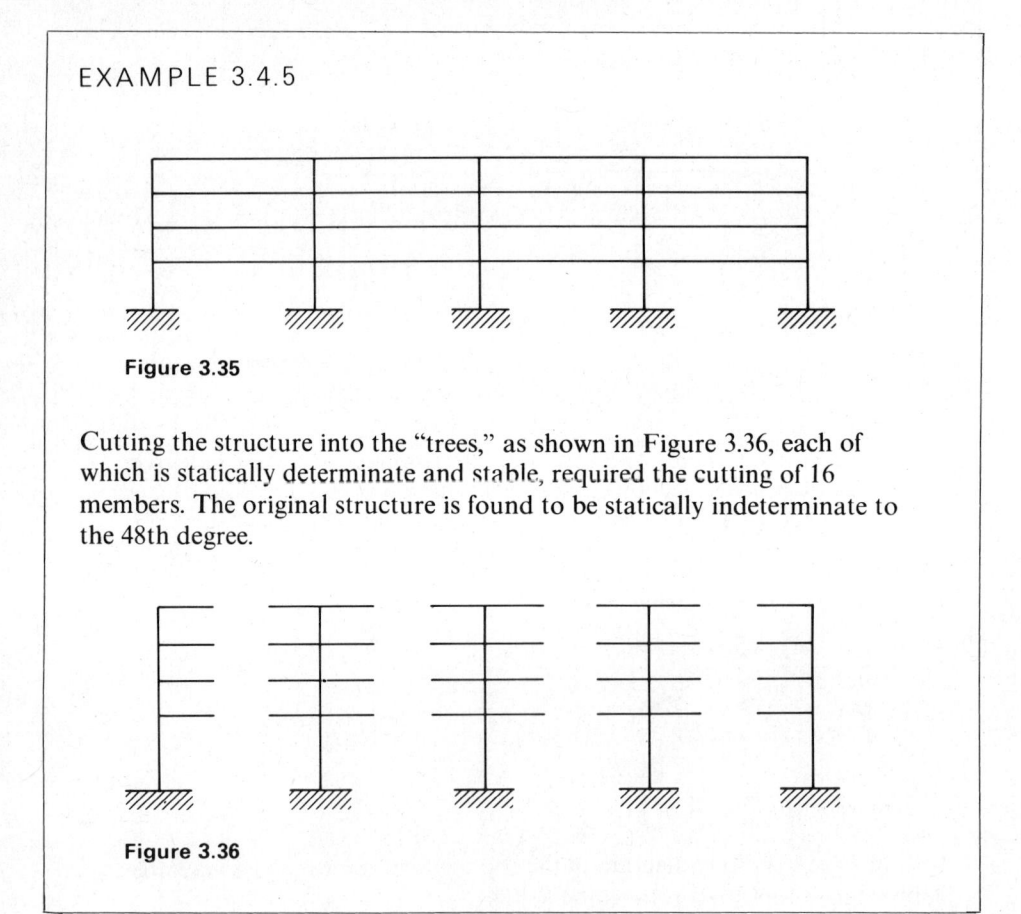

EXAMPLE 3.4.5

Figure 3.35

Cutting the structure into the "trees," as shown in Figure 3.36, each of which is statically determinate and stable, required the cutting of 16 members. The original structure is found to be statically indeterminate to the 48th degree.

Figure 3.36

PROBLEMS

3.1 Determine the stability and the static determinacy of the rigid bodies shown.

(a)

(b)

(c)

(d)

(e)

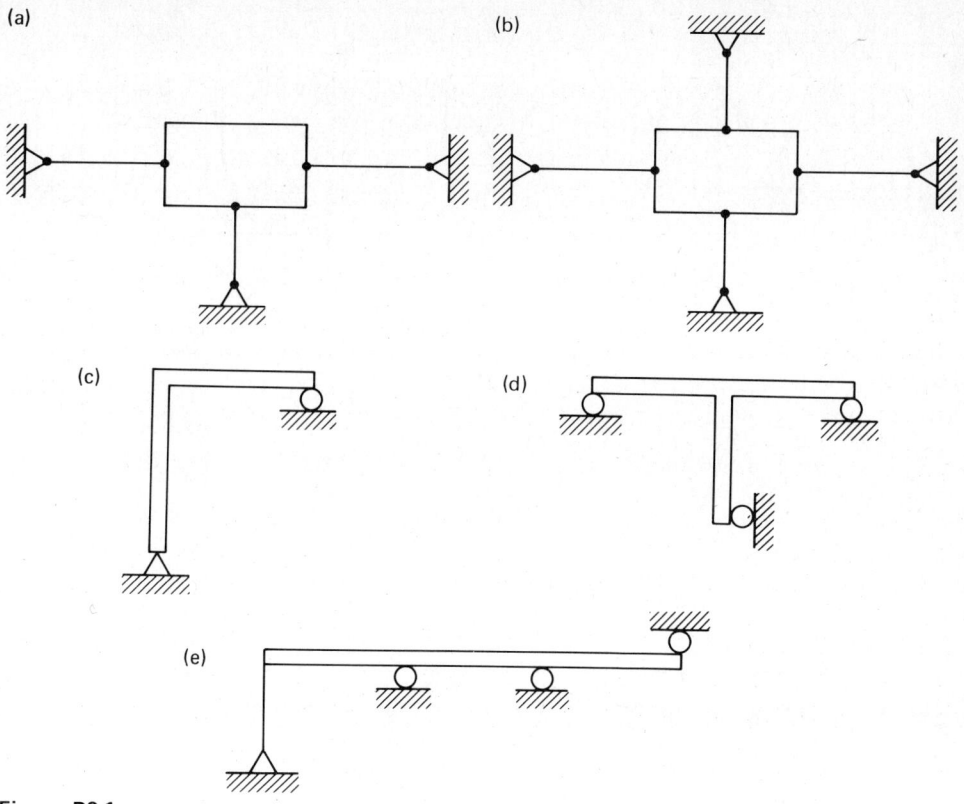

Figure P3.1

3.2 Determine the stability and the static determinacy of the trusses shown.

(a)

(b)

(c)

(d)

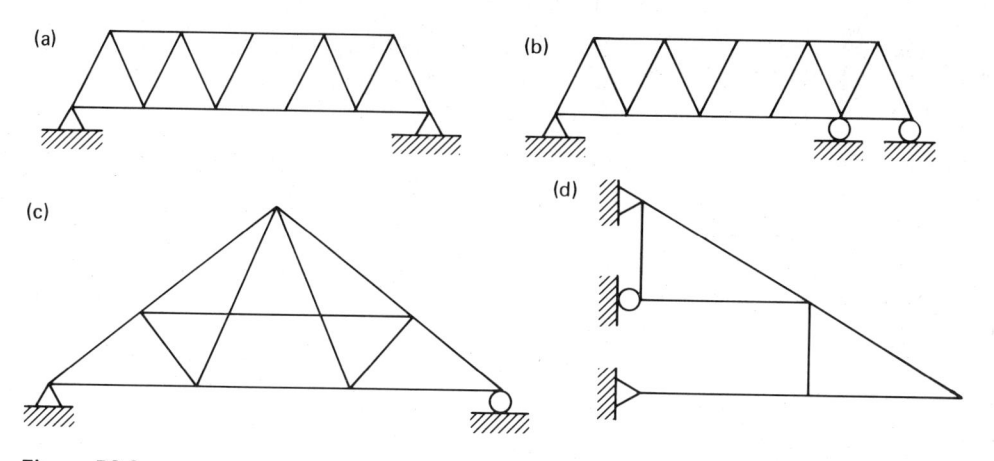

Figure P3.2

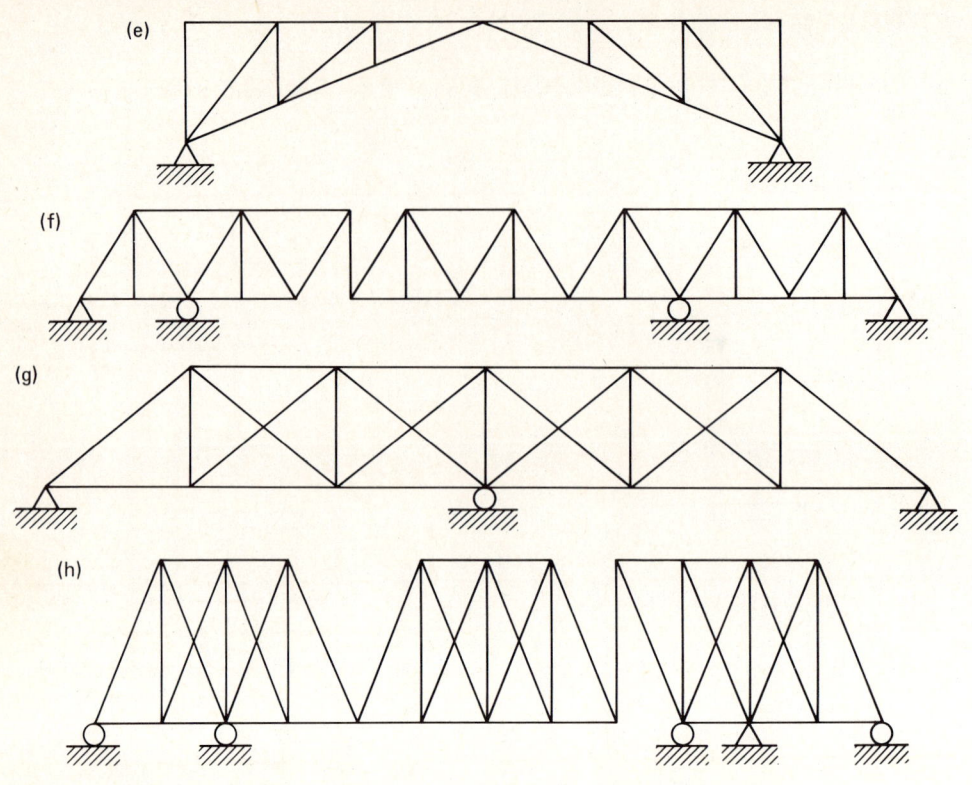

Figure P3.2 (continued)

3.3 Determine the stability and the static determinacy of the beam structures shown.

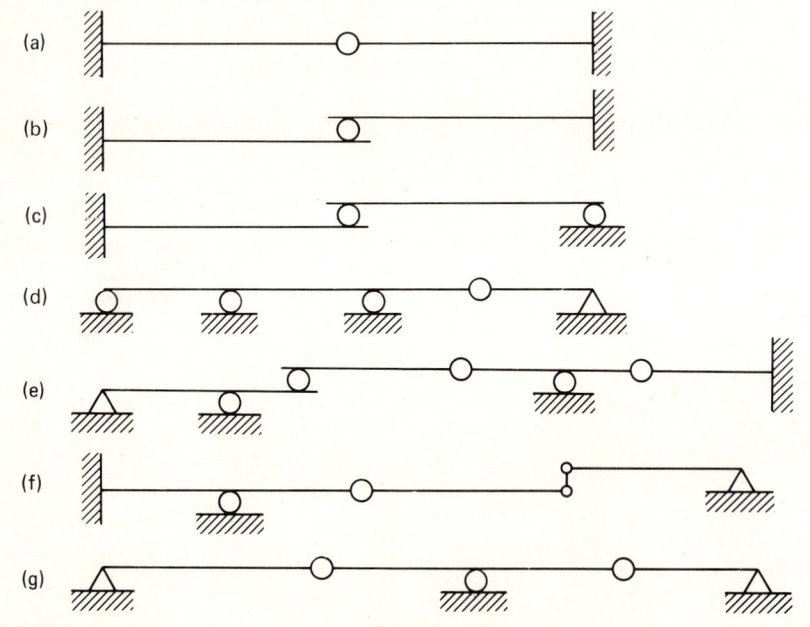

Figure P3.3

3.4 Determine the stability and the static determinacy of the rigid frames shown:

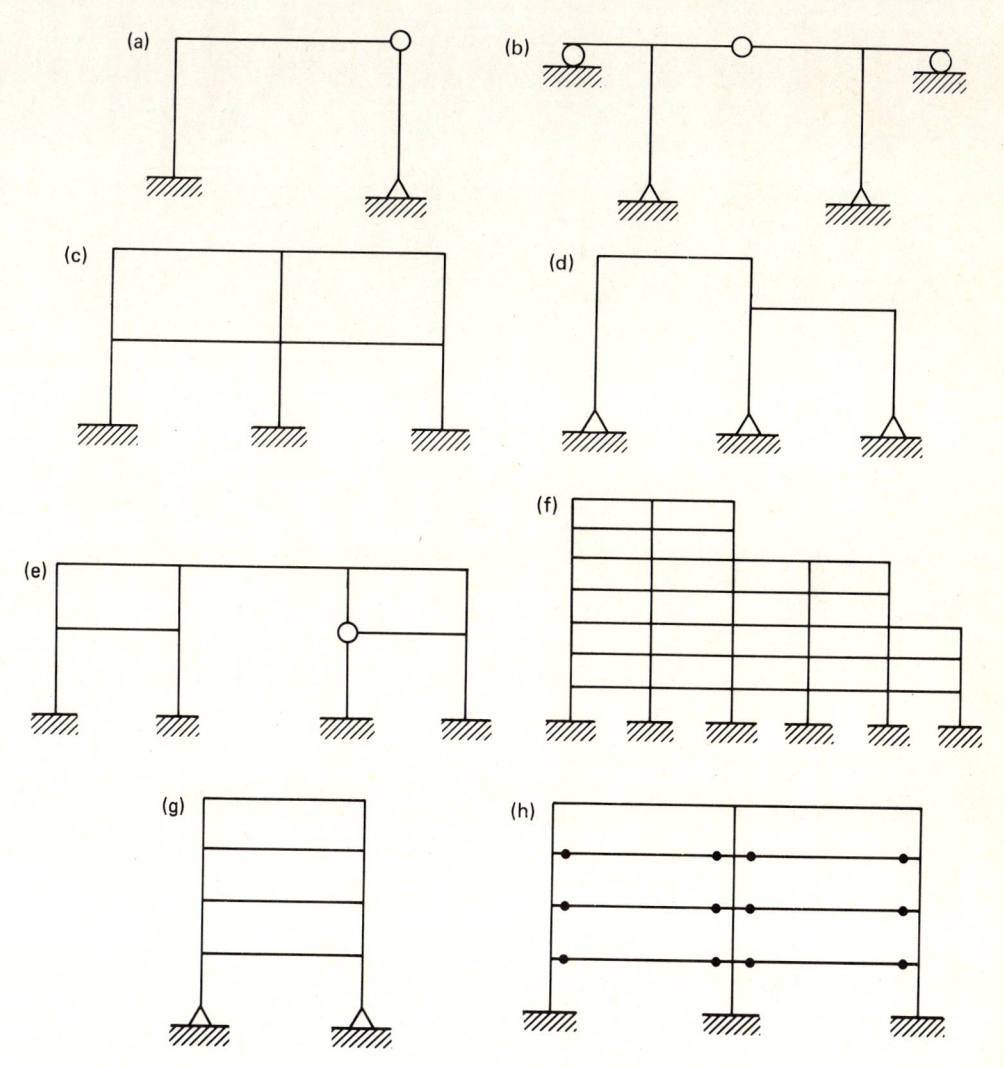

Figure P3.4

Stress and Force Analysis of Statically Determinate Structures

... and then, too late, he realized he was sitting on the very branch that he had been sawing ... and in quite the wrong place.

4.0 INTRODUCTION

If the woodsman were a structural analyst, he would deserve the fate that seems in store for him. A tree branch is a statically determinate member. The internal forces and the bending moments can be determined using static equilibrium principles alone. In many cases less obvious than that of the woodsman, it is important to know the forces and moments carried internally at all points in a member in order to "size it" properly. This chapter reviews some of the techniques you learned in statics and strength of materials courses and extends them to new methods and new applications.

4.1 TRUSSES

There are two fundamental methods of analysis for trusses: the method of joints and the method of sections. Both start with a free-body diagram of the truss as a whole, from which the equilibrium equations are written and solved for the support reactions.

4.1.1 Method of Joints

After the support reactions have been found, a joint is selected that has no more than two members connecting for which the axial forces are unknown. The free-body diagram of that joint is drawn, the forces are summed in two directions, and each sum is equated to zero. When drawing the free-body diagram, it is a good idea to assume that the unknown forces are tensions and to show them so on the free-body diagram by their exerting a pull on the joint. When this is assumed, the resulting sign of the unknowns when evaluated will match the conventional + for tension and − for compression.

Once a joint has been analyzed, its members become knowns, and adjacent joints, which might have had three or more unknowns, can then be solved since some of these unknowns have become knowns. This process continues from joint to joint, each time selecting a joint whose number of unknown members does not exceed 2.

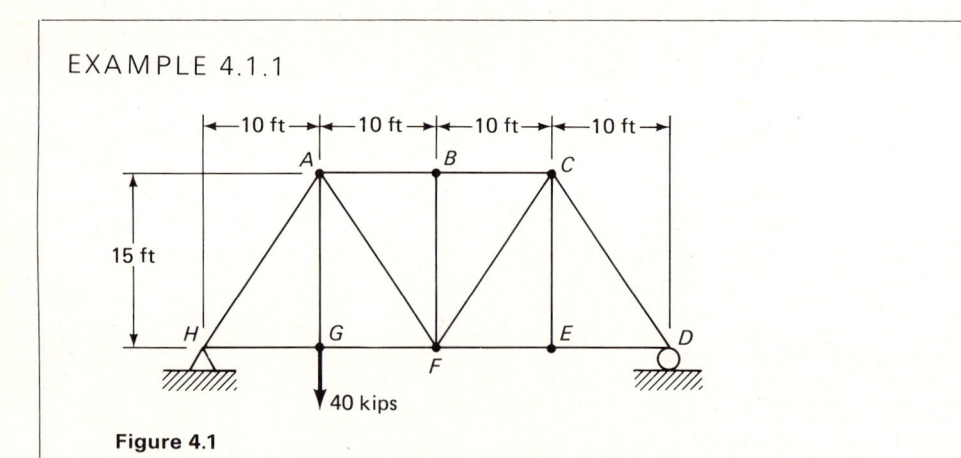

EXAMPLE 4.1.1

Figure 4.1

From the free-body diagram of the whole structure the equilibrium equations are written and solved for the support reactions.

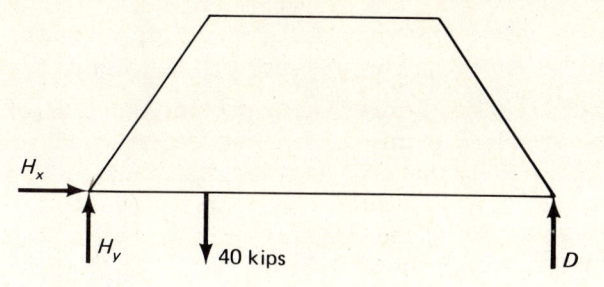

Figure 4.2

$$\pm\sum F_x = H_x = 0$$
$$+\circlearrowleft\sum M_H = -40(10) + 40D = 0$$

Hence

$$D = 10 \text{ kips}$$
$$\uparrow\sum F_y = H_y - 40 + D = 0$$
$$H_y = 40 - D = 40 - 10 = 30 \text{ kips}$$

Joint H was selected to start since there are only two unknown members that join there. Joint D would have been as good a choice. All of the other joints, however, have too many unknowns. All other forces applied by the support at H are known. The free-body diagram is

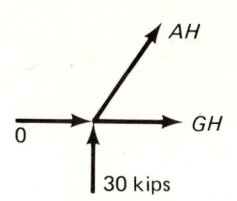

Figure 4.3

$$+\uparrow\sum F_y = \frac{15}{\sqrt{15^2 + 10^2}} AH + 30 = 0$$
$$AH = -36.1 \text{ kips}$$

$$\pm\sum F_x = GH + \frac{10}{\sqrt{15^2 + 10^2}} AH = 0$$

$$GH = \frac{-10}{\sqrt{15^2 + 10^2}}(-36.1) = +20 \text{ kips}$$

In these two equilibrium equations, the coefficients

$$\frac{15}{\sqrt{15^2 + 10^2}} \quad \text{and} \quad \frac{10}{\sqrt{15^2 + 10^2}}$$

are $\sin\theta$ and $\cos\theta$, respectively

Next, joint G may be considered since only AG and GF are unknowns. Joint A cannot be analyzed yet since, of the four members intersecting there, three remain as unknowns.

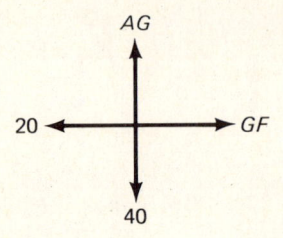

Figure 4.4

$+\uparrow \sum F_y = AG - 40 = 0 \qquad AG = 40 \text{ kips}$

$\pm \sum F_x = -20 + GF = 0 \qquad GF = 20 \text{ kips}$

Now joint A may be analyzed, since both AG and AH have become known forces and only two, AB and AF, are unknowns.

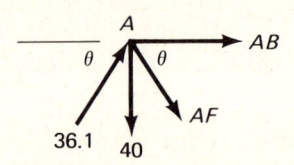

Figure 4.5

In this free-body diagram the forces in AG and AH were shown with their known values and directions. Applying the equilibrium equations,

$+\uparrow \sum F_y = 36.1 \dfrac{15}{\sqrt{15^2 + 10^2}} - 40 - \dfrac{15}{\sqrt{15^2 + 10^2}} AF = 0$

$AF = -12.0 \text{ kips}$

$\pm \sum F_x = AB + \dfrac{10}{\sqrt{15^2 + 10^2}} (36.1) + \dfrac{10}{\sqrt{15^2 + 10^2}} AF = 0$

$AB = -\dfrac{10}{\sqrt{15^2 + 10^2}} (36.1) - \dfrac{10}{\sqrt{15^2 + 10^2}} (-12.0)$

$= -13.3 \text{ kips}$

The solution of this example would continue, going in order through joints B, F, E, and C.

Finally, joint D would be analyzed. But by then all members intersecting there would be known, so its analysis serves as a check. The equilibrium equations must be satisfied for joint D. If they are not, you know that you have made an error in the calculations, and it is essential that this error be found and corrected.

Joints connecting four members that are arranged as two pairs of collinear members are often used in trusses. If there are no other forces applied externally to the joint, the joint can be considered equivalent to two separate members, one replacing one of the pairs of collinear members, and the second replacing the other pair. These are arranged to be in the same directions as the members replaced and not touching each other, as shown in the sample system. Members 1 and 2 and one pair of collinear members, and members 3 and 4 are the second pair.

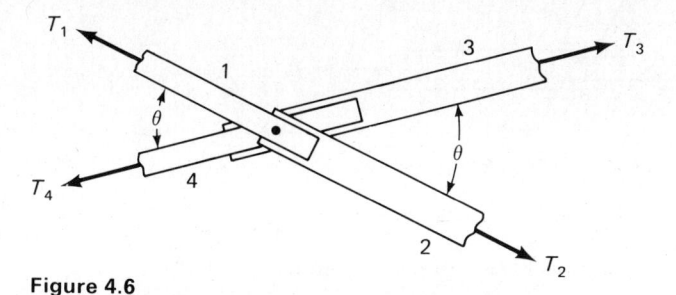

Figure 4.6

It could be replaced with the system

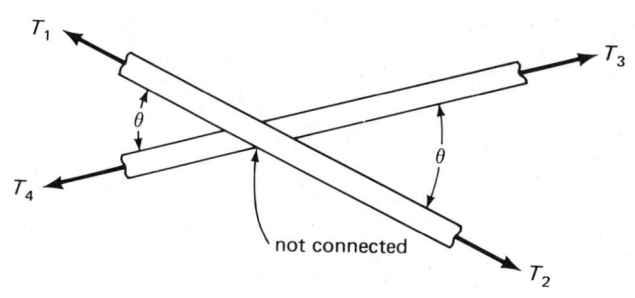

Figure 4.7

For a proof, take the summation of forces perpendicular to members 1 and 2 in the original system:

$$\sum F = T_3 \sin \theta - T_4 \sin \theta = 0$$

from which

$$T_3 = T_4$$

A similar summation of forces perpendicular to members 3 and 4 in the original system would show that

$$T_1 = T_2$$

In the equivalent system the identical results should be obvious. Also notice that if

(1) there is an externally applied force at the joint, or
(2) there is an additional member connected to the joint, or
(3) one or both pairs of members are not collinear,

then the results are invalidated.

52 STRESS AND FORCE ANALYSIS OF STATICALLY DETERMINATE STRUCTURES

An extension of the idea can be made where one of the four members is replaced with an externally applied force in its direction.

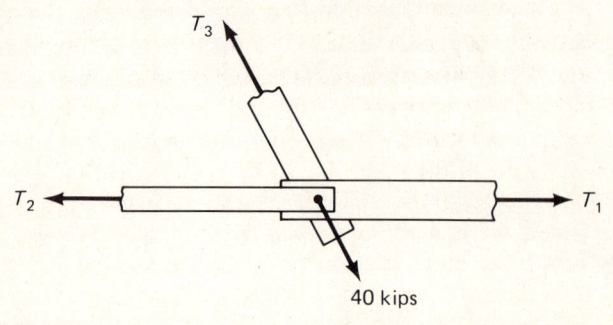

Figure 4.8

In this case we conclude that $T_3 = 40$ kips (tension) and that $T_1 = T_2$.
If the 40-kip force were removed, then $T_3 = 0$.

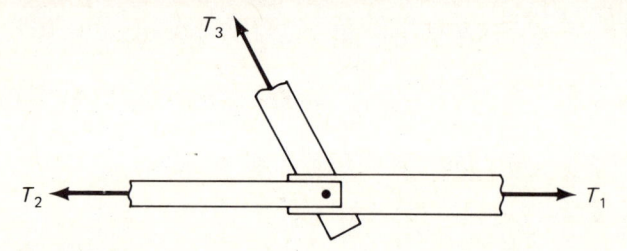

Figure 4.9

Thus in T connections, as shown, at which no external force is applied, the "leg" of the T is a zero force member.

This latter case can be extended even further to one where two members terminate at a joint, either with or without external forces.

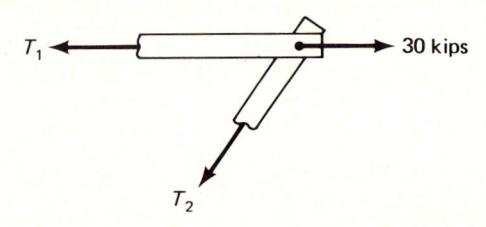

Figure 4.10

Inspection of this joint should show that $T_1 = 30$ kips (tension) and $T_2 = 0$.

4.1.2 Method of Sections

Almost all truss systems are configured so that analysis using the method of joints must begin at one end and proceed joint by joint toward the other end. If it is necessary to evaluate the forces carried by a member located some distance from the ends, the method of joints requires the calculation of the forces in many members before the desired one is reached. The method of sections provides a means for a direct calculation in these cases.

After the support reactions have been calculated, the truss is cut through (analytically) so that one part of the truss is completely severed from the rest. When this is done, no more than three unknown members should be cut. If possible the cut should pass through the member or members whose internal forces are to be found.

A free-body diagram of the part of the truss on one side of this section is drawn, and the internal forces are found through the equilibrium equations. Since the system of forces on the free-body diagram is a plane nonconcurrent force system, three equilibrium equations may be written and solved for the three unknowns.

EXAMPLE 4.1.2

Find the forces carried in *BC*, *FC*, and *FE* of the truss in Example 4.1.1.

A section is taken through the three unknown members to be analyzed, which completely cuts through the truss. The free-body diagram for the left-hand half is

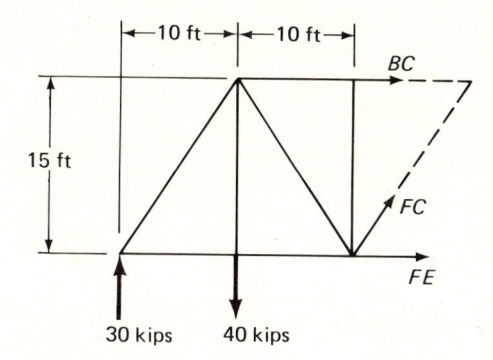

Figure 4.11

The 30-kip reaction force was found in Example 4.1.1. Notice that the three unknown forces all are assumed to be in tension, as was assumed in the method of joints:

$$+\uparrow \sum F_y = 30 - 40 + \frac{15}{\sqrt{10^2 + 15^2}} FC = 0 \qquad FC = +12 \text{ kips}$$

$$+\circlearrowleft \sum M_F = -30(20) + 40(10) - BC(15) = 0 \qquad BC = -13.3 \text{ kips}$$

$$+\circlearrowleft \sum M_c = -30(30) + 40(20) + 15\,FE = 0 \qquad FE = 6.67 \text{ kips}$$

Identical values would have been found had the section to the right of the cut been used for the free-body diagram, and had the equilibrium equations been written from it and solved.

4.1.3 Compound Truss Analysis

Most statically determinate compound trusses can be analyzed using the method of sections, the method of joints, or a combination of the two. Occasionally, however, these methods fail. When they do, analysis should begin with the separation of the compound truss into its constituent simple trusses, and the forces in the connecting members or pins should be found. Then each of the simple trusses may be analyzed by the methods of sections and joints.

EXAMPLE 4.1.3

Find the forces in members JK, KF, and FG of the truss.

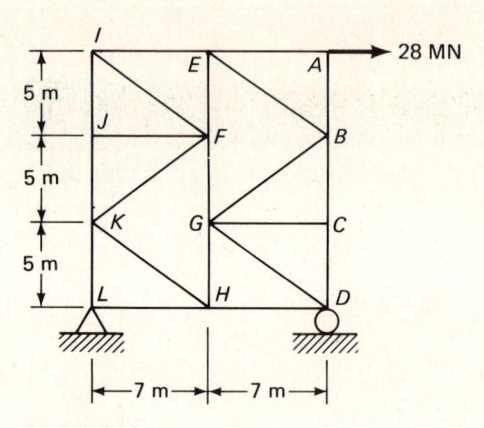

Figure 4.12

First we use equilibrium of the entire truss to find the support reactions.

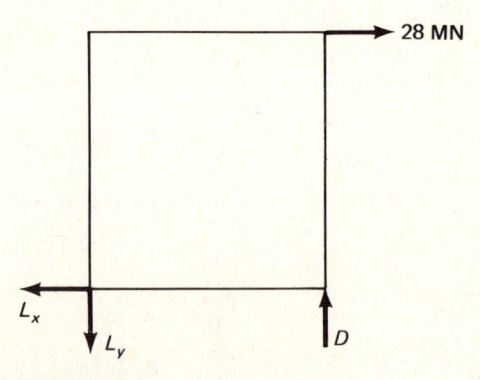

Figure 4.13

$$+\circlearrowright \sum M_L = 14D - 28(15) = 0 \qquad D = 30 \text{ MN}$$
$$\pm \sum F_x = -L_x + 28 = 0 \qquad L_x = 28 \text{ MN}$$
$$+\uparrow \sum F_y = -L_y + D = 0 \qquad L_y = 30 \text{ MN}$$

If joint A is analyzed first, AE and AB are seen to carry 28 MN and 0, respectively. And if joint L is analyzed, LK and LH are found to carry 30-MN and 28-MN tension, respectively. But if the method of joints is tried for the remainder of the analysis, no other joints in the truss are found that connect fewer than three unknown members. Further, the method of sections also fails since there is no way to cut through the truss without cutting more than three unknowns.

If this compound truss is separated into its constituent simple trusses, the following series of free-body diagrams results.

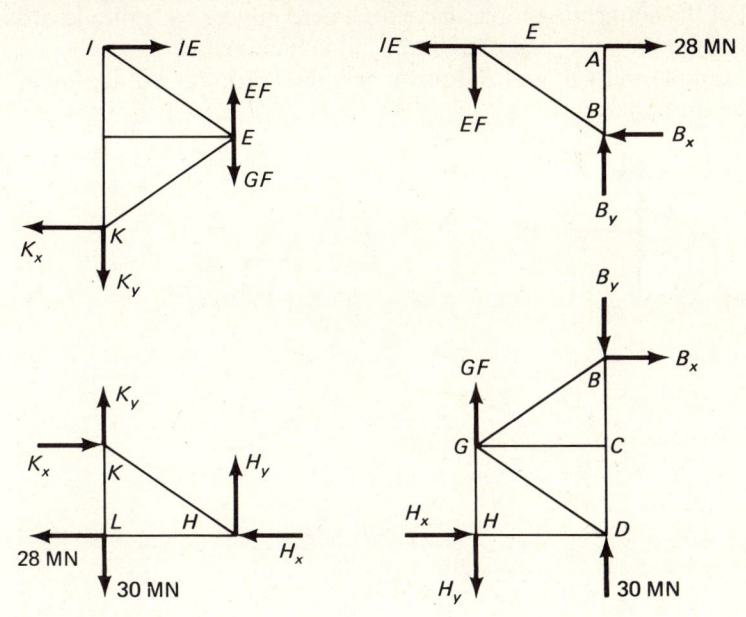

Figure 4.14

At first this series may seem to provide a formidable combination of unknowns, but in most situations like this a few relatively simple calculations will produce the unknown connecting forces.

We start with moments about H in the free-body diagram of $HGBD$ to get

$$+\!\!\circlearrowleft\sum M_H = 7 \times 30 - 7B_y - 10B_x = 0$$

so that

$$10B_x + 7B_y = 210$$

A second equation involving B_x and B_y can be written by taking moments about E on the free-body diagram of ABE,

$$+\!\!\circlearrowleft\sum M_E = -5B_x + 7B_y = 0$$

Solving these two equations simultaneously gives

$B_x = 14$ MN

$B_y = 10$ MN

Now the rest of the unknown forces on *HGBD* and *ABE* can be solved. For example, using *ABE*,

$$\pm \sum F_x = -IE - B_x + 28 = 0$$

$$IE = 28 - B_x = 28 - 14 = 14 \text{ MN}$$

$$+\uparrow \sum F_Y = -EF + B_y = 0$$

$$EF = B_y = 10 \text{ MN}$$

When all of the connecting forces have been determined, each simple truss may be analyzed to give the forces in the individual members.

For example, with B_y and B_x known, members *BG* and *BC* are found using joint *B*.

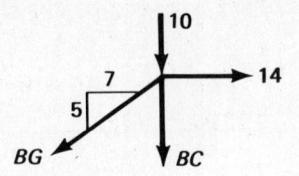

Figure 4.15

$$\pm \sum F_x = -\frac{7}{\sqrt{5^2 + 7^2}} BG + 14 = 0$$

$$BG = +17.2 \text{ MN}$$

$$+\uparrow \sum F_y = -BC - \frac{5}{\sqrt{5^2 + 7^2}} BG - 10 = 0$$

$$BC = -20 \text{ MN}$$

The remaining forces and moments are left for you to evaluate.

4.2 AXIAL FORCE, SHEAR FORCE AND BENDING MOMENT IN STATICALLY DETERMINATE RIGID FRAMES

Most rigid frames are statically indeterminate. There are many methods for their analysis, a few of which will be studied in later chapters of this book. Several of these require an analysis of one or more rigid frames which are similar in configuration to the statically indeterminate one, but altered in support or connections so that they are statically determinate. The primary reason then for our study of statically determinate rigid frames is directed toward their anticipated use in the analysis of statically indeterminate rigid frames.

Some of the methods to be used will require mathematical manipulations of the expressions for the internal forces and moments of the statically determinate frames. We will develop methods to get these expressions.

Statically determinate frame analysis is also used at a later stage of statically indeterminate analysis: when enough support reactions and internal forces have been found to permit the rest to be determined by using equilibrium relations alone.

4.2 AXIAL FORCE, SHEAR FORCE AND BENDING MOMENT IN RIGID FRAMES

Rigid frames are distinguished from other structures by the use of joints that prevent relative motion between the connected members. In a large proportion of rigid frames the members intersect at right angles. In these the relation of forces and moments between connected members is simple. Example 4.2.1 provides a numerical example.

EXAMPLE 4.2.1

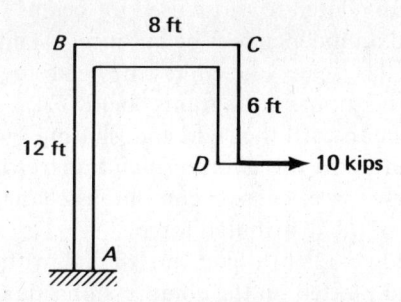

Figure 4.16

The free-body diagram is drawn and reactions are calculated, giving

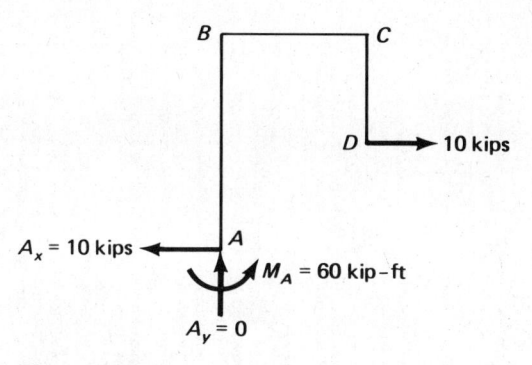

$A_x = 10$ kips
$M_A = 60$ kip-ft
$A_y = 0$

Figure 4.17

Next the members are separated at the joints and their free-body diagrams drawn.

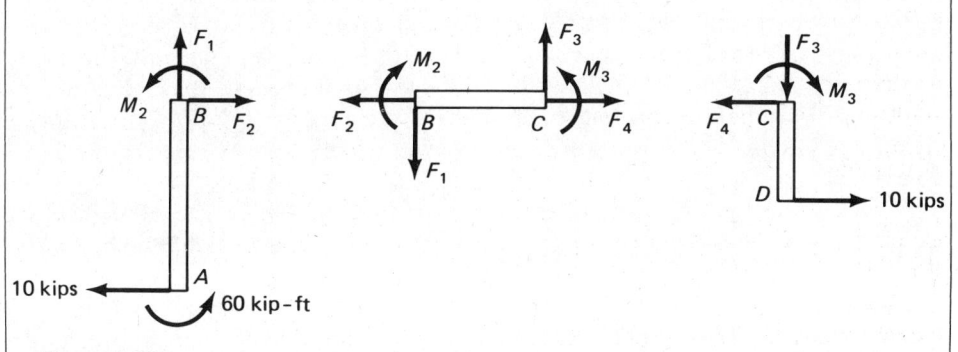

Figure 4.18

The reader can verify easily that

$F_1 = 0$ $F_3 = 0$ $M_2 = 60$ kip-ft

$F_2 = 30$ kips $F_4 = 10$ kips $M_3 = 60$ kip-ft

Notice that force F_2 is an end shear force on member AB, but is an axial force on member BC. A similar relation should be obvious regarding force F_4. Moment M_2 is an end moment on both AB and BC.

Shear and moment diagrams for the frame can be developed now. The techniques for their construction are related to those used for beams. In fact each member may be considered an independent beam once its end forces and moments are known. To do this, select one end of the beam to be the "left" end and sketch the shear and moment diagrams "below" it by starting at the left end and proceeding toward the right end, shifting the shear diagram by the amount of and in the direction of each transverse force encountered. Distributed transverse forces cause a continuous change in shear at a rate equal to the intensity of the distributed force.

The moment diagram is constructed next, starting at the left end, with a moment equal to the end moment and plotted on the compression side of the member. The moment diagram changes at a rate equal to the shear magnitude, and in a direction corresponding to the shear plot. Example 4.2.2 extends Example 4.2.1 by sketching the shear and moment diagrams of the elements.

The Newport Bridge, New England's longest suspension bridge spans Narragansett Bay, linking Jamestown and Newport, RI (Courtesy of Bethelem Steel Corporation).

4.2 AXIAL FORCE, SHEAR FORCE AND BENDING MOMENT IN RIGID FRAMES

EXAMPLE 4.2.2

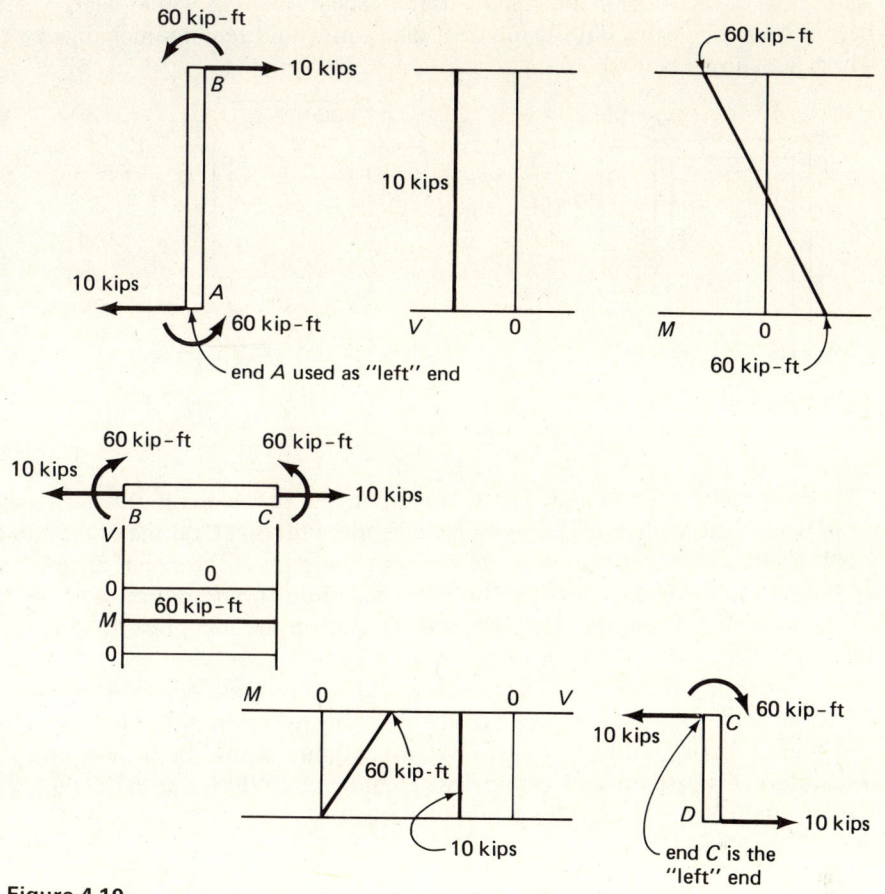

Figure 4.19

The construction of these may be more recognizable if you turn your book 90° clockwise for *AB*, restore it to its usual orientation for *BC*, and rotate it 90° counterclockwise for *CD*.

Usually the shear and moment diagrams for a rigid frame are drawn on sketches of the structure itself, and are merely the assemblage of the individual diagrams.

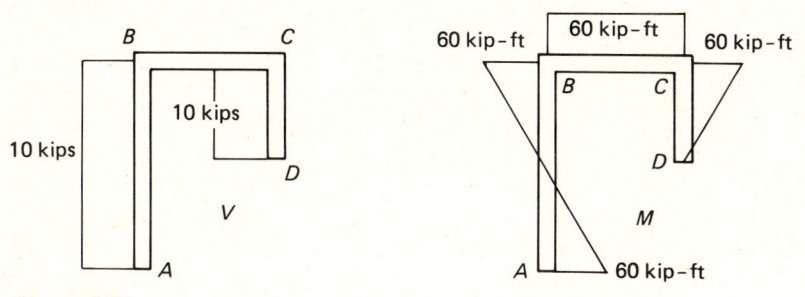

Figure 4.20

It is interesting to observe that had ends *D*, *C*, and *B* been chosen for the "left" ends of the three members, the frame's shear diagram would have been reversed in signs, but the moment diagram would remain unchanged. These are shown below.

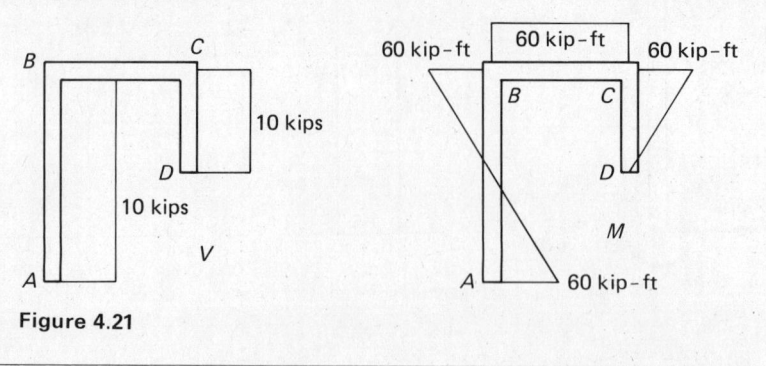

Figure 4.21

The signs of shear forces usually are not important, so the possibility of two shear diagrams causes no concern. The sign of the bending moment, on the other hand, is very important. It is fortunate that the moment diagram is the same, regardless of which end of the member is used as the "left" end. You should notice carefully that the curve for the moment diagram will always plot on the compression side of the members.

Expressions for the internal shear force and bending moment will be used in later sections. They may be developed easily by taking a section through the structure in each region for which a different free-body diagram would be drawn, and then writing and solving appropriate equilibrium equations. For the frame in Example 4.2.1 sections through *AB*, *BC*, and *CD* would be necessary.

EXAMPLE 4.2.3

Develop an expression for the internal bending moments for the frame of Example 4.2.1.

There are no critical sign conventions to be used here. The reason for this casual use of conventions will become apparent when the use of the internal bending moment expressions is studied.

Cut a section through *CD*, and draw the free-body diagram of a part to one side of the cut.

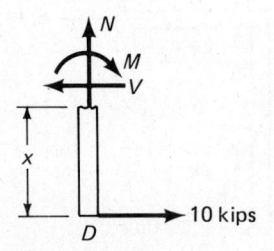

Figure 4.22

4.2 AXIAL FORCE, SHEAR FORCE AND BENDING MOMENT IN RIGID FRAMES

$$\pm \sum F_x = -V + 10 = 0$$
$$V = 10 \text{ kips}$$
$$+\circlearrowleft \sum M_{cut} = -M + 10x = 0$$
$$M = 10x \text{ kip-ft}$$

These expressions are valid for $0 < x < 6$ ft with $x = 0$ at D.

Next cut BC and draw the free-body diagram of a part to one side of the cut.

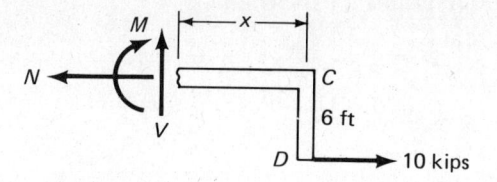

Figure 4.23

$$\updownarrow \sum F_y = V = 0$$
$$\pm \sum F_x = -N + 10 = 0$$
$$N = 10 \text{ kips}$$
$$+\circlearrowleft \sum M_{cut} = -M + 6(10) = 0$$
$$M = 60 \text{ kip-ft}$$

for $0 < x < 6$ ft with $x = 0$ at C.

Finally cut AB,

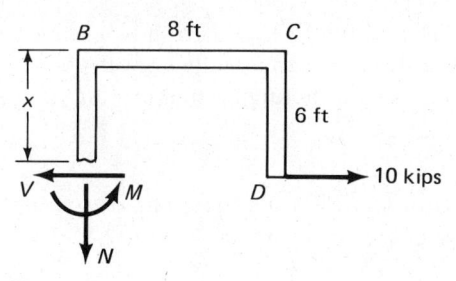

Figure 4.24

$$\pm \sum F_x = -V + 10 + 0$$
$$V = 10 \text{ kips}$$
$$+\uparrow \sum F_y = -N = 0$$
$$+\circlearrowleft \sum M_{cut} = M - (x - 6)10 = 0$$
$$M = 10x - 60 \text{ kip-ft}$$

for $0 < x < 12$ ft with $x = 0$ at B.

62 STRESS AND FORCE ANALYSIS OF STATICALLY DETERMINATE STRUCTURES

Occasionally more complicated structures are encountered, such as those with members that are not perpendicular, those with distributed loading not perpendicular to the member, or those in which externally applied forces or couples act at joints. In these a much more careful analytic procedure is necessary. Example 4.2.4 illustrates all of these.

EXAMPLE 4.2.4

Draw the shear and moment diagrams for the frame.

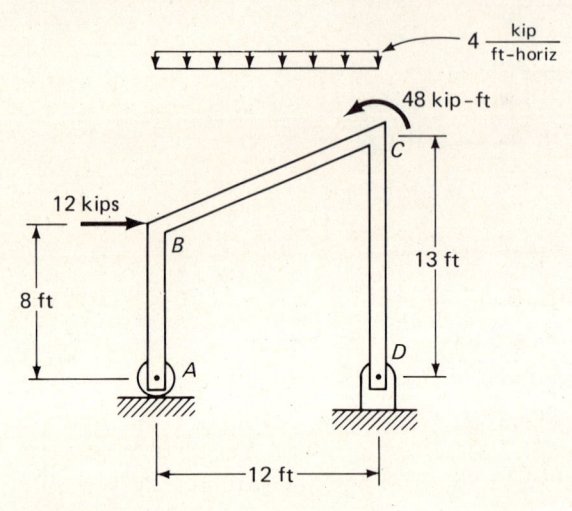

Figure 4.25

In addition to the 48-kip-ft couple and the 12-kip force shown, there is a 4-kip/ft downward distributed force shown schematically above BC. This distributed force should be interpreted as acting on member BC, and its

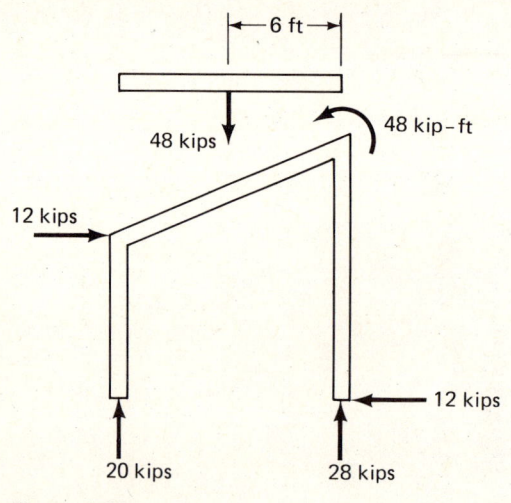

Figure 4.26

intensity of 4 kip/ft should be interpreted as a downward force per unit *horizontal* dimension.

The support reactions are found in the usual way and are shown on a free-body diagram of the frame.

First we consider AB, taking a section a distance x above A, drawing the free body diagram and writing the equilibrium equation to get

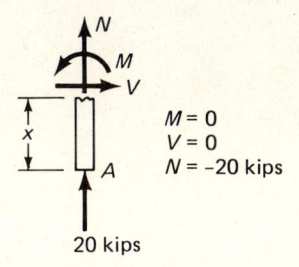

$M = 0$
$V = 0$
$N = -20$ kips

Figure 4.27

$$M = 0$$

$$V = 0$$

$$N = -20 \text{ kips}$$

for all locations between A and B.

Next we examine member BC, taking a section a distance x *measured along the member* from B. We draw a free body diagram for the left hand part of the structure.

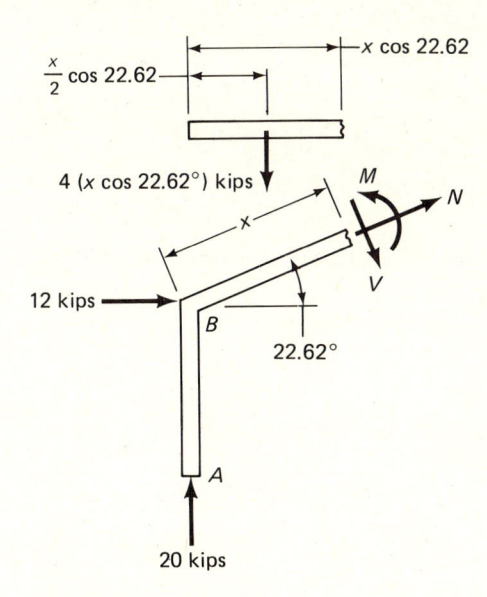

Figure 4.28

We avoid simultaneous equations by summing forces in directions perpendicular to and parallel to member BC.

$+\searrow \sum F = 20 \cos 22.62° - 12 \sin 22.62° - 4(x \cos 22.62°) \cos 22.62° - V = 0$

$V = -3.41x + 13.85$ kips

$+\nearrow \sum F = 20 \sin 22.62° + 12 \cos 22.62° - 4(x \cos 22.62°) \sin 22.62° + N = 0$

$N = 1.42x - 18.77$ kips

Then we sum moments about the sectioned end.

$+\circlearrowleft \sum M_{cut} = M + 4(x \cos 22.62)\left(\dfrac{x}{2} \cos 22.62\right) + 12x \sin 22.62°$

$\qquad - 20x \cos 22.62° = 0$

$M = -1.704x^2 + 13.85x$ kip-ft $\quad (x-\text{ft})$

Finally we consider CD. For this member, evaluation of M and V is easier if a section is taken a distance x measured upward from D, and a free body diagram drawn for the part below the section. Thus

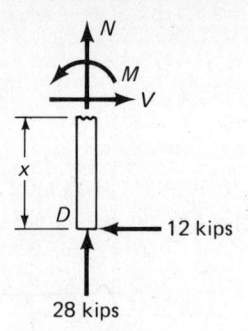

Figure 4.29

From equilibrium

$V = 12$ kips

$M = 12x$ kip-ft

$N = -28$ kips

The shear and moment diagrams are sketched for the complete frame. They are

Figure 4.30

The peak value in BC, M = 28.1 kip-ft was found by determining the value of x where V = 0 (x = 4.06 ft) and evaluating M at that location.

It is recommended that a table be used to record the part, the origin and range of variable, and the expression for M. Such a table follows:

Table 4.1

Member	Origin	Range of Variable (ft)	M (Kip-ft)
AB	A	0–8	0
BC	B	0–13	$-1.704x^2 + 13.85x$
CD	D	0–13	$12x$

PROBLEMS

Sequenced Problem 1, parts (a), (b), (c), (d), (e), and (f).
Sequenced Problem 2, parts (a), (b), (c), (d), and (e).
Sequenced Problem 3, parts (a), (b), and (d).
Sequenced Problem 4, parts (a), (b), and (d).

4.1

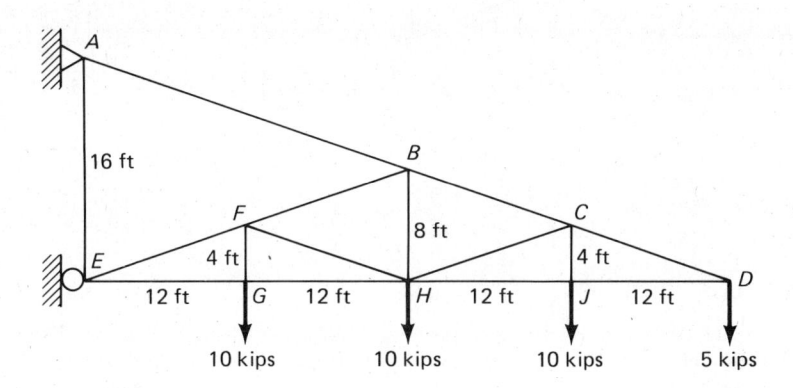

Figure P4.1

Determine the forces in the following members, indicating tension (+) or compression (−).

(a) BC, CH, and HJ.
(b) AB, BF, FH, and GH.
(c) FG, BH, CJ, and AE.

4.2

Determine all the member forces, indicating tension (+) or compression (−).

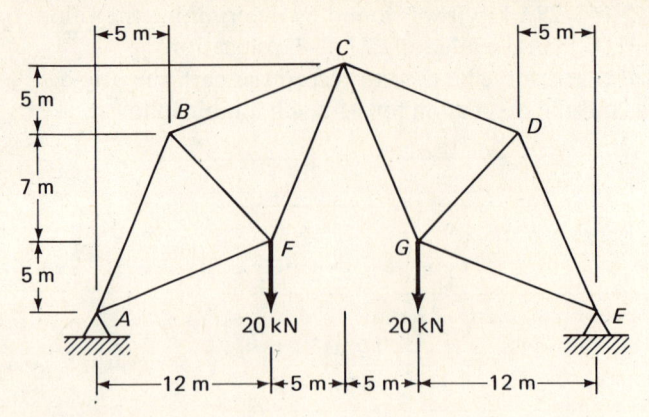

Figure P4.2

4.3

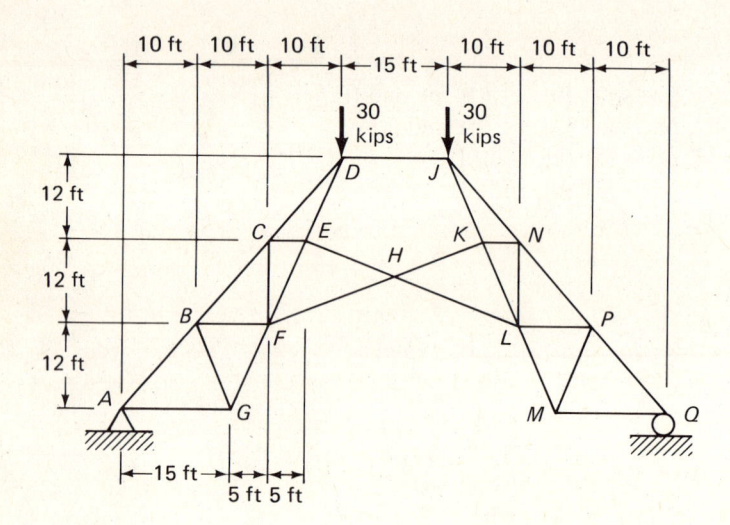

Figure P4.3

The truss is symmetric. Determine the following member forces, indicating tension (+) or compression (−).
(a) *DJ*, *EH*, and *FH*.
(b) *BC*, *CF*, and *EF*.
(c) *AG*, *AB*, *FG*, and *BG*.

4.4

(a) Draw the shear and moment diagrams for *ABCD*.
(b) Write expressions for the shear and the moment in *ABCD*. Write separate expressions for each region, as necessary, indicating the region and the range of variables for each expression.

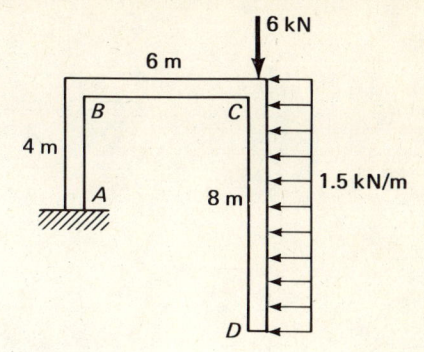

Figure P4.4

4.5

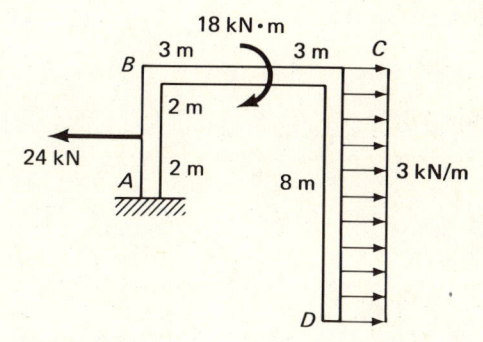

Figure P4.5

(a) Draw the shear and moment diagrams for *ABCD*.
(b) Write expressions for the shear and the moment in *ABCD*. Write separate expressions for each region, as necessary, indicating the region and the range of variables for each expression.

4.6

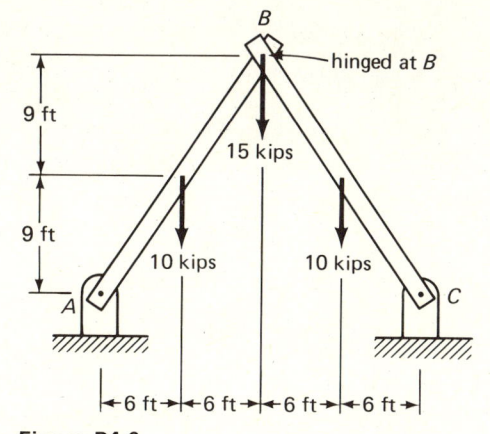

Figure P4.6

Draw the shear and moment diagrams for *ABC*.

4.7

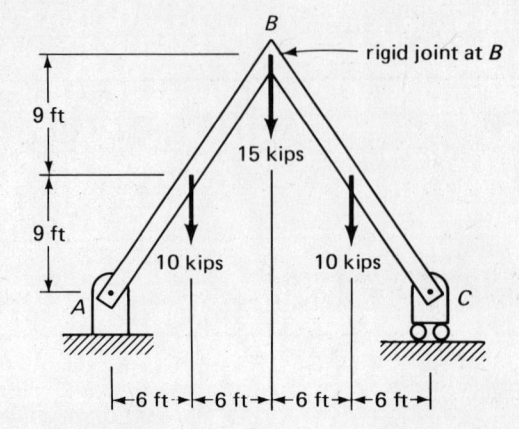

Figure P4.7

Draw the shear and moment diagrams for ABC.

Structural Loads

As he stepped off the scale he read aloud from the "Weight and Fortune" card it had dispensed: "You are resourceful and intelligent. You will make many valuable contributions to society."

His wife grabbed the card from him. "Let me see that!" she said. "Aha! Look here! See? It's got the weight wrong, too!"

5.0 INTRODUCTION

A structure must support its own weight and withstand the forces from its use and environment. Once the structure has been designed, its weight can be calculated with accuracy, almost as if it were placed on a scale. Use and environmental loads related to future conditions are more elusive.

The complexities of load analysis are subtle. It is not enough to assume huge, perhaps unrealistic loads just to be "on the safe side." Resulting structures would be excessive and expensive. Nor is it enough to presume that the most severe loading condition occurs when each member is loaded with all of its possible loads, all at their maximum. Strangely, a combination of loads in which some members or parts of members are not loaded at all may be the worst loading for one member, while a different combination may be the worst for another. We must, therefore, explore a nearly limitless number of load combinations; indeed, we may need to do so for each member. Thus at first glance, load analysis may seem such an immense task and so uncertain that you might believe you could do as well with a good guess or even by reading the load from a "Weight and Fortune" card.

Our knowledge of weather, earthquakes, static and dynamic use loads, and so on, although far from complete, is extensive and increasing. Furthermore, by using influence lines, to be examined in Section 5.2, the worst of the many possible loading arrangements can be determined easily. Thus load analysis need not overwhelm you.

5.1 LOAD CLASSIFICATION AND DETERMINATION

There are many approaches to load classification. Two will be presented in this section: classification by geometry and classification by load source.

Classification by geometry has three members: concentrated forces, line loads, and surface loads.

A force applied over an area small in extent compared to the member or structure may be considered a *concentrated force* and assumed to act at a point. Examples are cable tension forces, individual vehicle wheel loads on a road surface, and column forces.

A force applied over a long, narrow area such that only the long dimension is significant compared to the member or structure may be considered a *line load* and assumed to be distributed along the long dimension. Line loads have units of force per length and often vary with position along the line. Distributed forces on beams and the weight of a curtain wall on a floor slab are two examples.

A force applied over an area of significant size in both directions is a *surface load*. This type of load has units of force per area (pressure) and may vary with position. Examples are water pressure on a dam face and floor loads from stored material in a warehouse.

The second classification system has eight members based on load source. They are dead load; building occupany and use live loads; bridge live loads; snow, ice, and rain loads; wind loads; earthquake loads; earth pressure loads; and shrinkage, settlement, and misalignment loads. The following will show you how these different loads are estimated and how they are treated in some of the codes. A lot of detail has been omitted for brevity, so you are cautioned not to use this section as the basis of a real design. Instead, use the complete provisions of the applicable local codes. If there are none, take guidance from one of the recognized national codes such as ANSI[1] or UBC[2].

5.1.1 Dead Load

This is the weight of the structure itself. Its most prominent feature is its permanency. Dead load does not change in magnitude, location, or direction. It includes the weight of the structural members and may include nonstructural items that are permanently fastened to the structure. Thus heating ducts, lighting fixtures, and curtain walls often are included in the dead load.

5.1.2 Building Use and Occupancy Live Loads

These relate to the planned use of the building. They are movable loads,[3] that is, changeable in location within the structure. Minimum occupancy loads for different uses are listed in most building codes, usually as surface loads acting on floor areas, but occasionally as concentrated forces, or as concentrated forces in combination with surface loads. Several are listed in Table 5.1.

Many codes permit the calculated occupancy live load to be reduced by 0.08% for each square foot of live load surface supported if that live load surface exceeds 150 ft^2. Underlying the reduction is the low probability that such an extensive area would be completely loaded. The codes do not permit live load reductions for places of public assembly, live loads greater than 100 lb/ft^2, roof loads, or storage facilities.

Table 5.1

Use	Uniform Load (lb/ft^2)	Concentrated Load* (lb)
Auditorium seating areas		
Fixed	50	0
Movable	100	0
Library reading rooms	60	1000
Library stacks	125	1500
Offices	50	2000
School classrooms	40	1000
Storage warehouses, light	125	
Storage warehouses, heavy	250	

*Concentrated load may be considered as acting on a surface 2½ ft^2 wherever on an otherwise unloaded floor it produces stresses greater than those caused by the uniform load. From E. H. Gaylord and C. N. Gaylord, *Structural Engineering Handbook*, 2nd ed., McGraw-Hill, New York, 1979.

---

[1] "Building Code Requirements for Minimum Design Loads in Buildings and Other Structures," American National Standards Institute, New York, ANSI A58.1—1972.
[2] "Uniform Building Code," International Conference of Building Officials, Whittier, CA, 1979.
[3] Movable loads are presumed to be relocatable gradually. There is no implication of dynamic load effects.

The codes also place upper bounds on the permitted reduction. Typically, reductions may be:

(1) Not more than 40% for horizontal members or for vertical members supporting the live load for a single floor
(2) Not more than 60% for vertical members supporting the live load from more than one floor
(3) Not more than 23.1 $(1 + D/L)$%, where D and L are the live loads supported by the member

5.1.3 Bridge Live Loads

Bridge live loads are of two types: highway and railroad. Highway bridge live loads are specified in the AASHTO specification.[4] In it, design loads are associated with five different standard trucks. The selection of which one to design for depends on the intended use of the bridge, for example, as part of an interstate highway as contrasted with one for a rural road.

There are three H(M) and two HS(MS) loadings, with H(M) representing two-axle trucks and HS(MS) representing semitractors with trailers. We illustrate these with the one designated HS20-44(MS18), applicable to interstate highway bridges. Each wheel carries half of its axle load, with a wheel spacing of 6 ft (1.8 m).

A bridge designed to meet this standard loading will use as design load the axle loads prescribed for a *single* truck placed where it will cause the most severe loading, and with its trailer length selected to maximize that severity.

More than one truck could be on the bridge at one time, but the probability of many such trucks in a row is small. More likely, one truck will be interspersed with other vehicular traffic. The AASHTO specification handles lines of traffic by specifying an equivalent line load w and a single force P for each classification. For the HS20-44(MS18), P is taken as 18 kips when evaluating moments and 26 kips when evaluating shears, while w is 0.64 kip/ft for both.

To determine the most severe loading condition, segments of the line load w are

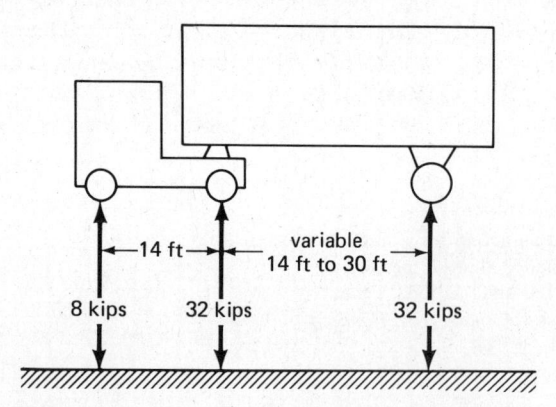

Figure 5.1

---

[4]"Standard Specification for Highway Bridges," 12th ed., The American Association of State Highway and Transportation Officials, Washington, DC, 1977.

placed on parts of a bridge traffic lane selected to cause the most severe stresses. Then the single concentrated force *P* is added and placed for the greatest increase in severity. These steps are taken for each traffic lane for multilane bridges, taking into account any allowed live load reductions.

The worst loading is the more severe of the two cases examined: the single truck versus the line load and concentrated force combination.

Live load reductions may be used for multiple lane bridges. They are based on probability theory and assume that it is unlikely that a multiple lane bridge will be loaded bumper to bumper in all lanes with such heavy trucks. AASHTO permits a 10% reduction for three lanes and a 25% reduction for four or more lanes.

Bridge traffic is not only movable but moving, so that dynamic and impact effects must be considered for every part of a structure likely to be affected by impact loading: superstructures, legs of rigid frames, parts of structure down to the main foundation, and above-ground parts of steel and concrete piles.

The increase, expressed as a fraction of the live load, is determined using the formula

$$I = \frac{50}{l + 125}$$

in which

I = fractional multiplier of live load

l = length (feet) of the part of the span that is loaded to produce maximum loading for the member

The calculated live load increase is the live load multiplied by I.

The AASHTO specification has much more detail and other considerations than could be included in a textbook. Highway bridge designers will find the AASHTO specification an essential reference.

Railroad bridge live loads, similar to highway bridge live loads, are published in the AREA[5] specification. The loadings correspond to a train with two locomotives pulling a string of freight cars. Known by two different names, Cooper loadings and E loadings, they consist of 18 axle loads (9 per engine) followed by a line load representing the freight cars. Relative values for the individual axle loads and the line load are identical for all the Cooper loadings, and the axle spacings are the same for all. A number following the E, such as E40, represents the axle load in kips at each of the ten drive wheels. The E80 loading is shown below. It is typical of contemporary railroad equipment.

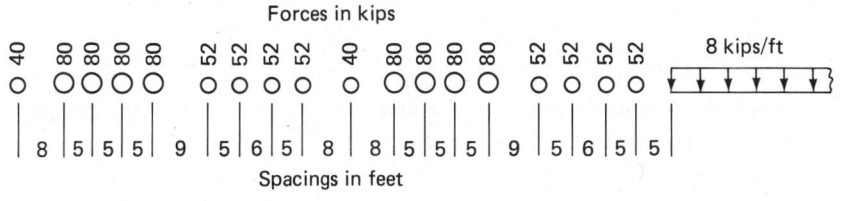

Figure 5.2

[5]*Manual for Railway Engineering*, American Railway Engineering Association. 59 E. Van Buren St., Chicago, Illinois 60605, 1979.

The loading is applied at that location along the bridge which produces the most severe stresses for the member under consideration. All 18 axle loads must be used, spaced as prescribed. The line load must be continuous and connected to the engines, and of a length that maximizes the severity of loading.

The AREA code has extensive provisions for dynamic loading, for live load reductions, and for lateral and longitudinal forces. These will not be treated in this book.

5.1.4 Snow, Ice, and Rain Live Loads

Snow loads are given in most local codes. They reflect local conditions and phenomena and, therefore, are likely to be reliable. Other codes, such as ANSI A58.1,[6] may be used if there is no local code. The ANSI code has maps of snow loading in the United States for 25-, 50-, and 100-year mean recurrences. Most buildings are designed for 50-year recurrence loads, but for very high or very low potentials of hazard to life and property, ANSI recommends use of the 100-year and 25-year mean recurrence loads, respectively. A 50-year mean recurrence implies that there is an average interval of 50 years between such snow accumulations. But it is only an average interval; so there could be a much shorter period between any two individual occurrences, perhaps only a few weeks.

Snow loads are calculated by applying the 25-, 50-, or 100-year recurrence surface load for the region, multiplied by a constant, C_S, as a uniform surface load on the horizontal projected area of the roof. The constant C_S, with a basic value of 0.8, is adjusted by ANSI to account for such things as

(1) Nonuniform accumulations on pitched, curved, and multiple series roofs
(2) Extra accumulations sliding from adjacent higher roofs
(3) Accentuated drifting adjacent to upward projections, such as parapets, penthouses, or adjacent buildings
(4) Slide-off from pitched roofs
(5) Wind removal of snow

As an example of the adjustments and application of C_S, consider a simple gabled roof with slopes of 35° on a building 25 ft tall to be located in a region with a 50-year mean recurrence surface snow load of 25 lb/ft². The building will occupy a site 100 ft from a building 45 ft tall.

For gabled roofs with slopes greater than 20° ANSI requires evaluations of two different C_S. One is used for uniform loading over the entire roof:

$$C_S = 0.8 - \frac{\lambda - 30}{50}$$

where λ is the slope in degrees.

The other C_S evaluation is used for a uniform loading applied to just one of the sloping surfaces while the other remains unloaded:

$$C_S = 1.25 \left(0.8 - \frac{\lambda - 30}{50} \right)$$

---

[6]"Building Code Requirements for Minimum Design Loads in Buildings and Other Structures," American National Standards Institute, New York, ANSI A58, 1—1972.

In our example

$$C_S = 0.8 - \frac{35-30}{50} = 0.7$$

over the entire roof and

$$C_S = 1.25\left(0.8 - \frac{35-30}{50}\right) = 0.875$$

over one sloping surface. The surface loadings are

$$w = 25(0.7) = 17.5 \text{ lb/ft}^2$$

over the entire roof and

$$w = 25(0.875) = 21.9 \text{ lb/ft}^2$$

acting on one slope only.

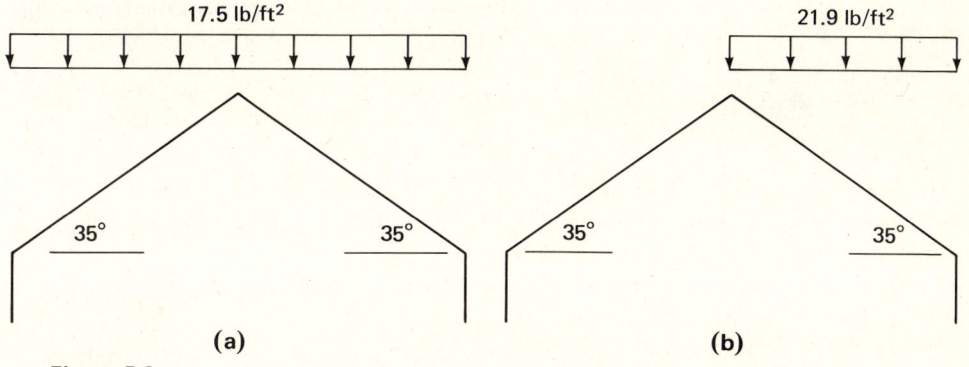

Figure 5.3

The structure must be designed to accommodate both loading conditions.

If the 45-ft adjacent building were not there, or if it were more than 200 ft away (10 times the height differential), the ANSI code would reduce both values of C_S by 25%, in effect reducing the loading on the entire roof from 17.5 to 13.1 lb/ft² and the one-slope loading from 21.9 to 16.4 lb/ft².

Ice loads can be significant. Mariners and aviators understand this. Ships have capsized from heavy accumulations of topside ice; airplanes have crashed from the deterioration of the lift of ice-altered air foils concomitant with the added weight of the ice itself. In some regions there are frequent accumulations of "blue" ice up to 2 in thick and with a density near that of water.

Ice accumulations may alter wind lift and drag coefficients for any structure. Usually the changes are insignificant, but for light truss structures and cable or wire structures they may be important. They should be regarded carefully for tall towers such as those used in radio transmitters. The phenomenon known as electric transmission line gallop is attributed to changes in the lift coefficient from ice buildup.

Rain loads rarely are an important consideration. Occasionally clogged roof drains on flat roofs lead to excess loads. This alone is reason for not designing a flat roof. Designers should be alert to "ponding" where a flat roof, lacking adequate stiffness and being a little slow to drain, sags under the weight of accumulated water.

This permits more water accumulation, which leads to more sag, and so on. The AISC code[7] addresses ponding by prescribing a minimum stiffness for roof structures.

5.1.5 Wind Loads

Wind loads can be very important. They have caused many serious structural failures, including the spectacular collapse of the Tacoma Narrows Bridge. At best, wind load analysis is inexact, replete with subtleties and uncertainties. Helpful references include the ASCE report,[8] the ANSI code,[9] and a paper by Vellozzi and Cohen.[10] There are many more.

The ASCE report, in discussing many subtleties, alerts its readers to potential wind-related dangers. Also, because it is a principal source of the material on wind loads in various codes, it gives an excellent background for understanding and applying those codes.

Wind forces, induced by a moving fluid, are inherently dynamic forces. Yet, with few exceptions, they are treated as static forces. In most cases this is satisfactory, but if the structure

(1) Has a relatively low stiffness, or
(2) Has combinations of cross-sectional shapes that are likely to generate *and respond* to a Karmen vortex street, or
(3) Is likely to have simultaneous torsional and bending oscillations,

then a more detailed dynamic analysis should be made, possibly including wind tunnel studies.

The ANSI code does include an appendix on gust effects using a dynamic analysis, but the appendix is not officially part of the code.

The ANSI code gives contour maps of 25-, 50-, and 100-year mean recurrence wind speeds for the United States. They are for winds at an elevation of 30 ft, where small surface variations in otherwise open country cause only negligible effects in wind speed. They are for the "fastest mile of wind," which means it is the average speed in a mile of moving air and ignores gusts or short-distance variations. The maps ignore winds from tornadoes; structures generally are not designed to survive them. Some local codes may have provisions for tornadoes, but these generally relate to minimizing personal injury from flying objects, for example, rather than to prescribing structural strength. A structure designed to survive a tornado would be very expensive.

The ANSI code equates wind pressure to the dynamic pressure found by applying appropriate terms of Bernoulli's equation so that

$$q = 0.00256 V^2$$

where q is the dynamic pressure in lb/ft^2 when V is the wind speed in mi/h. Since wind

---

[7] *Manual for Steel Construction*, 8th ed., American Institute of Steel Construction, Chicago, 1980.
[8] "Wind Forces on Structures," Final Report of the Task Committee on Wind Forces of the Committee on Loads and Stresses, Structural Division, ASCE; *Transactions of the American Society of Civil Engineers*, pt. II, vol. 126, pp. 1124–1198, 1961.
[9] "Building Code Requirements for Minimum Design Loads in Buildings and Other Structures," American National Standards Institute, New York, ANSI A58.1—1972.
[10] J. W. Vellozzi and E. Cohen, "Gust Response Factors," *Journal of the Structural Division, American Society of Civil Engineers*, vol. 94, no. ST6, pp. 1295–1313, June 1968.

speed varies with both elevation and exposure, it is necessary to make adjustments in the velocity pressure. The ANSI code provides tables of *effective velocity pressure* q_F for "ordinary buildings and structures," which include these adjustments and, for convenience, make an additional adjustment for the structure's dynamic response to gusts.

There are separate tables for each of three classes of exposure:

(1) Centers of cities or very rough or hilly terrain
(2) Suburban areas, towns, city outskirts, wooded areas, or rolling hills
(3) Flat open country, flat coastal belts, or grasslands

Values of q_F are found for the site by entering the applicable table with the region's basic wind speed for the 25-, 50-, or 100-year mean recurrence wind and reading q_F for the desired elevation.

There are two similar sets of tables for effective velocity pressures. One is for q_P, the pressure acting on *parts* (such as girts or purlins) or *portions* of the structure that have wind force tributary areas less than 200 ft^2. The other is for q_M, the pressure assumed to act on internal surfaces at the given elevation.

Each q (q_P, q_F, q_M) is multiplied by a coefficient to give the effective pressure acting on a surface area. A coefficient C_p is used to multiply both q_P and q_F to determine effective external pressures, while C_{pi} is used to multiply q_M to find effective internal pressure. ANSI gives values of C_p for many cases; a few listed in Table 5.2.

Table 5.2

Description	C_p
Windward exterior wall	0.8
Leeward exterior wall with height/width >2.5	-0.6
Side walls	-0.7
Flat roofs with	
wall height/least width <2.5	-0.7
wall height/least width >2.5	-0.8

The negative sign is interpreted as a "suction." ANSI lists many more C_p, some of which are complicated functions or tables such as those for sloping or arched roofs.

Values of C_{pi}, also given in ANSI, depend on the ratio of open to closed surface areas, and the relative direction of the wind. C_{pi} may be $+$ or $-$, with $-$ indicating a suction on an inside surface. The negative values are associated with openings on side walls (walls parallel to the wind) and leeward walls.

There are several additional values for C_p given for special small portions of the structure, such as at corners of walls, ridges, and eaves of roofs. For example, $C_p = -2.4$ for the part of a sloping roof from its windward edge to a location one-tenth of the distance to the ridge. These are treated as independent wind loads, not acting in concert with other wind loads.

For other structures, such as signs, water tanks, chimneys, or guy wires, ANSI gives coefficients C_f, which multiply q_F and are assumed to act on the projected area of the surface or member.

5.1.6 Earthquake Loads

An earthquake is a sudden local motion of the ground. It may be horizontal or vertical and may have amplitudes of several inches. Newmark[11] cites records from the 1940 El Centro, California, earthquake that indicate a peak-to-peak ground motion of more than 1 ft, a maximum speed of nearly 14 in/s, and a maximum acceleration of about 10 ft/s^2, or 0.3g. A perfectly rigid building with sufficient strength would respond with matching motions. But buildings are not perfectly rigid and may lack the strength to remain whole during these violent motions. Buildings deform instead, sometimes elastically, sometimes plastically, and sometimes catastrophically.

Most codes consider only the *horizontal* ground motions. The vertical components are not less violent, but they are less destructive because typical design and construction practices give structures a greater reserve strength for their vertical loads than for their horizontal ones. Although earthquakes are an obvious dynamic phenomenon, most codes treat them by using equivalent static forces. Depending on the specific code, the total horizontal shear force on the structure is given as

$$V = ZKCW \quad \text{or} \quad V = ZIKCSW$$

and is distributed as a set of horizontal forces applied at each floor.

Z is a zone factor that depends on the geographic location of the structure. Local codes give the local value of Z, while national codes give maps that show regional values. Values of Z range from 0 to 1, with the latter assigned to regions with a history of damaging earthquakes. There are only a few areas with $Z=0$ and many with $Z=1$. Some of these are not noted for having a history of damaging earthquakes, but have experienced them nevertheless. Two of these are coastal Massachusetts and southeastern Missouri/western Kentucky. In 1811 an earthquake demolished the town of New Madrid, Missouri. Its intensity ranks with those of the more notorious 1906 San Francisco and 1964 Prince William Sound, Alaska, earthquakes.

K is a factor that accounts for the inelastic energy absorption capacity of the structure. Values depend on the type of structural system in use and are assigned according to the description of the structural type listed in the codes. Most framing systems have $K=1.0$, but values can be as big as 3.0 for elevated tanks supported by four or more cross-braced legs that rest directly on the ground. Values of K may be as small as 0.67 for buildings with moment resisting frames that are capable of resisting the entire lateral earthquake force. Box systems that resist the lateral force through shear walls or through axially loaded bracing systems usually have a low capacity for inelastically absorbing energy, bringing an assignment of a larger K, $K=1.33$. Structural systems employing combinations of shear walls and moment resisting space frames as described in the codes may qualify for $K=0.80$. Careful consultation of the codes for correct values of K is important. A valuable detailed discussion of K values is given by Smith.[12]

C is called the seismic coefficient. It considers the structure's vibratory response

---

[11] N. M. Newmark, "Relation Between Wind and Earthquake Response of Tall Buildings," Proceedings of 1966 Illinois Structural Engineering Conference, University of Illinois Engineering Experiment Station, Urbana, February 1969, pp. 137–156.

[12] H. C. Smith, "Lateral-Force Design," in *Structural Steel Designer's Handbook*, F. S. Merritt, Ed., McGraw-Hill, New York, 1972, Sec. 9.

to transient lateral motion at its base and, depending on the specific code, is given by

$$C = 0.05/T^{1/3} \quad \text{or} \quad C = 1/15T^{1/2}$$

where T is the fundamental period of vibration of the structure, in seconds, for the direction of motion under consideration. The codes give several methods for determining T, and prescribe upper bounds for C or combinations of CS.

I is an occupancy importance factor ranging in values from 1 to 1.5. The larger values are for essential facilities such as hospitals, fire and police stations, and disaster relief centers.

S is a site-structure resonance factor. It relates T, the fundamental period of natural vibration of the structure, to T_S, the characteristic period of ground vibration at the site.

The codes using the factor S give formulas for its calculation. S may have values between 1 and 1.5; and in cases where T_S is unknown, exact resonance is assumed, giving $S = 1.5$.

W is the dead load in kips, except for warehouses or other heavy storage facilities for which 25% of the live load is added to the dead load.

The force V is distributed to the different floors in the following way:

$$\text{top floor } F_t = 0.004V \left(\frac{h_n}{D_s}\right)^2 \leq 0.15V$$

where D_s is the horizontal dimension in feet of the lateral force resisting system in the direction of the force (motion) under consideration.

For other floors, such as the xth floor,

$$F_x = \frac{(V - F_t)w_x h_x}{\sum_{i=1}^{n} (w_i h_i)}$$

where w_x and w_i are the parts of W previously described for the xth and ith floors, respectively, and h_x and h_i are the height in feet above the base to floor x and floor i, respectively.

The codes also prescribe lateral forces on parts or portions of buildings or other structures according to

$$F_p = ZC_p W_p$$

where Z is the zone factor described previously, W_p is the weight of the part or portion, and C_p is a horizontal force factor that is assigned according to a description of the part and the direction of the force. For example, ANSI lists $C_p = 1.00$ for cantilever parapets and cantilever walls, except retaining walls with force perpendicular to the plane of the wall. $C_p = 0.20$ for exterior bearing and nonbearing walls, interior bearing walls and partitions, interior nonbearing walls over 10 ft high, and masonry fences over 6 ft high, all loaded with the force perpendicular to the plane of the wall. There are others listed.

Additional requirements pertaining to overturning moment, relative drift between stories, separation between buildings, and provisions for motion of the exterior wall panels will not be addressed in this section.

5.1.7 Earth Pressure Loads

Building codes are not very detailed in treating loads that may be exerted by the soil on basement walls or floors. They require the design of below grade vertical (or nearly so) surfaces to account for the lateral pressure of the adjacent soil, including suitable allowances for surcharge. If below a waterfree surface, this pressure must include the full hydrostatic pressure plus the weight of the soil as diminished by buoyancy. For the uplift (buoyant force) on a basement floor, only the hydrostatic pressure need be considered.

5.1.8 Shrinkage, Settlement, and Misalignment Loads

Forces from the shrinking or swelling of materials or components, the settling of footings on foundations, or the force-fitting of structural components that "don't quite fit" are rarely addressed in codes. Yet because they can and do cause failures, they must be considered whenever there is a potential for them. When they can be predicted during design, there should be provisions for them. For example, the change in length of a bridge, which accompanies the change in temperature between winter and summer, can be estimated. Then the number of expansion joints needed and their range of movement can be determined for use in the bridge.

Unexpected settlements, misfits, and shrinkages can happen during construction, requiring field engineers to respond quickly with well conceived analyses. These may show that no corrective action is necessary, or they may show that some structural damage would result if left uncorrected. Occasionally they show that the remanufacture of a component is necessary, or that additional shoring should be used, or even that some completed work must be razed and redone. Such costly steps are not done casually.

Forces from causes such as misalignments and misfits occur only in statically indeterminate structures, never in statically determinate ones. For this reason their evaluation requires the analysis of a statically indeterminate structure. Techniques for this are developed in later chapters.

5.1.9 Load Combinations

It would seem desirable to design a structure so that it could support the *simultaneous* application of the maximum value of every one of its load sources. The laws of probability do tell us that, *given enough time*, that very severe combination will occur. The codes recognize, however, that the probability is small that such an event will occur *in the life* of the structure. Accordingly, the codes permit some reduction in the total load for certain combinations of source. For example, ANSI[13] permits a 25% reduction in total load resulting from the simultaneous maximum load from combinations of:

---

[13] "Building Code Requirements for Minimum Design Loads in Buildings and Other Structures," American National Standards Institute, New York, A58.1—1972.

dead, live, snow/rain, and wind

or

dead, live, snow/rain, and earthquake

or

dead, live, snow/rain, and thermal

or

dead, wind, and/or thermal

or

dead, earthquake, and thermal

It permits a 33% reduction in total load for the combinations of

dead, live, snow/rain, wind, and thermal

or

dead, live, snow/rain, earthquake, and thermal

No reductions are permitted for other combinations, nor does the code permit increases in allowable stresses as a consequence of these reductions.

5.2 INFLUENCE LINES

We have examined loads from a variety of sources so that they can be used as design forces. The magnitude of force is not our only concern. Position is also important. In a design load a movable force must be considered in the position of its severest effect. If there are several movable forces, their combination must be the severest. Sometimes some forces should be excluded if that will produce a more serious design load.

Imagine that you are designing the two supports for beam ABC that must carry a 4-kN/m downward distributed force that could act over the entire beam or just portions of it.

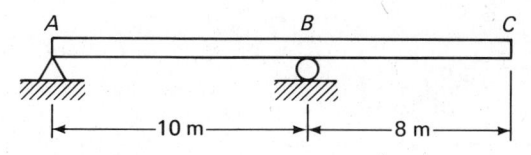

Figure 5.4

If you assume that the worst loading occurs when the whole beam is loaded, you will design support A to carry a compressive 7.2 kN. But support A could be subjected to greater forces if only AB or only BC were loaded: 20 kN compressive for AB, 12.8 kN tensile for BC. Clearly designing support A on the basis of ABC being entirely loaded could lead to its failure, either by exceeding its capacity or by the beam lifting from the knife edge if the potential of a tensile requirement were overlooked.

82 STRUCTURAL LOADS

The situation is different for support B. Its maximum possible compressive force occurs when all of ABC is loaded and no arrangement of downward forces is possible that will produce a tension. The conclusion is important. *"Worst loading" arrangements vary from component to component of a structure and must, therefore, be determined individually.*

If there are several potential failure modes for a single component, each must be examined for its worst load. A beam carrying both shear and moment must be analyzed for the worst loading for each.

In simple structures the worst loading may be obvious or easily found by trial and error, using a few short calculations. In more complex structures that approach could require almost exhaustive calculations and still leave concern that there may be a worse loading arrangement that has been overlooked. In these cases influence lines become important. They show the worst loading scheme almost directly and do so with a minimum of calculation.

Since influence lines are used almost exclusively in connection with gravity-induced live loads, we will limit our work to those for downward directed forces. Other applications are possible and used occasionally. The ideas that follow can be extended to them without difficulty.

Consider beam AB with a unit downward force that can be located anywhere along the beam.

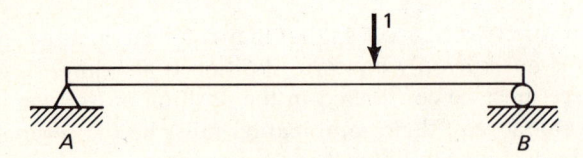

Figure 5.5

First we will develop the influence lines for the support reactions at A and B, that is, we will develop and plot the relations between the support reactions and *the position* of the unit force. From equilibrium,

$$r_A(x) = 1 - \frac{x}{L} \qquad r_B(x) = \frac{x}{L}$$

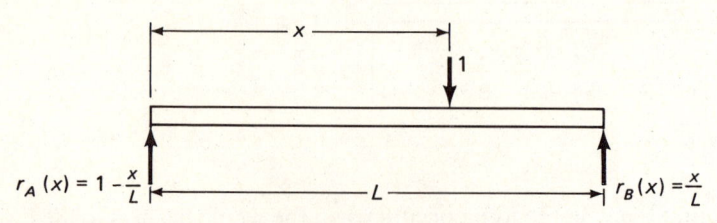

Figure 5.6

It is customary to draw influence lines directly below a diagram of the structure so that the influence line (IL) values are related visually to positions of the unit force along the structure. (Due to page make-up for books that is not possible here.)

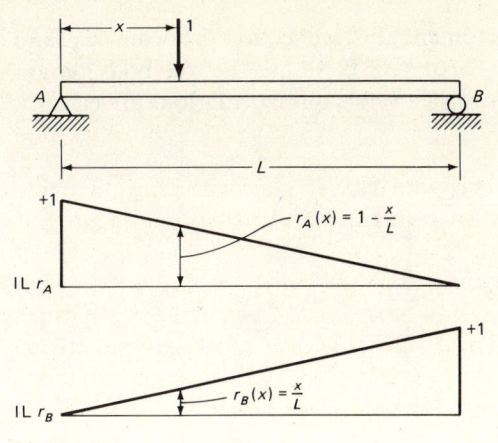

Figure 5.7

IL r_A shows that the support reaction at A has a value of 1 when the unit force is applied at A, and decreases linearly to zero as the unit force moves to B. Notice that the ordinate values are for the support reaction *at A*, and that the *independent variable* is the *position* of the unit force. The interpretation of IL r_B by its analogy with IL r_A should be obvious.

It may seem peculiar, but no dimensions are assigned to the unit force. The force is 1, not 1 kip nor 1 kN, just 1. Influence lines display the relationship between the *position* of the unit force and a response function such as a support reaction. Use of a unit dimensionless force suppresses features dependent on the magnitude of the force or its dimensions, and leaves its position as the sole consideration. When an influence line is used in connection with an applied force having a specific magnitude and dimension, the response is found by multiplying the influence line ordinate at the point of application of the force by the force's magnitude *and dimension*.

Now let us develop the influence line for the shear force and the bending moment at point C located $0.25L$ from A on the same beam. The beam loading is the same as before, the unit force. Clearly the support reactions $r_A(x)$ and $r_B(x)$ are the same.

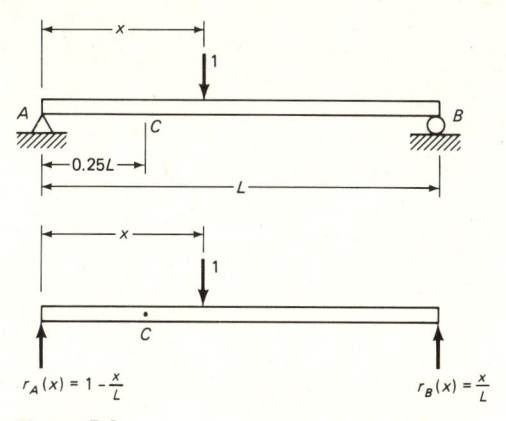

Figure 5.8

84 STRUCTURAL LOADS

To determine the shear and the moment at C, we section the beam at C and draw a free-body diagram of either part AC or CB. If we use the free-body diagram for CB when the unit force acts on AC and vice versa, our calculations are easier.

When the force is on AC

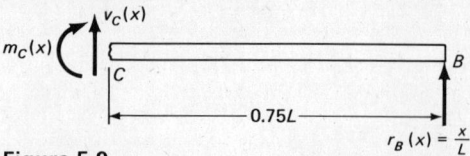

Figure 5.9

From equilibrium,

$$v_C(x) = -r_B(x) = -\frac{x}{L}$$

$$m_C(x) = 0.75L[r_B(x)] = 0.75x$$

for $0 < x < 0.25L$.

When the force is on CB

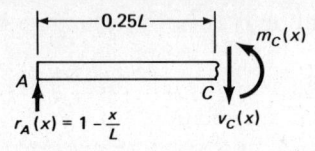

Figure 5.10

Again from equilibrium,

$$v_C(x) = r_A(x) = 1 - \frac{x}{L}$$

$$m_C(x) = 0.25L[r_A(x)] = 0.25L - 0.25x$$

for $0.25L < x < L$.

They are plotted as follows:

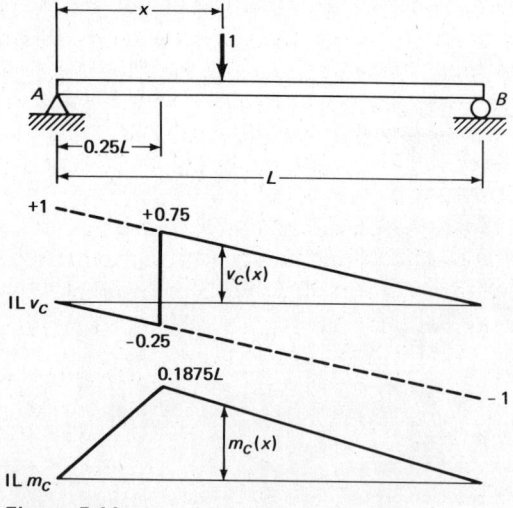

Figure 5.11

5.2 INFLUENCE LINES

IL v_C shows that the shear force at C is zero when the unit force acts at A, and that the shear force at C changes linearly to -0.25 as the unit force is moved to C. The shear force at C shifts abruptly to $+0.75$ and linearly decreases to zero as the unit force is moved to B.

We could write a formal definition of an influence line now, but before doing so, let us make two comparisons that may help you interpret the definition. First, using *traditional shear and moment diagrams*, we determine the shear and the moment at C caused by a unit downward force at the *midpoint* of AB.

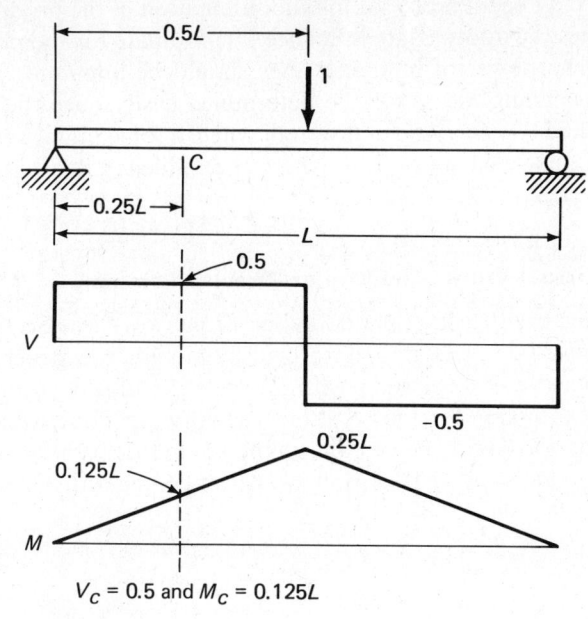

$V_C = 0.5$ and $M_C = 0.125L$

Figure 5.12

Notice that these are the ordinate values of the *influence lines* for v_C and m_C for $x = 0.5L$, the point of application of the unit force. Keep that comparison in mind as you study the following definition: *An influence line for a particular function (for example, a support reaction, beam shear at some point, and so on) is a curve for which the ordinate value at an abscissa point is the value of that function caused by a unit downward force applied to the structure at a position corresponding to that abscissa point.*

Although the use of influence lines will be considered in detail in Section 5.2.3, it may be helpful to glance at their use now.

If beam AB were 12 m long and loaded with a 20-kN force at $x = 8$ m, we could evaluate the resulting support reaction at A and B and the resulting shear and moment at C by taking the ordinate values for their influence lines at $x = 8$ m and multiplying them by 20 kN. For $x = 8$ m and $L = 12$ m, the influence line ordinates are

$r_A = 0.333$

$r_B = 0.667$

$v_C = 0.333$

$m_C = 1.00$ m

so that for a 20-kN force,

$$R_A = (20 \text{ kN})(0.333) = 6.67 \text{ kN}$$
$$R_B = (20 \text{ kN})(0.667) = 13.33 \text{ kN}$$
$$V_C = (20 \text{ kN})(0.333) = 6.67 \text{ kN}$$
$$M_C = (20 \text{ kN})(1.00 \text{ m}) = 20 \text{ kN} \cdot \text{m}$$

If several forces act on the beam simultaneously, the support reactions and the shear and moment at C each may be found by taking the summation of the products of each force individually times the appropriate influence line ordinate for the point of application of the force. The power of influence lines should be apparent. The response to a variety of load arrangements can be determined easily once the influence line has been constructed. Further, the question of where a force might act to produce its greatest effect is answered by finding where the influence line has its largest ordinate.

5.2.1 Influence Lines by Virtual Work: The Müller–Breslau Principle

Influence lines can be generated easily using the method of virtual work (see Section 2.4). Support reactions and internal forces are determined by moving the structure through a virtual displacement that is consistent with all the structural conditions and constraints, writing the expression for the resulting virtual work done, which, upon equating to zero, may be solved for the desired unknown. Let us try it for the support reaction at A of beam AB loaded by a downward unit force a distance x from A.

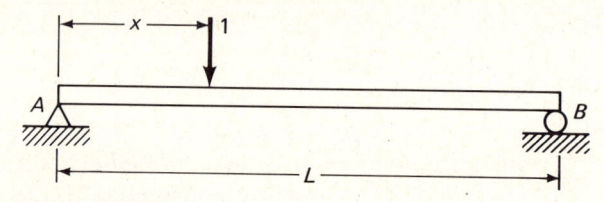

Figure 5.13

The vertical support at A is removed and replaced by the support reaction r_A. The structure is given a vertical displacement so that r_A moves a distance δ.

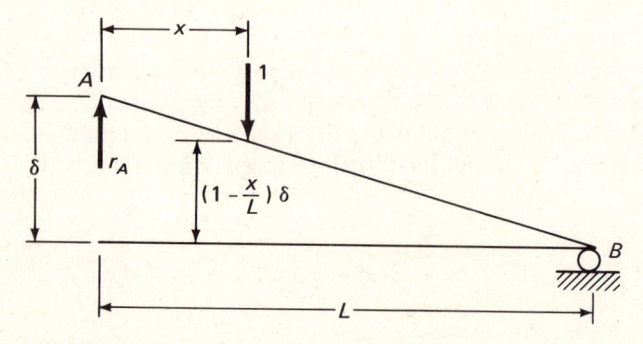

Figure 5.14

5.2 INFLUENCE LINES

The virtual work is

$$\delta Wk = r_A \delta - 1\left(1 - \frac{x}{L}\right)\delta = 0$$

If we were to follow the procedures of Section 2.4, we would divide by δ and solve for r_A. Instead, let us accomplish the same thing by setting $\delta = 1$. The result is

$$1(r_A) = r_A = 1 - \frac{x}{L}$$

matching the previous result. But something more important has been done here: the *structure itself*, when given a *unit displacement* of the support reaction r_A, assumed a geometry that matches IL r_A directly. Compare the beam, displaced a unit distance at A, to IL r_A found earlier. From that observation we state the Müller–Breslau principle.

> The influence line for any response function (support reaction, beam shear at a point, and so on) is identical to the geometry of the displaced structure when the restraint provided by the response function itself is removed and a unit displacement in the direction of the response function is imposed there.

Let us apply the Müller–Breslau principle to get the influence line for the beam shear force at C of beam AB.

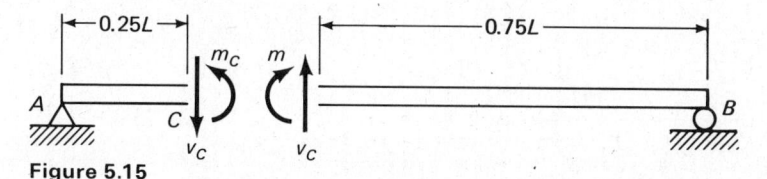

Figure 5.15

Here v_C acts on both pieces. In applying the Müller–Breslau principle, the two v_C must have a total displacement of 1, with the downward directed v_C on AC moving downward and the upward directed v_C on CB moving upward. How that total displacement is shared depends on the requirement that any other forces or couples present (in this case m_C and the two support reactions) can do no net work when making the displacement. Thus we do not permit displacements at the supports, and we require pieces AC and CB to have identical rotations so that the work done by the two equal, opposite m_C will exactly cancel.

Only one displaced shape satisfies these conditions. It is identical to the IL v_C developed in Section 5.2.

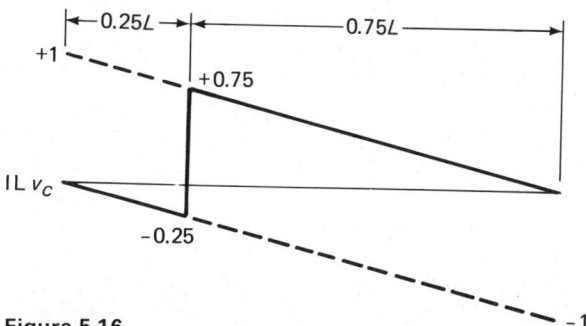

Figure 5.16

88 STRUCTURAL LOADS

The same method can be used to draw the influence lines for the shear at other points of the same beam. All are similar. As an example, IL v_D, where D is $0.7L$ from A, is

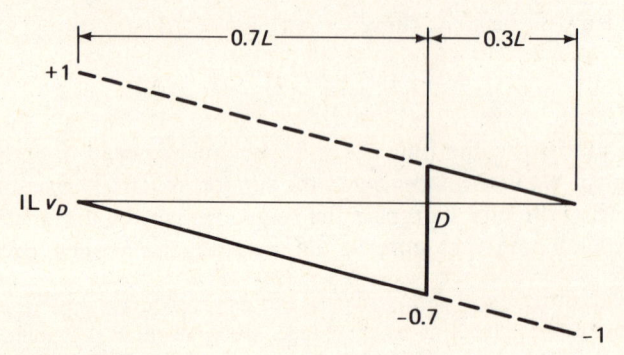

Figure 5.17

Note that in this case the unit jump occurs at D and that the slopes of AD and DB are the same, so that extended they have values of -1 and $+1$ at B and A, respectively. Now let us construct the influence line for the moment at C of the same beam.

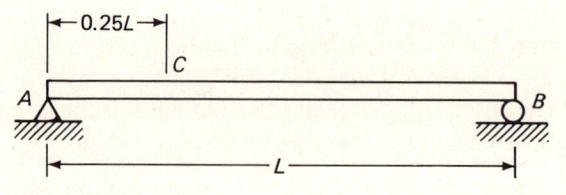

Figure 5.18

We introduce a hinge at C and cause a unit relative rotation between pieces AC and CB. AC rotates counterclockwise θ_A, and BC rotates clockwise θ_B, so that each bending moment couple does positive work when they are assumed to act in directions for the positive bending moment.

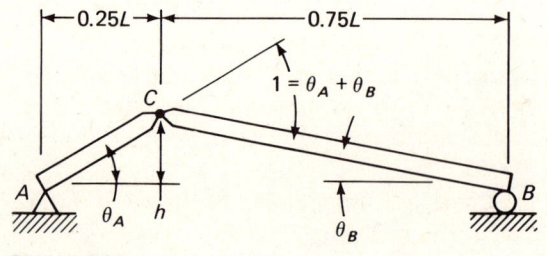

Figure 5.19

No net work is done by stationary supports A and B and no net work is done by the internal shear force at C.

From geometry,

$$\theta_A + \theta_B = 1$$
$$h = 0.25L \cdot \theta_A = 75L \cdot \theta_B$$

Thus $\theta_B = \theta_A/3 = 0.25$ and $h = 0.75L(0.25) = 0.1875L$, giving the influence line that matches the IL m_C found previously.

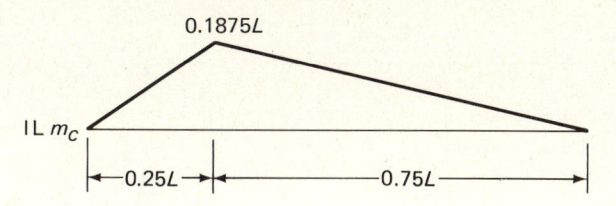

Figure 5.20

You probably interpreted θ_A and θ_B as having radians for dimensions, but that is not necessarily the case. In applying the principle of virtual work, the virtual displacement imposed must be small. If $\theta_A + \theta_B = 1$ meant that $\theta_A + \theta_B = 1$ rad, then the virtual displacement would place a kink of nearly 60° in the beam at C. That's no small displacement. A better interpretation is to presume that θ is sufficiently small that when it is imposed, the resulting motion of any point on the beam is vertical, and its horizontal displacement is negligible. The vertical motions can then be calculated by multiplying an angle by a member length, such as $h = 0.25L \cdot \theta_A$. The values of the influence line ordinates are found by assigning a value of 1 for the appropriate displacement of the member, θ in our example. This deforms the structure to a geometry that matches the influence line, but implies nothing in the sense of a real displacement.

We have looked only at influence lines for a simple beam. Let us look at some influence lines for a more complicated structure.

EXAMPLE 5.2.1

For the structure of Figure 5.21 construct influence lines for the three support reactions, the bending moments at C and G, and the shear forces at G and H.

Each influence line was drawn by applying the Müller–Breslau principle. A release was made of the constraint that corresponds to the function for the particular influence line. The released point was given a unit displacement in the positive direction of the function. The displaced configuration satisfied all remaining constraints and was such that no net work would be done by other functions.

For IL r_A, end A was moved upward through a unit displacement with no displacements made at C and D. The hinge at B made that combination possible.

IL r_C and IL r_D had upward unit displacements at C and D, respectively. Again hinge B permitted these while the beam remained undisplaced at the remaining supports. Notice that these lines have regions with

Figure 5.21

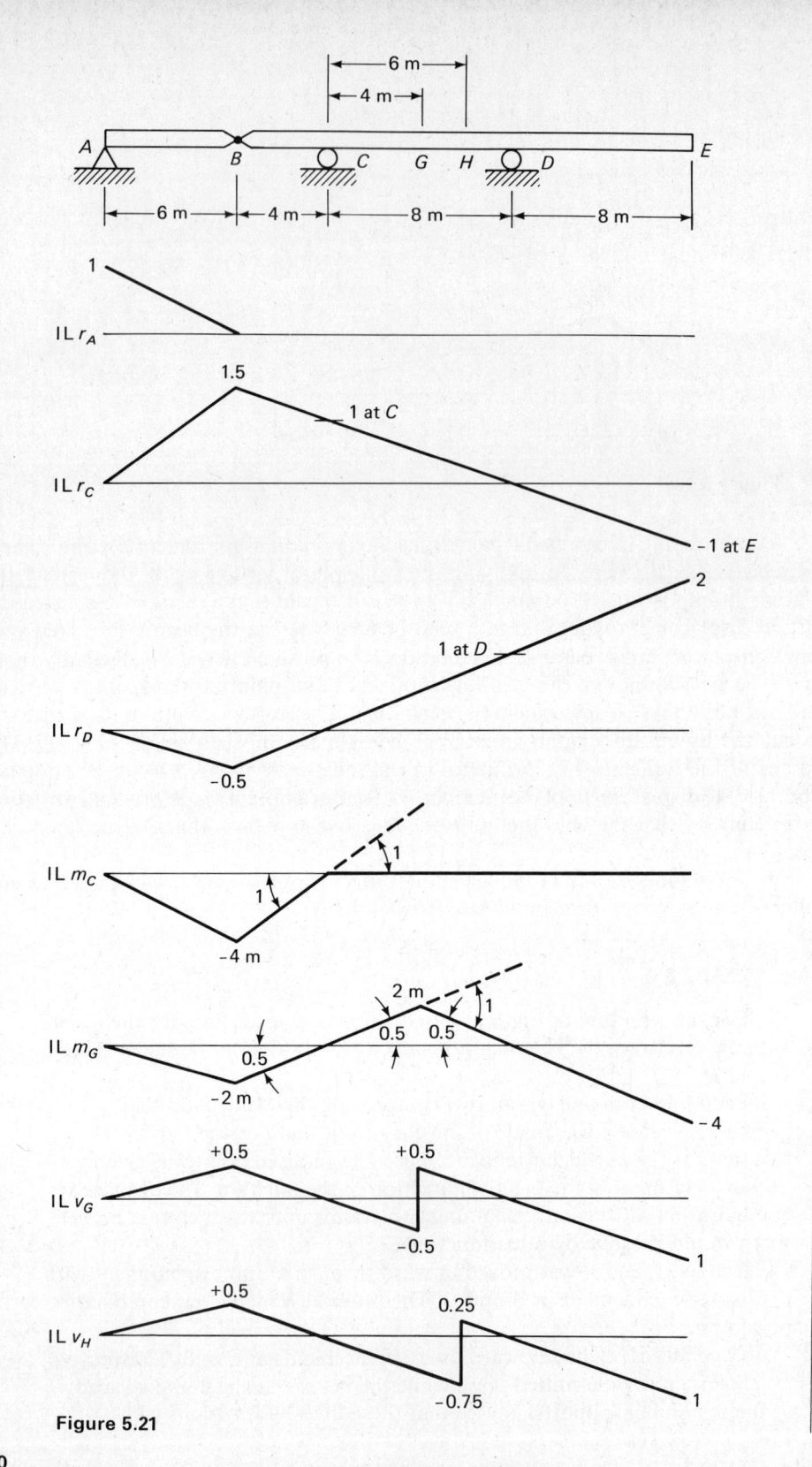

negative ordinates. A unit downward force applied in these regions causes a reversal in the assumed direction of the support reaction.

In IL m_C a hinge was added at C to permit a unit relative rotation of the parts on opposite sides of C. Since supports C and D prevent a rotation of part CD, all of that relative rotation must occur as a rotation of BC. The hinge at B permits this while preserving a zero displacement at A. The ordinate value, -4 m, found at B is the product of the unit rotation of BC about B and the 4-m length of BC.

The parts between C and D for both IL v_G and IL v_H are similar to the influence lines for shear for a simple beam considered earlier. The portions beyond C and D are extensions of those lines across C and D, preserving their slope. Likewise, the part between C and D for IL m_G is similar to that for the influence line for the moment in a simple beam. The lines extend across C and D without a change in slope until a hinge or internal roller permits one.

5.2.2 Influence Lines for Girders and Members of Trusses

In most structures loads are carried to main members through systems of smaller members. In an arrangement often used in buildings, a floor slab rests on beams called joists, which are supported at their ends by floor beams. These, in turn, are carried by main girders, which are supported by columns.

A similar arrangement is used in bridges. The roadway rests on simple beams called stringers, which are supported by floor beams. The floor beams may be framed to the joints of a main truss for their support or framed to the main girders. The trusses or main girders rest on the bridge piers.

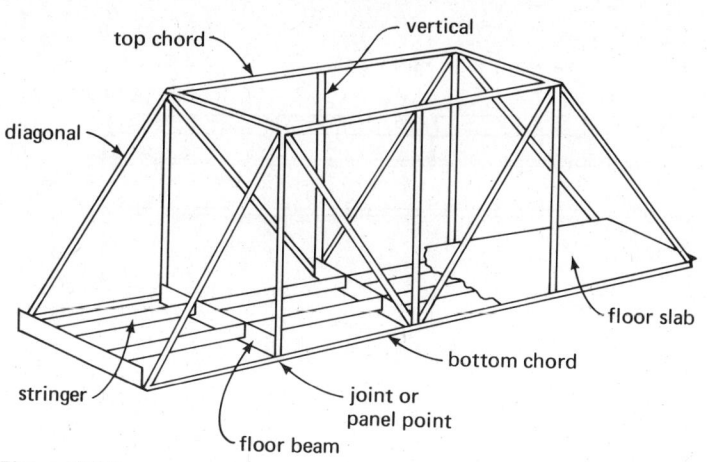

Figure 5.22

Influence lines for the shear and the moment at a point in the main girder are invaluable in determining girder design loads. They may be drawn easily using the Müller–Breslau principle, but because they may seem difficult to interpret at first, we will use an equilibrium approach to start the first example.

EXAMPLE 5.2.2

Draw the influence line for the shear force and the bending moment at point G of the main girder.

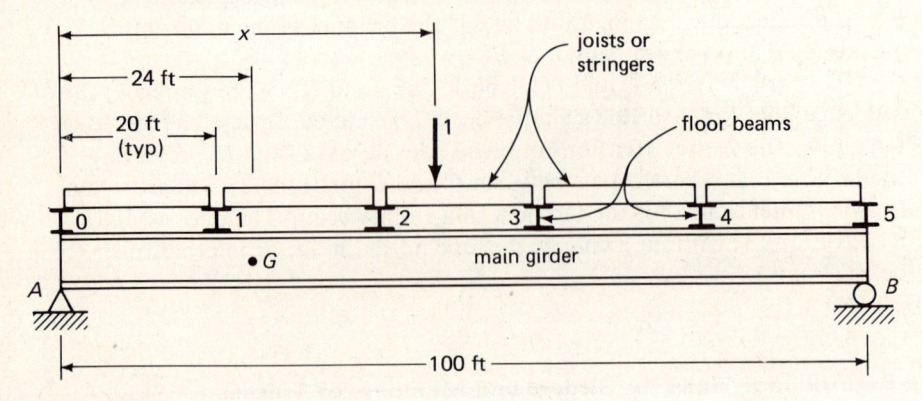

Figure 5.23

The support reactions at A and B are the same as for a simply supported beam,

$$r_A = 1 - \frac{x}{100} \quad \text{and} \quad r_B = \frac{x}{100}$$

When the unit force acts on stringer 0–1, it is convenient to use the free-body diagram of part GB of the girder.

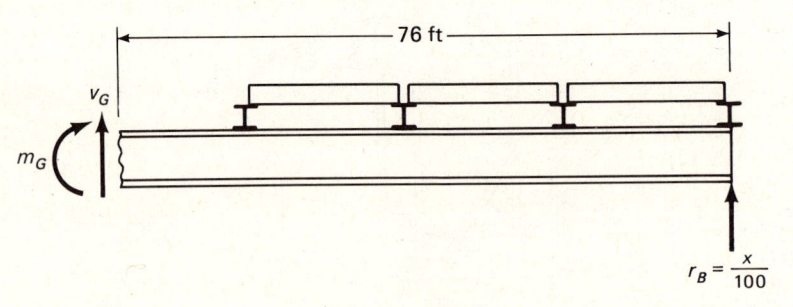

Figure 5.24

From equilibrium,

$$v_G = -r_B = -\frac{x}{100}$$

$$m_G = 76 r_B = \frac{76x}{100} \text{ ft}$$

for $0 < x < 20$ ft.

When the unit force acts on stringers 2–3, 3–4, or 4–5, we use the free-body diagram for AG.

$$r_A = 1 - \frac{x}{100}$$

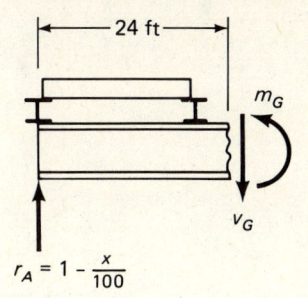

Figure 5.25

$$r_A = -\frac{x}{100}$$

From equilibrium

$$v_G = r_A = 1 - \frac{x}{100}$$

$$m_G = 24r_A = 24 - \frac{24x}{100} \text{ ft}$$

for 40 ft $< x <$ 100 ft.

When the unit force acts on stringer 1–2, a slightly more complicated situation arises which requires consideration of the support reactions of the stringer itself. Its free-body diagram is

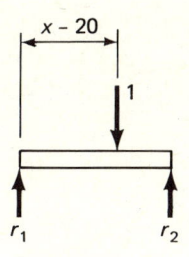

Figure 5.26

We obtain

$$r_1 = 1 - \frac{x - 20}{20} \quad \text{and} \quad r_2 = \frac{x - 20}{20}$$

for 20 ft $< x <$ 40 ft.

The free-body diagram of AG must include r_1.

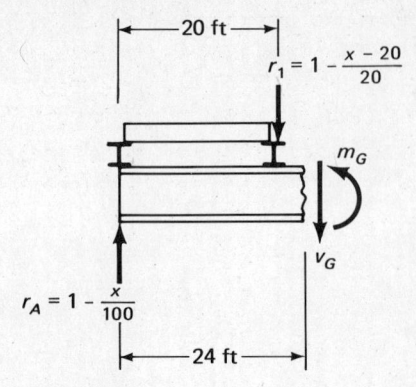

Figure 5.27

From equilibrium,

$$v_G = \frac{4x}{100} - 1$$

$$m_G = 16 - \frac{4x}{100} \quad \text{ft}$$

for 20 ft $< x <$ 40 ft.

Both are linear functions of x, as were the expressions in the other regions. Thus the influence lines plot as continuous linear functions.

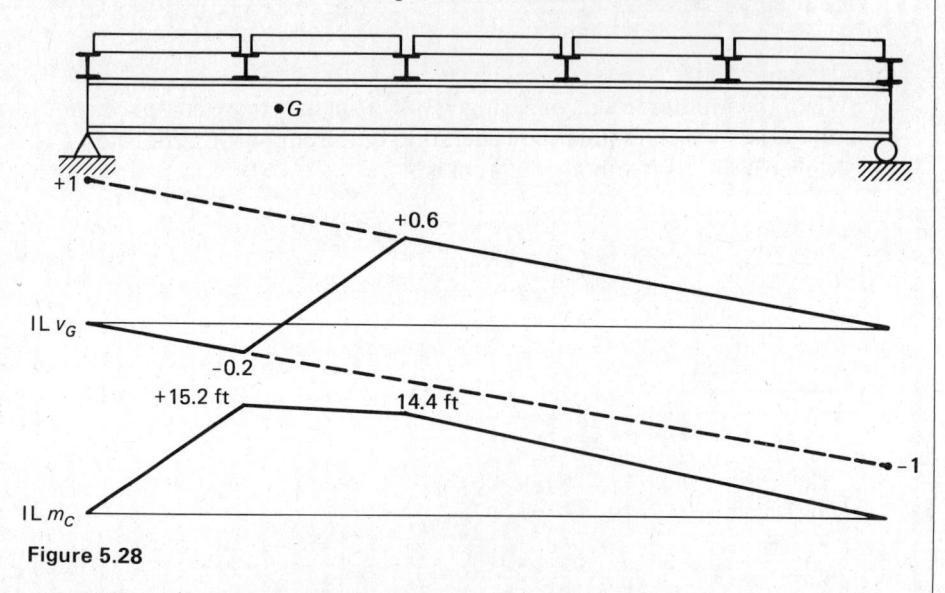

Figure 5.28

EXAMPLE 5.2.3

Develop the same influence lines of Example 5.2.2 using the Müller–Breslau principle.

For IL v_G we sever the girder at G and move the separated parts through a unit relative shear displacement in the directions for the positive shear forces. Of course we must satisfy all the other restrictions: no displacements at the supports and no relative rotation between AG and GB (AG and GB remain parallel) to avoid net work by support reactions or the bending moments at G. The similarity to the use of the Müller–Breslau principle for beam shear described in Section 5.2.1 should be obvious. The stringers, maintaining contact with the girder through the floor beams, are displaced. Since *they* carry the unit force, it is *their* displaced positions that form the influence line.

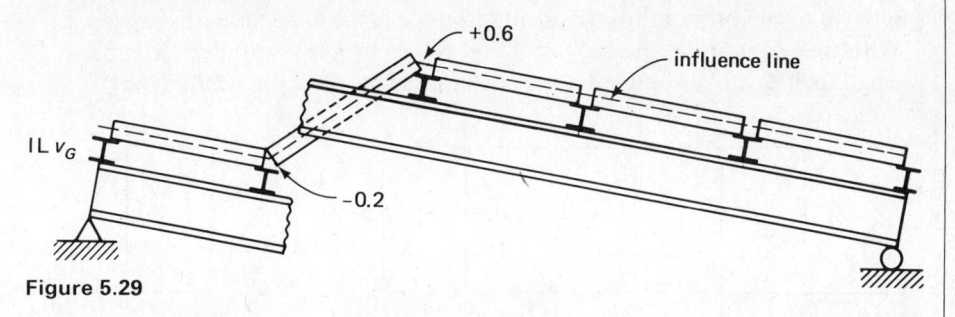

Figure 5.29

A similar sequence is used for IL m_C. A unit relative rotation is introduced at G between parts AG and GB with directions consistent with positive bending moments at G. The supports are undisturbed. The displaced positions of the stringers form IL m_C.

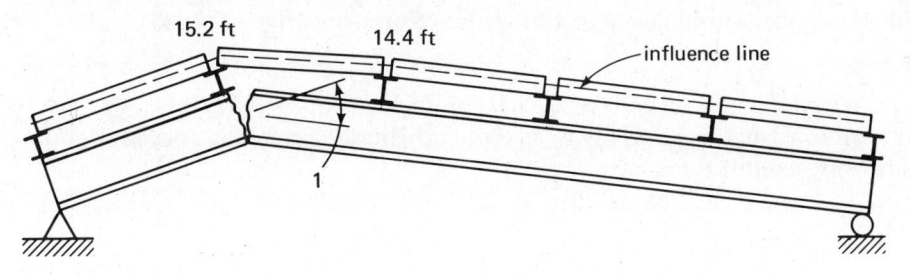

Figure 5.30

The Müller–Breslau principle can be applied to trusses as well as to girders. Unfortunately the geometric constructions usually are more difficult. There are some useful exceptions, as will be illustrated.

EXAMPLE 5.2.4

Determine the influence line for the force in member GQ of the truss shown.

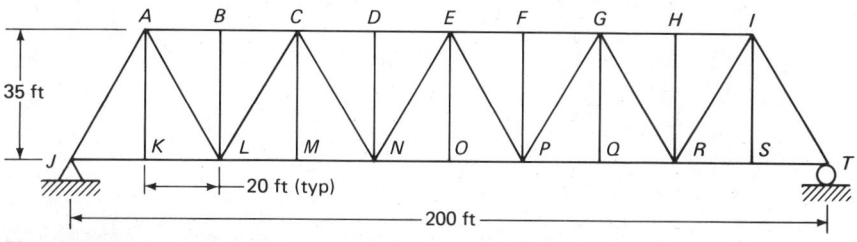

Figure 5.31

96 STRUCTURAL LOADS

The roadway is supported by 20-ft stringers, which rest on floor beams at the joints of the bottom chord.

To apply the Müller–Breslau principle we sever GQ and apply a unit displacement of the severed ends, causing the assumed member tensions to do positive work. In essence member GQ undergoes a unit decrease in length while all the other member lengths remain unchanged. Because the displacements must be small, joints Q and G cannot move horizontally relative to the other joints, requiring triangle GPR to remain unchanged. With that constraint, the only way that GQ can have a unit decrease in length is if Q moves upward a unit distance. The resulting geometry is

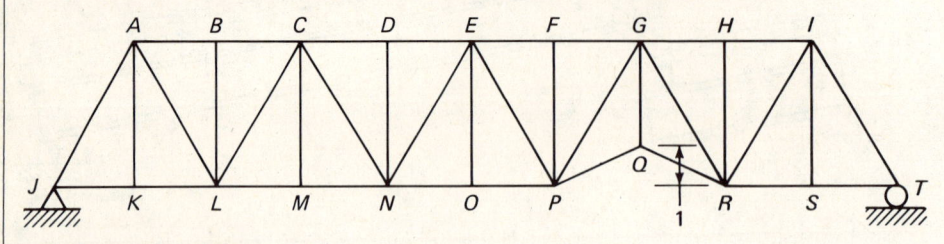

Figure 5.32

The bottom chord, coincident with the stringers, forms the influence line. Evidently it has a value of zero everywhere except from P to R, where it rises linearly from 0 at P to 1 at Q and returns linearly to 0 at R.

The development of influence lines for such members as PQ and PG involves awkward constructions in applying the Müller–Breslau principle. They are developed easily using equilibrium.

EXAMPLE 5.2.5

Construct influence lines for the member forces in PQ and PG of the structure of Example 5.2.4.

The support reactions are the same as for a simple beam,

$$r_J = 1 - \frac{x}{200} \quad \text{and} \quad r_T = \frac{x}{200}$$

A section is taken through FG, PQ, and PG, and free-body diagrams are drawn. As before, we use the free-body diagram for that part of the structure *not* carrying the unit force. Thus when the force acts between J and P, we use

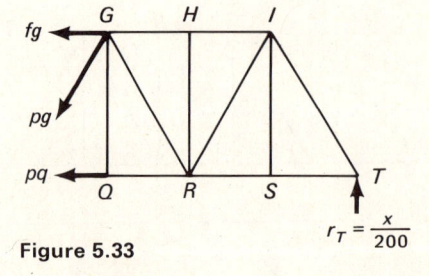

Figure 5.33

We find

$$pq = 0.00857x \quad \text{and} \quad pg = +0.00576x$$

for $0 < x < 120$ ft.

When the unit force acts between Q and T, we use

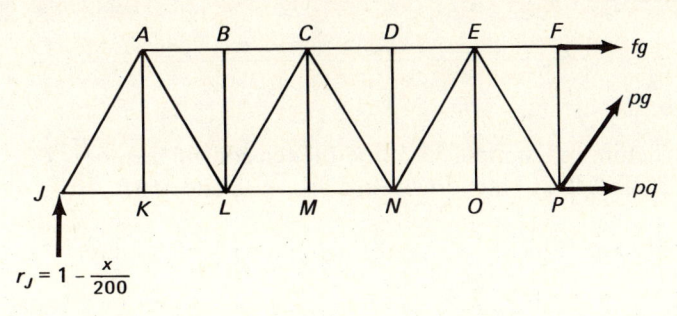

Figure 5.34

We obtain

$$pq = 4 - 0.02x \quad \text{and} \quad pg = -1.152 + 0.00576x$$

for $140 \text{ ft} < x < 200 \text{ ft}$.

Analogous to the influence lines for shear and moment in a main girder, the part of these influence lines for $120 \text{ ft} < x < 140 \text{ ft}$ is a straight line that joins the other regions.

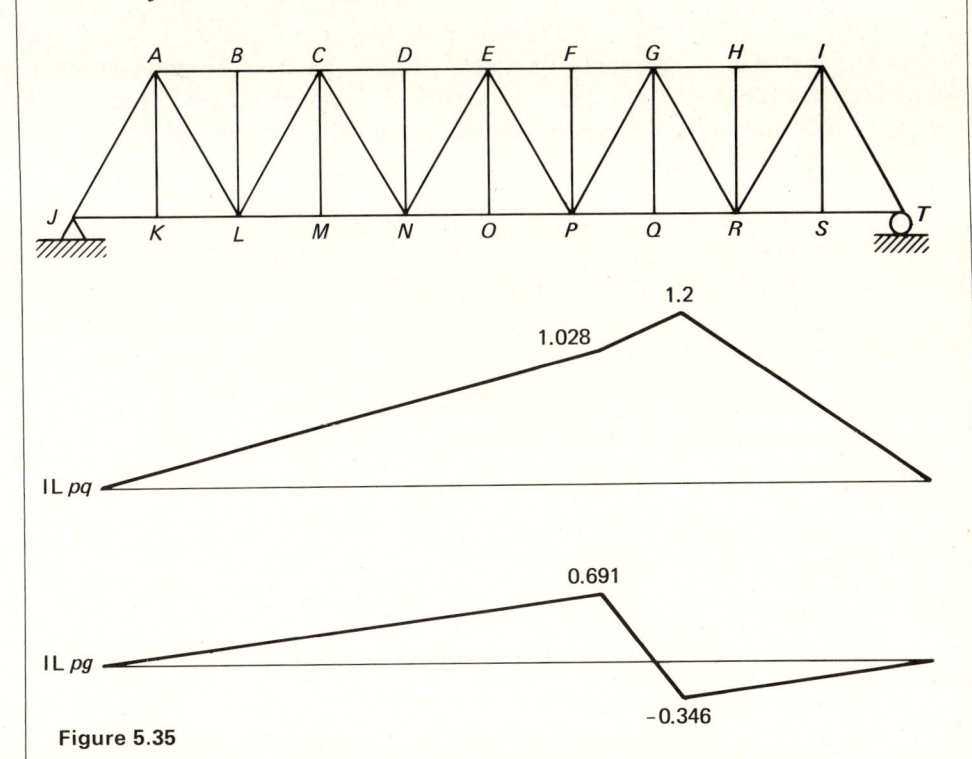

Figure 5.35

5.2.3 Use of Influence Lines

Values of support reactions, beam shears, and so on, caused by a single force may be found by multiplying the force by the influence line ordinate at the force's point of application. If there are several forces, the sum of these products gives the support reaction, beam shear, and so on.

EXAMPLE 5.2.6

For the structure in Example 5.2.5 find the tensions in members PQ and PG when 20 and 40 kips act downward at $x = 35$ and 160 ft, respectively.

The influence line values $x = 35$ and 130 ft, taken from Example 5.2.4, are

$pq]_{x=35 \text{ ft}} = 0.300 \qquad pq]_{x=160 \text{ ft}} = 0.800$

$pg]_{x=35 \text{ ft}} = 0.2016 \qquad pg]_{x=160 \text{ ft}} = -0.2304$

so that

$PQ = (20 \text{ kips})(0.300) + (40 \text{ kips})(0.800) = 38 \text{ kips}$

$PG = (20 \text{ kips})(0.2016) + (40 \text{ kips})(-0.2304) = -5.18 \text{ kips}$

where the negative sign means compression.

A distributed force $w(x)$ acting on the structure can be considered a collection of concentrated forces $w(x)\,dx$. Response to the distributed force is found by multiplying each $w(x)\,dx$ by the influence line ordinate at x, summed by integration. For a support reaction at A, for example,

$$R_A = \int_0^L w(x)(\text{IL } r_A)\,dx$$

If $w(x)$ is constant, that is, a uniformly distributed force, then

$$R_A = w \int_0^L (\text{IL } r_A)\,dx$$

$= w \times$ area under the "A reaction" influence line

EXAMPLE 5.2.7

Determine the reactions at A, C, and D if 40 kN/m is applied over the whole structure of Example 5.2.1.

$R_A = (40 \text{ kN/m})(1 \times 6 \text{ m} \times 0.5) = 120 \text{ kN}$

$R_C = (40 \text{ kN/m})(1.5 \times 18 \text{ m} \times 0.5 - 1 \times 8 \text{ m} \times 0.5) = 380 \text{ kN}$

$R_D = (40 \text{ kN/m})(-0.5 \times 10 \text{ m} \times 0.5 + 2 \times 16 \text{ m} \times 0.5) = 540 \text{ kN}$

In each, the second factor is the area of the influence line.
If only the part from A to D had been loaded,

$$R_C = (40 \text{ kN/m})(1.5 \times 18 \text{ m} \times 0.5) = 540 \text{ kN}$$

and if only C to E had been loaded,

$$R_D = (40 \text{ kN/m})(2 \times 16 \text{ m} \times 0.5) = 640 \text{ kN}$$

These two are the maximum loading cases for C and D. They are the result of loading the structure selectively in regions that have a single sign for their influence line ordinates, either an all positive region or an all negative one.

It is important to realize that every beam has a family of bending moment influence lines, a separate line for each point of the beam. In using influence lines to determine the largest bending moment, you must select the one line from the family that will reveal the largest moment and use it to establish the worst loading arrangement and its resulting maximum moment.

EXAMPLE 5.2.8

Determine the worst loading arrangement for the bending moment in member AB if the structure could be loaded with a movable 36 kips/ft.

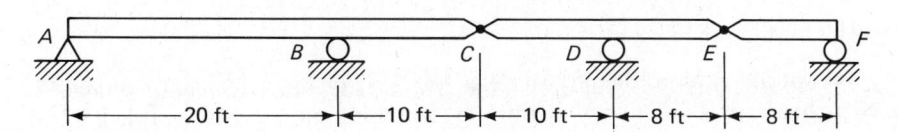

Figure 5.36

The bending moment influence line for an arbitrary point G between A and B, a distance l from A, is

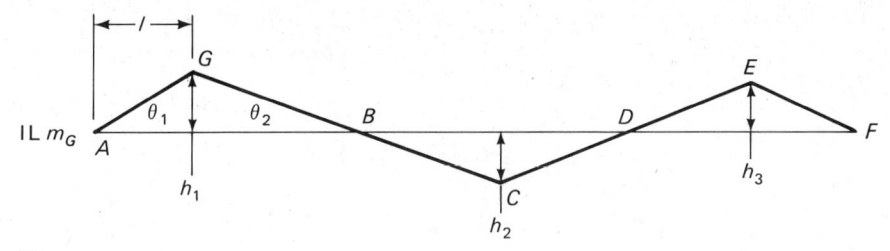

Figure 5.37

We have

$$\theta_1 + \theta_2 = 1 \qquad h_1 = l\theta_1 = (20-l)\theta_2 \qquad \text{or} \qquad \theta_1 = \frac{20-l}{l}\theta_2$$

$$\frac{20-l}{l}\theta_2 + \theta_2 = 1 \qquad \theta_2 = \frac{l}{20}$$

so

$$h_1 = (20-l)\frac{l}{20}$$

$$\frac{h_2}{10} = \frac{h_1}{20-l} \quad \text{or} \quad h_2 = \frac{10}{20-l}(20-l)\frac{l}{20} = \frac{l}{2}$$

$$\frac{h_3}{8} = \frac{h_2}{10} \quad \text{or} \quad h_3 = \frac{8}{10}\frac{l}{2} = \frac{4l}{10}$$

If both *AB* and *DEF* were loaded while *BCD* remained unloaded, the largest positive bending moment would result. It would be

$$M_G = +(36 \text{ kips/ft})\left[\frac{1}{2}(20-l)\left(\frac{l}{20}\right)(20) + \frac{1}{2}\left(\frac{4l}{10}\right)(16 \text{ ft}^2)\right]$$

$$= -18l^2 + 363.2l$$

Clearly the location of point *G* depends on *l*.

To find the maximum + moment for M_G we take the derivative with respect to *l* and equate it to zero,

$$\frac{dM_G}{dl} = -36l + 363.2 = 0 \qquad l = \frac{363.2}{36} = 10.09 \text{ ft}$$

for which $M_G = 1832$ kip-ft.

We should explore the largest negative value of M_G also. That would occur when only *BCD* is loaded and the remaining parts are not loaded.

$$M_G = -(36 \text{ kips/ft})(20h_2 \times \tfrac{1}{2} \text{ ft}^2)$$

$$= -36\left(10 \times \frac{l}{2}\right) = -180l$$

Taking a derivative of this linear function will not provide the maximum. Its maximum and minimum will occur at the extremes of *l*, that is, 0 and 20 ft. Thus

$$M_G]_{l=0} = 0$$
$$M_G]_{l=20} = -180(20) = -3600 \text{ kip-ft}$$

Evidently the largest bending moment will occur for $l = 20$ ft, that is, when *G* and *B* are coincident. It is a negative bending moment and happens only when span *BCD* alone is loaded.

If the loading is a set of concentrated forces with some prescribed spacing, such as a set of standard truck wheel loads, the worst loading is usually found using a few trial calculations.

EXAMPLE 5.2.9

Determine the maximum force in member *PG* of the truss of Example 5.2.5 that would result from the HS20-44(MS18) wheel loads of Section 5.1.3.

The influence line found in Example 5.2.5 is

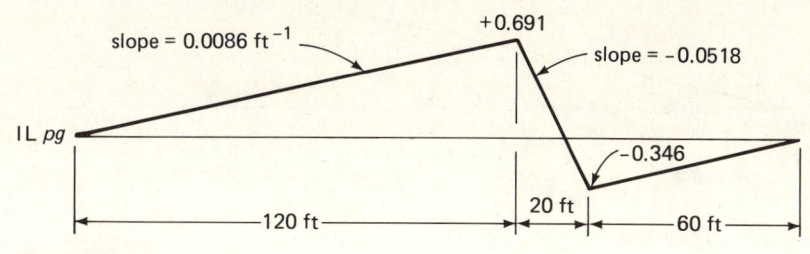

Figure 5.38

The HS20−44(MS18) wheel loads are

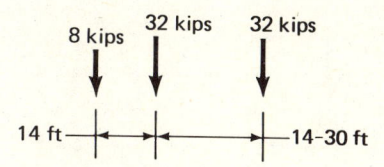

Figure 5.39

In seeking the maximum *PG*, these forces should be located as close to the influence line maximum as possible. The optional 14–30 ft length is assigned 14 ft in this case.

There are two load arrangements that are candidates for maximum tension in *PG*. They should be compared and the larger of these selected.

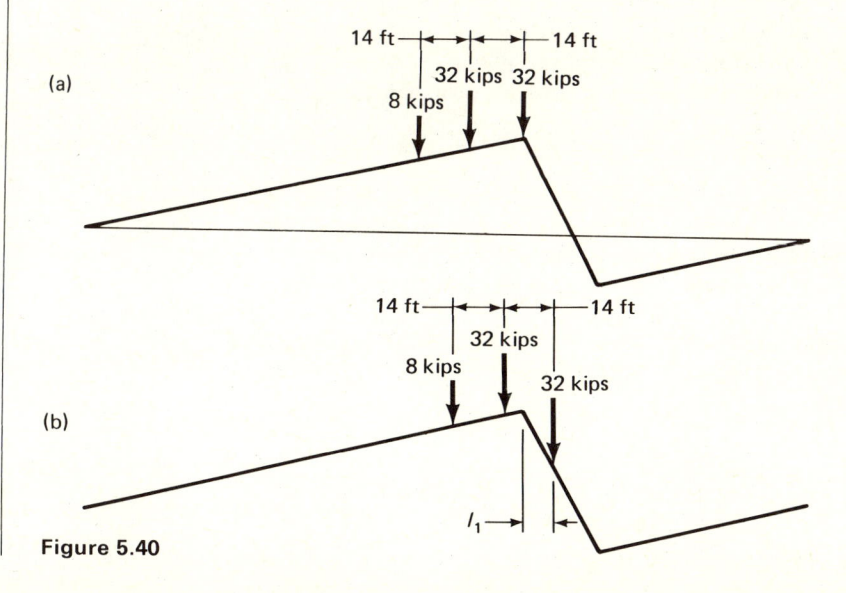

(a)

(b)

Figure 5.40

Forces acting on a single straight line segment of an influence line can be replaced by their total, acting at their center of gravity. For that approach, case (a) above becomes

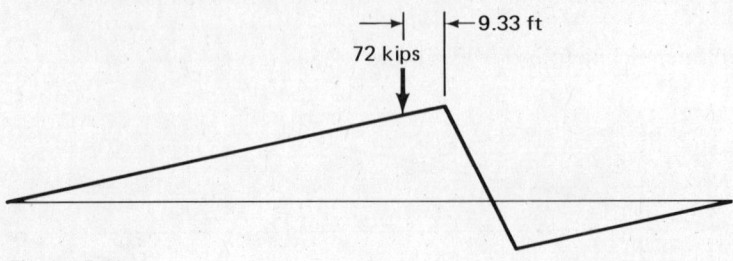

Figure 5.41

and case (b) becomes

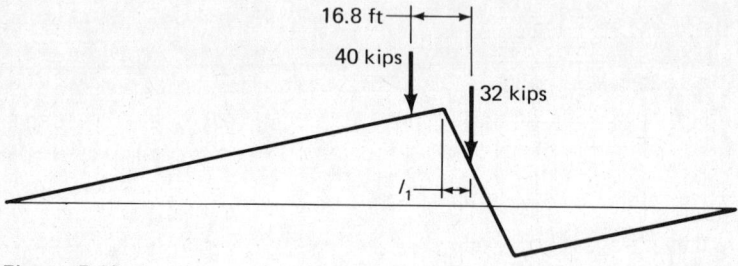

Figure 5.42

We examine case (b) to evaluate changes in PG as l_1 is varied:

$$\frac{\Delta PG}{\Delta l_1} = 32(-0.0518) + 40(0.0086) = -1.314 \text{ kips/ft}$$

Evidently as l_1 increases, PG decreases, suggesting that l_1 be minimized. Its smallest possible value is 0, for which case (b) is the equivalent to case (a). So for maximum tension in PG we need only evaluate case (a):

$$PG_{\max} = (72 \text{ kips})[0.691 - 9.33(0.0086)] = 44.0 \text{ kips}$$

Recall that in addition to the wheel loads, a movable line load, $w = 0.64$ kip/ft, and a single concentrated force, $P = 18$ kips, must be considered with placements to cause a maximum for PG. The location of the zero value for IL pg between 120 and 140 ft is at 133.3 ft, so that

$$PG_{\max} = 0.64(0.691 \times 133.3 \times 0.5) + 18(0.691)$$
$$= 29.5 + 12.4 = 41.9 \text{ kips}$$

The wheel loads, producing a larger PG, govern; thus PG must be designed to carry 44.0 kips in tension.

Because IL pg has negative ordinate regions, there can be loading situations that produce compression in PG. Although not shown in detail here, you are encouraged to show that the wheel loading gives a maximum compressive force of 21 kips and the line load/single concentrated force combination 13.6 kips. Clearly *PG* must be capable of carrying both 44 kips in tension and 21 kips in compression.

PROBLEMS

5.1 A series of 40-ft long roof beams, spaced 16 ft centerline to centerline, support purlins weighing 8 lb/ft and spaced 5 ft centerline to centerline. The purlins support a roof weighing 5.5 lb/ft^2. The pitch of the roof is slight, so the roof is considered flat. For 50-year recurrence snow loads of 26 lb/ft^2 and $C_S = 0.8$, determine the roof beam loading.

5.2 Assume that the roof beams for Problem 5.1 are part of a building 40 ft × 160 ft overall and 18 ft high, occupying an exposed site for which 50-year mean recurrence winds are 90 mi/h. For these conditions ANSI gives $q_F = 20$ lb/ft^2. Determine the roof beam loading. Assume *no* snow loading.

5.3 A column supports vertical loading from a total dead weight of 24 kips. In addition it supports live load for an area 18 ft × 20 ft from each of three floors, planned for office use. The total floor space so loaded is 5000 ft^2 on each floor. In addition, the column supports an area 18 ft × 20 ft of the flat roof on which a 50-year mean recurrence snow loading is 30 lb/ft^2. What is the total column load, considering any live load reductions?

5.4 Consider two walls 24 ft high. One is a windward wall and the other a leeward wall of a building. For a design wind pressure $q_F = 32$ lb/ft^2, determine the horizontal loading on one frame for which the frame spacing is 16 ft. Assume $C_p = 0.8$ and -0.7 for the windward and leeward walls, respectively.

5.5

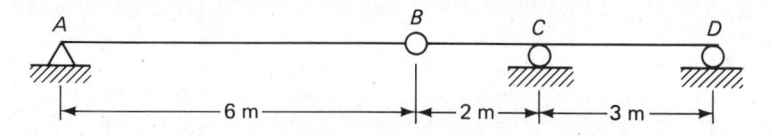

Figure P5.5

Draw the influence lines for the support reactions at *A* and *D*, and for the shear at a point midway between *A* and *B*.

5.6

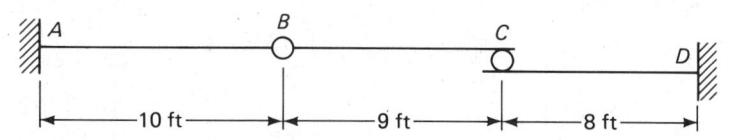

Figure P5.6

Draw the influence lines for the moment at *A* and *D*, the shear at *A* and *C*, and the moment midway between *B* and *C*. Use the virtual work method. Evaluate maximum shear at *A* and *C* due to a moving uniform load of 2.7 kips/ft.

5.7

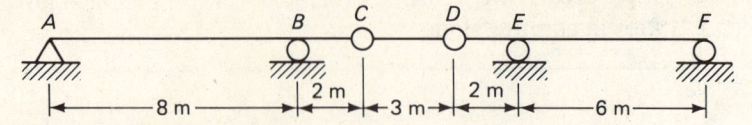

Figure P5.7

Draw the family of influence lines for bending moments for all points between E and F and the family of influence lines for shear for all points between A and B. Use each to determine the maximum moment between E and F and the maximum shear between A and B due to a moving uniform load of 80 kN/m. Indicate in each where on the beam these maximum values occur.

5.8

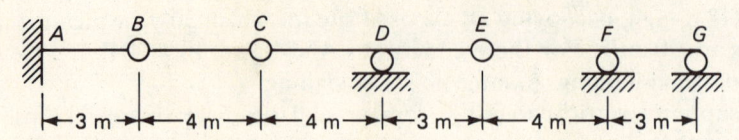

Figure P5.8

Draw the influence lines listed below ONLY and ~~evaluate the maximum responses~~ for a moving ~~150-kN/m uniform load~~.
(a) Vertical supports at A, D, F, G.
(b) Moment at A.
(c) Shear midway between B and C. — MIDWAY
(d) Moment at any general location between E and F.
(e) Shear at any general location between E and F.

5.9 Draw influence lines for the internal force for the truss members listed. Roadway loads are transferred to the truss through stringers to floor beams connected at the bottom chord joints.
(a) Members BH and BC.

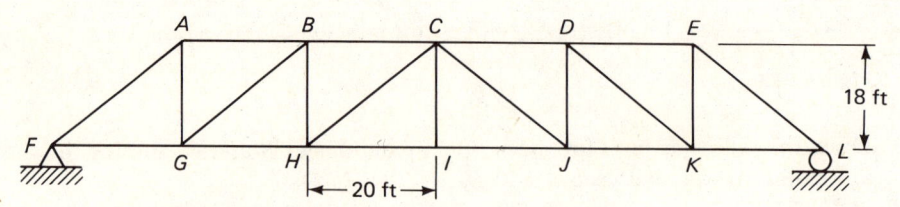

(b) Members BF, BC, and CG.

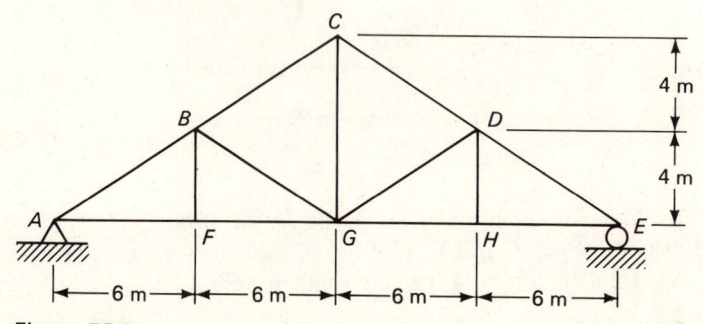

Figure P5.9

Elastic Deflections of Plane Statically Determinate Structures

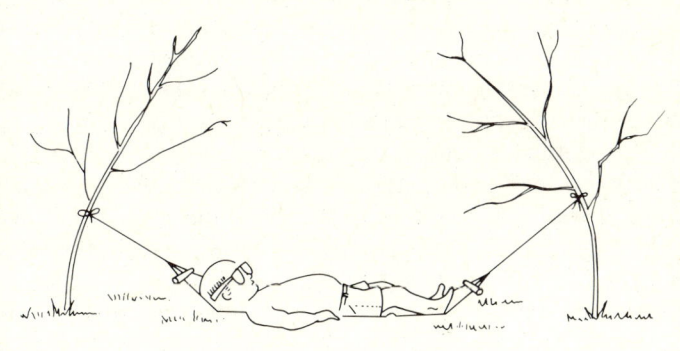

Maybe I should have tied it higher.

6.0 INTRODUCTION

The portly gentleman has hit bottom. His design of the support structure is poor; his remedy will be no better.

Deflections occur in all structures. If they are excessive, the structures may be useless. In some situations even small deflections can be excessive. For example, harmful aiming errors of a radio telescope can arise from almost immeasurable deflections of its supporting structure.

Avoiding excessive deflections, or evaluating the size of unavoidable deflections, is one use of deflection analysis, but not its only use. Deflection analysis is the focus of the method of consistent deformations, an important technique for the analysis of statically indeterminate structures. It is used to determine how much camber (upward bow) must be used in bridges to offset the deflection caused by the bridge's own weight. It is used to determine the shapes of fascia pieces to be manufactured to fit closely with deflected members of a structure. It is used to verify the validity of assumptions and calculations of a structural design by comparing deflections measured during construction to those predicted for the structure by its designer.

6.1 BEAM DEFLECTION

There are many methods for calculating slopes and deflections of beams. Several popular methods will be examined in detail in this section. They should serve the needs of most analysts. Two others, the method of successive integrations and the method of superposition, will be mentioned only briefly here. Your knowledge of them is presumed.

The method of successive integrations often is the first basic deflection analysis technique studied in a strength of materials course. It is the method generally used to develop the beam deflection formulas found in handbooks. The method has limited application. It may be used to analyze the deflections of beams for which no more than two expressions are needed to define the deflection for all points of the beam. Some other method should be used to analyze beams for which more than two such expressions are needed, since the complications in evaluating the constants of integration for these cases become formidable.

The method of superposition is valuable and powerful. There are very few support and load combinations that cannot be analyzed easily by a clever analyst equipped with a good set of beam deflection formulas. The analyst breaks up the loading of the beam into parts and combines the deflection equations for the different loadings by adding them algebraically, taking care that the expressions added apply to appropriate regions of the beam, and that the support conditions are satisfied.

Unquestionably both methods are valuable to the analyst. If you feel insecure with either of them, you should review them in your strength of materials textbook.

6.1.1 Conjugate Beams

Most structural analysts become proficient at determining shears and moments in beams. The conjugate beam method takes advantage of this proficiency to calculate slopes and deflections of beams by using an analogy that will be developed in the next few paragraphs.

Recall the method of successive integrations. The steps used in it to develop a beam's shear and moment are identical to those used to develop its slope and deflection. The method of conjugate beams substitutes a fictitious beam for the real one. The fictitious beam, called the conjugate beam, is loaded with a distributed "force" equal to M/EI of the real beam and is supported so that its shear and moment at its supports and ends are equal to the slope and deflection of the real beam.

The following notation relates the conjugate beam to the real beam.

Table 6.1

Real Beam		Conjugate Beam	
Slope θ	=	Shear $\bar{V}$	
Deflection y	=	Moment $\bar{M}$	
Shear V	=	Slope $\bar{\theta}$	(6.1)
Moment M	=	Deflection $\bar{y}$	
Curvature M/EI	=	Load $\bar{w}$	

A bar over a variable indicates that the variable is for the conjugate beam. It also refers implicitly to its equivalent variable for the real beam as defined above. Thus $\bar{V}$ and θ are the shear of the conjugate beam and the slope of the real beam, respectively, and they are equal.

Starting with the basic beam relations

$$w = \frac{dV}{dx} = \frac{d^2M}{dx^2} \qquad (6.2a)$$

and

$$\frac{M}{EI} = \frac{d\theta}{dx} = \frac{d^2y}{dx^2} \qquad (6.2b)$$

and substituting Eqs. (6.1) into Eq. (6.2b) gives

$$\bar{w} = \frac{d\bar{V}}{dx} = \frac{d^2\bar{M}}{dx^2} \qquad (6.3)$$

Except for the bar over the variables, Eq. (6.3) is identical to Eq. (6.2a). It is on this comparison that the method of the conjugate beam is built.

The support conditions for the conjugate beam must provide an appropriate match in the support conditions for the real beam, based on relations Eq. (6.1). The goal is to find supports for the conjugate beam which have $\bar{V}$, $\bar{M}$, $\bar{\theta}$, and $\bar{y}$ identical to θ, y, V, and M, respectively, for the real beam.

Strictly speaking, Eq. (6.3) gives no formal basis for assigning $V = \bar{\theta}$ and $M = \bar{y}$. It is done for the convenience of establishing conjugate ends where the real beam has $M = 0$ or $V = 0$, or both. For these cases the real beam will also have values of θ or y, or both, that are unspecified, and hence their conjugates $\bar{V}$ or $\bar{M}$, or both, must also be unspecified. By assigning $V = \bar{\theta}$ and $M = \bar{y}$, suitable end conditions giving an unspecified V or M, or both, are produced.

(1) Real beam, fixed end support:

$$y = 0 \quad \text{and} \quad \theta = 0$$

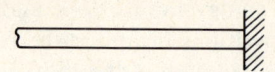

To satisfy Eqs. (6.1), its conjugate must have $\bar{M}=0$ and $\bar{V}=0$. A free end provides these, so the conjugate support of a fixed end support is a free end.
(2) Real beam, free end:

$$M=0 \quad \text{and} \quad V=0$$

Figure 6.2

Its conjugate must have $\bar{y}=0$ and $\bar{\theta}=0$. Since a fixed end would give these, the conjugate of a free end is a fixed support.
(3) Real beam pin, knife edge, or roller support at end:

$$M=0 \quad \text{and} \quad Y=0$$

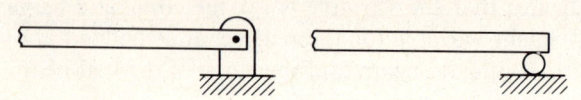

Figure 6.3

Its conjugate must have $\bar{y}=0$ and $\bar{M}=0$. Another pin, knife edge, or roller at the end would give these. Thus knife edges, pins, and rollers at ends are the conjugates of themselves.
(4) Interior hinge connection in a real beam

Figure 6.4

The conditions for this are

$y_L = y_R$
$\theta_L \neq \theta_R$
$V_L = V_R$
$M_L = M_R = 0$

where the subscripts L and R refer to the variables immediately to the left or right of the hinge. According to Eqs. (6.1), its conjugate must have the following conditions:

$\bar{M}_L = \bar{M}_R$
$\bar{V}_L \neq \bar{V}_R$
$\bar{\theta}_L = \bar{\theta}_R$
$\bar{y}_L = \bar{y}_R = 0$

which all would be provided by an interior pin or roller support.

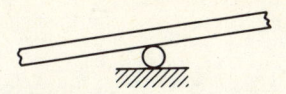

Figure 6.5

(5) Interior pin, knife edge, or roller support in a real beam

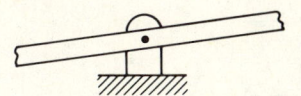

Figure 6.6

The conditions for these are

$y_L = y_R = 0$

$\theta_L = \theta_R$

$V_L \neq V_R$

$M_L = M_R$

The conjugate for these must have the following conditions:

$\bar{M}_L = \bar{M}_R = 0$

$\bar{V}_L = \bar{V}_R$

$\bar{\theta}_L \neq \bar{\theta}_R$

$\bar{y}_L = \bar{y}_R$

all of which would be satisfied by an interior hinge or roller connection.

Poplar Street Bridge, St. Louis, Mo., with the St. Louis Gateway Arch in the background. (Courtesy of Bethlehem Steel Corporation).

Summarizing the conjugate supports,

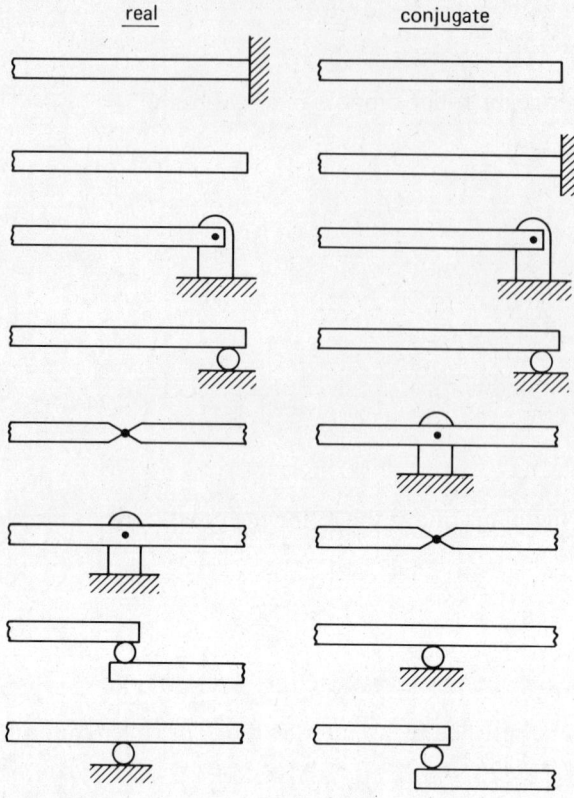

Figure 6.7

The procedure for the conjugate beam is simple.

(1) Draw the M/EI diagram for the real beam, loaded with its real loads and supported by its real supports.
(2) Draw the conjugate beam by replacing each support, interior condition, or end condition with its conjugate.
(3) Apply M/EI to the conjugate beam as its loading. At this stage the procedure may take one of two directions.
(4a) Draw the shear and moment diagrams of the conjugate beam loaded with M/EI, using the usual procedures for drawing shear and moment diagrams. Then replace the shear $\bar{V}$ with its equivalent θ and the moment $\bar{M}$ with its equivalent y. These diagrams then become the slope and deflection diagrams of the real beam.
(4b) Otherwise, if the slope and the deflection at a specific location are to be found, use equilibrium procedures to determine the internal shear $\bar{V}$ and the bending moment $\bar{M}$ there for the conjugate beam, and then substitute θ for $\bar{V}$ and y for $\bar{M}$.

The recommended sign conventions for the conjugate beam method are consistent with the notations introduced in Section 4.2.

6.1 BEAM DEFLECTION 111

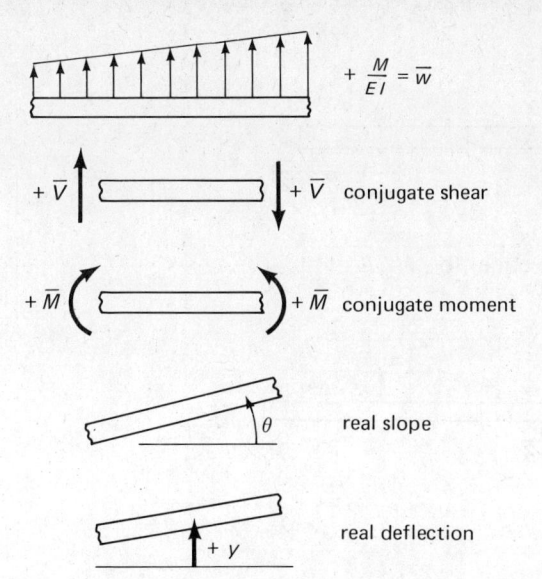

$+\dfrac{M}{EI} = \bar{w}$

$+\bar{V}$ conjugate shear

$+\bar{M}$ conjugate moment

real slope

real deflection

Figure 6.8

EXAMPLE 6.1.1

Sketch the slope and deflection curves for the beam supported and loaded as shown, and calculate the critical values for these curves.

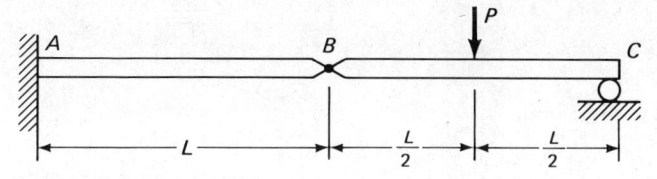

Figure 6.9

First the free-body, shear, and moment diagrams are drawn.

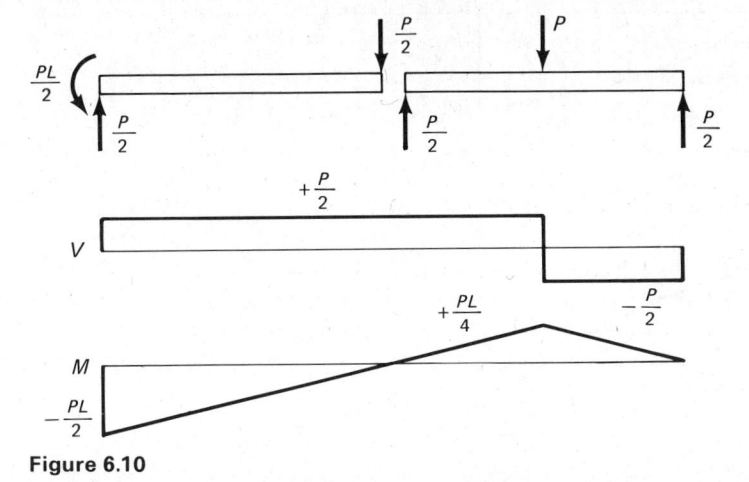

Figure 6.10

112 ELASTIC DEFLECTIONS OF PLANE STATICALLY DETERMINATE STRUCTURES

Then the conjugate beam is drawn

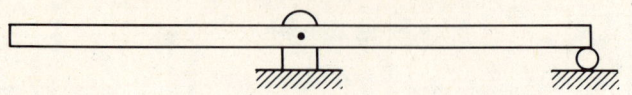

Figure 6.11

and loaded with a distributed force equal to M/EI.

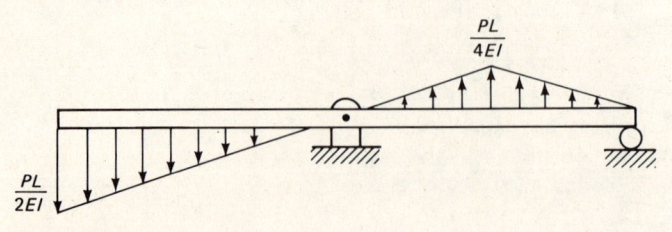

Figure 6.12

Its free-body diagram is drawn, and support reactions are determined for equilibrium.

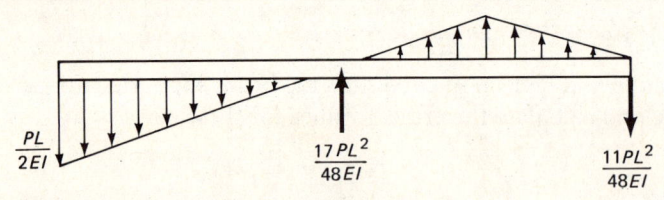

Figure 6.13

Its shear diagram is sketched.

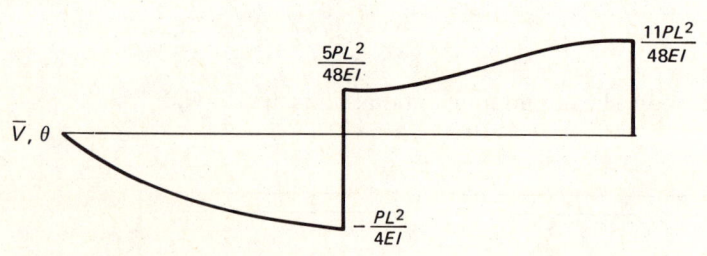

Figure 6.14

Its moment diagram is sketched.

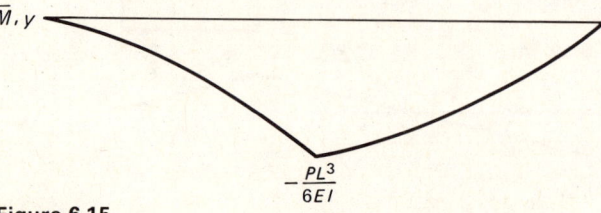

Figure 6.15

The values for the shear and moment diagrams were found using areas under their preceding diagrams, and by accounting for abrupt shifts caused by concentrated "forces" from the conjugate supports. For example, the moment $-PL^3/6EI$ was found by taking the area under the parabolic curve for the shear diagram. This was

$$\frac{2}{3} \times L \times \left(-\frac{PL^2}{4EI}\right) = -\frac{PL^2}{6EI}$$

If the slope or deflection is to be found at only one or two specific points, usually it is easier to take a section of the conjugate beam at that point, use equilibrium to determine its shear and moment there, and relate these to the desired slope and deflection. For example, if only the slope and deflection just to the left of point B were to be found, we take a section of the conjugate beam just to the left of B and solve for the shear and moment.

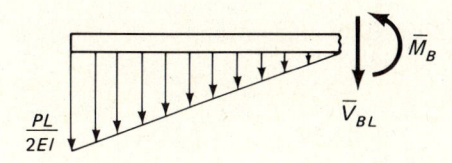

Figure 6.16

$$\sum F_y = -\frac{PL}{2EI}\left(\frac{L}{2}\right) - \bar{V}_{BL} = 0$$

$$\bar{V}_{BL} = -\frac{PL^2}{4EI}$$

$$\sum M_B = \bar{M}_B + \frac{PL^2}{4EI}\left(\frac{2L}{3}\right) = 0$$

$$\bar{M}_B = -\frac{PL^3}{6EI}$$

so the slope and deflection just left of B are

$$\theta_{BL} = -\frac{PL^2}{4EI}$$

$$y_B = -\frac{PL^3}{6EI}$$

the values that were found before.

There are interesting characteristics of conjugate beams related to their stability and determinacy. The conjugate of a stable, statically determinate beam is also a stable, statically determinate beam. But there is no such match in stability and determinacy between the real and conjugate beams if the real beam is either unstable

114 ELASTIC DEFLECTIONS OF PLANE STATICALLY DETERMINATE STRUCTURES

or statically indeterminate. Unstable real beams have statically indeterminate conjugate beams. Statically indeterminate real beams have unstable conjugate beams.

If a conjugate beam is found to be statically indeterminate, the conclusion must be made that the real beam is unstable and further analysis is not appropriate. If a conjugate beam is unstable, the assumption must be made that it remains in equilibrium through its conjugate loading, that is, the M/EI function. Thus the method of conjugate beams provides a useful method for solving some statically indeterminate beams. This will be developed in Section 9.1.

6.1.2 Moment-Area Methods

Moment-area methods are another useful approach for determining beam slopes and deflections. There are two theorems. Both are based on

$$\frac{d^2y}{dx^2} = \frac{d\theta}{dx} = \frac{M}{EI}$$

The first moment-area theorem is established by integrating

$$\frac{d\theta}{dx} = \frac{M}{EI}$$

between points A and B on the beam,

$$\int_{\theta_A}^{\theta_B} d\theta = \int_{x_A}^{x_B} \left(\frac{M}{EI}\right) dx$$

so that

$$\theta_B - \theta_A = \int_{x_A}^{x_B} \left(\frac{M}{EI}\right) dx$$

Thus the first moment-area theorem states:

> The change in the slope of the elastic curve between two points on a continuous beam is equal to the area under the M/EI diagram of the beam between those points.

EXAMPLE 6.1.2

Find the slope of the left-hand end of the beam loaded and supported as shown.

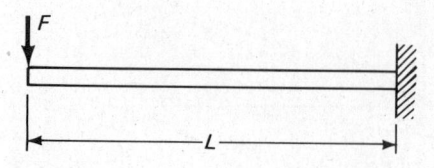

Figure 6.17

Its M/EI is constructed in the usual way.

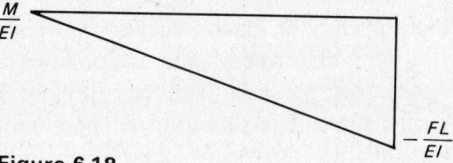

Figure 6.18

Since the beam is cantilevered, the right end has zero slope, that is, $\theta_R = 0$. θ_L is the slope at the left end to be determined. $\theta_R - \theta_L$ is the area under the M/EI diagram,

$$-\theta_L = \frac{L}{2}\left(-\frac{FL}{EI}\right)$$

$$\theta_L = +\frac{FL^2}{2EI}$$

The second moment-area theorem is developed using the M/EI and y diagrams.

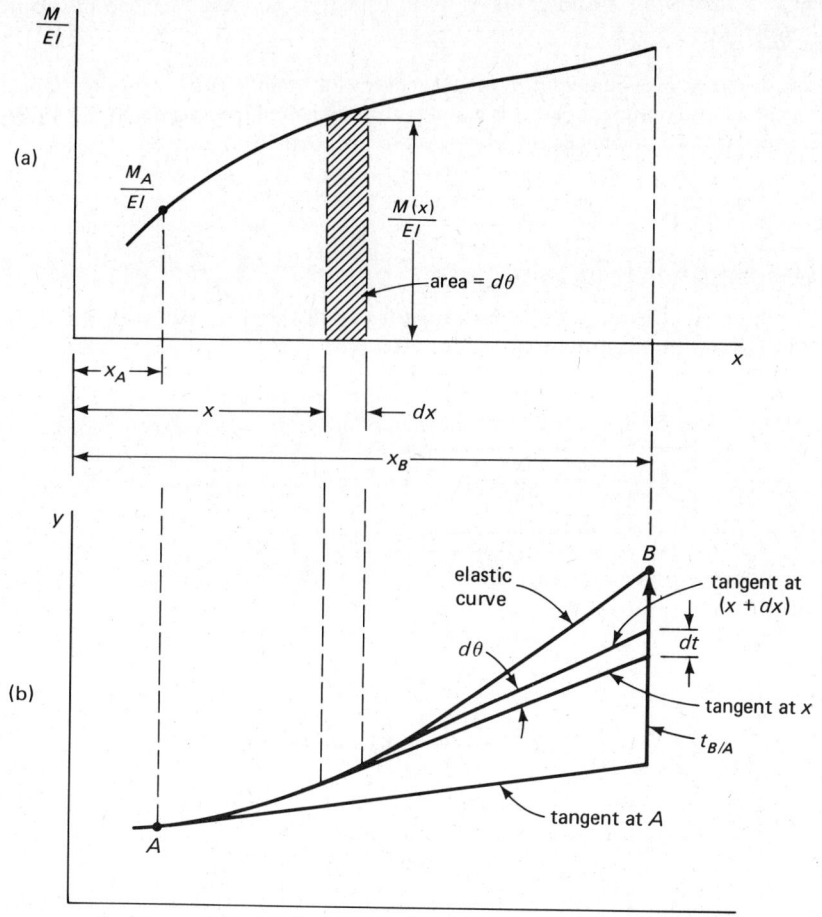

Figure 6.19

116 ELASTIC DEFLECTIONS OF PLANE STATICALLY DETERMINATE STRUCTURES

Figure 6.19 shows the M/EI diagram and the elastic curve for a beam. The x axis of Figure 6.19(b) is coincident with the undeformed beam. Three lines are drawn tangent to the elastic curve. One, tangent at A, passes a distance $t_{B/A}$ from B, a point on the elastic curve. Distance $t_{B/A}$ is measured in a direction perpendicular to the undeformed beam. The other two tangents are drawn at x and $x+dx$, two points that are dx apart and on the elastic curve between A and B. The distance dt at B between these two tangents is a portion of $t_{B/A}$. It is equal to the change in slope between these tangents times their distance to B. Since the difference in slope is $(M/EI)\,dx$ and the distance to B is $x_B - x$,

$$dt = (x_B - x)\frac{M}{EI}dx$$

Integration from A to B sums all of the dt, giving $t_{B/A}$,

$$t_{B/A} = \int_{x_A}^{x_B} (x_B - x)\frac{M}{EI}dx \qquad (6.4)$$

Close examination of Eq. (6.4) shows that the integral is the moment of the area under M/EI between A and B taken with respect to B. Thus the second moment-area theorem states:

> The distance of a point B on the elastic curve of a beam from a line drawn tangent to the elastic curve at another point A is equal to the moment of the area under the M/EI diagram between the two points taken with respect to the point B.

EXAMPLE 6.1.3

Find the distance from the tangent to the elastic curve at point A to point B on the elastic curve.

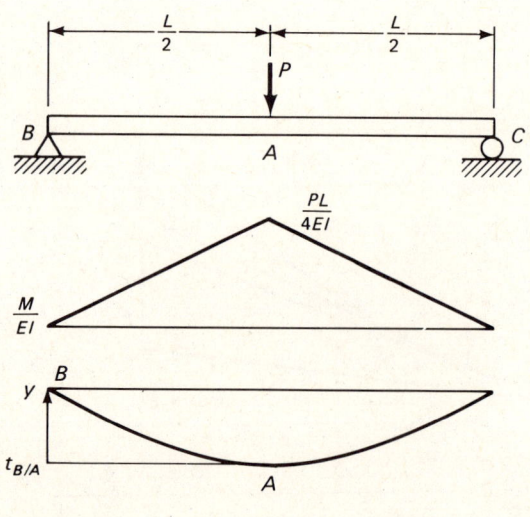

Figure 6.20

$$t_{B/A} = \frac{PL}{4EI} \times \frac{L}{2} \times \frac{1}{2} \times \frac{L}{2} \times \frac{2}{3} = +\frac{PL^3}{48EI}$$

Thus the distance of point B from the tangent at A is $PL^3/48EI$.

At first, knowing that $t_{B/A} = +PL^3/48EI$ may not seem useful. Its value comes through the use of some simple geometric reasoning from which the deflection of point A may be inferred.

Because of symmetry, the tangent at A is a horizontal line. Since $t_{B/A}$ is positive, point B must be above this horizontal tangent at A at a distance $PL^3/48EI$. Since point B cannot move, the horizontal tangent at A must be *below* a distance $PL^3/48EI$. So the deflection of A is found as $y_A = -PL^3/48EI$.

EXAMPLE 6.1.4

Find the slope of end A of the beam loaded and supported as shown.

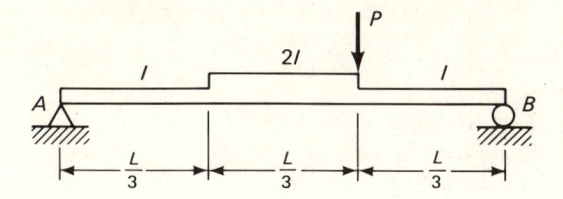

Figure 6.21

The moment diagram is

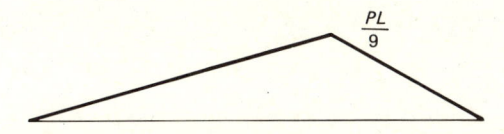

Figure 6.22

Because the stiffness changes from EI to $2EI$ at $L/3$ from either end, the M/EI diagram is

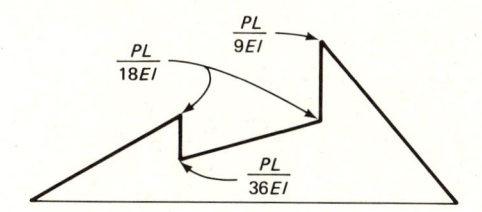

Figure 6.23

The elastic curve is assumed to have the approximate shape shown.

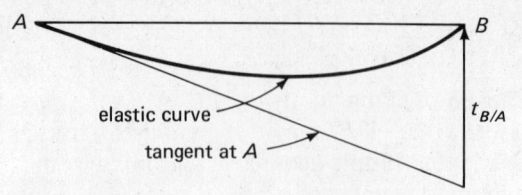

Figure 6.24

For this assumed elastic curve point B, which does not move, should be a distance $t_{B/A}$ above the tangent at A. Further, the slope θ_A is negative for the assumed shape. It must be concluded then that

$$\theta_A = -\tan^{-1}\left(\frac{t_{B/A}}{L}\right)$$

and since the structural deformations are assumed to be small, the tangent and its argument may be assumed to be equal. Thus

$$\theta_A = -\frac{t_{B/A}}{L}$$

It remains only to calculate $t_{B/A}$:

$$t_{B/A} = \frac{PL}{18EI} \times \frac{L}{3} \times \frac{1}{2} \times \left(\frac{2L}{3} + \frac{1}{3} \times \frac{L}{3}\right)$$

$$+ \frac{PL}{36EI} \times \frac{L}{3} \times \frac{1}{2} \times \left(\frac{L}{3} + \frac{1}{3} \times \frac{L}{3}\right)$$

$$+ \frac{PL}{36EI} \times \frac{L}{3} \times \frac{L}{2}$$

$$+ \frac{PL}{9EI} \times \frac{L}{3} \times \frac{1}{2} \times \left(\frac{2}{3} \times \frac{L}{3}\right)$$

$$= \frac{35PL^3}{1944EI}$$

Hence

$$\theta_A = -\frac{35PL^2}{1944EI} = -0.00180\frac{PL^2}{EI}$$

6.2 THE METHOD OF VIRTUAL WORK (UNIT LOAD METHOD)

The method of virtual work is a powerful means for calculating deflections in structures. As presented here, it is also known as the unit load method. Section 2.4 gave a brief venture into virtual work, but it was limited to rigid bodies. We will extend that study here to include deformable bodies and structures. This will require us to include in the work and energy equations some terms that involve the changes in the internal energy of the structure arising from changes in its size and shape. All of the assumptions made in Section 2.4 apply here except, of course, those related to a *rigid* body.

6.2 THE METHOD OF VIRTUAL WORK (UNIT LOAD METHOD)

Thus we assume that the structure is stable, in equilibrium, on rigid supports, and that all the connections between members are frictionless.

Consider an elastic body in static equilibrium under load P that causes the body to deform as shown.

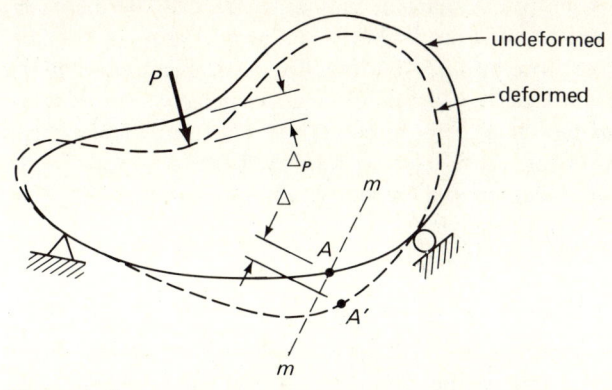

Figure 6.25

During loading the point of application of P is displaced Δ_P along P's line of action, thereby doing work equal to $\frac{1}{2}P\Delta_P$.[1] Internally, stresses and strains develop as P is applied. Consider, for example, the differential element with stresses σ_R that act on opposite faces, each with areas dA.

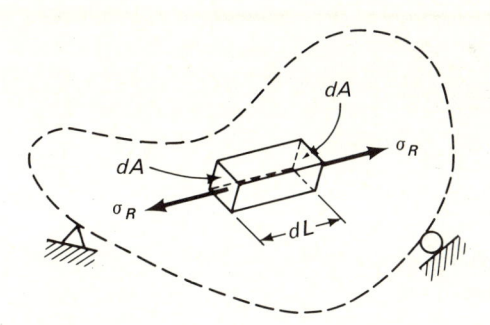

Figure 6.26

The corresponding strain ε_R causes the element to change its length by $\varepsilon_R \, dL$, and hence work is done on the element equal to $\frac{1}{2}\sigma_R \, dA \times \varepsilon_R \, dL$ or $\frac{1}{2}\sigma_R \, dV$. Every element in the body has similar stresses and strains so that the total work on all the elements is

$$\frac{1}{2} \int_{\substack{\text{whole} \\ \text{body}}} \sigma_R \varepsilon_R \, dV$$

This work is stored in the body as internal energy, and on the basis of the law of

---

[1] The factor $\frac{1}{2}$ often engenders confusion. It arises because as the force is increased from zero to P, the placement increases from zero to Δ_P, that is, the *full* force P is present only in the last increments of Δ_P, not through the entire Δ_P. The derivation of Eq. (6.11) in Section 6.3 may clarify this.

conservation of energy must equal the work done by the external force P. Hence

$$\tfrac{1}{2}P\Delta_P = \tfrac{1}{2}\int_{\substack{\text{whole}\\\text{body}}} \sigma_R \varepsilon_R \, dV$$

Let us assume that we wish to calculate Δ, the component of the displacement along line m–m of an arbitrary point A on the body as it moves to A' as P is applied. *Before applying P, let us make a preliminary loading. Let us apply a unit virtual force at A in the direction m–m, that is, a unit virtual force at the place and in the direction of the displacement to be determined.* This force causes a virtual displacement Δ_u at A and thus does virtual work of $\tfrac{1}{2}(1)(\Delta_u)$. It also induces virtual stresses σ_u and their corresponding strains ε_u to the elements of the body. Thus analogous to the case of the real force P, conservation of energy gives

$$\tfrac{1}{2}(1)(\Delta_u) = \tfrac{1}{2}\int_{\substack{\text{whole}\\\text{body}}} \sigma_u \varepsilon_u \, dV$$

Now *without removing the unit virtual force*, let us apply P. The same real external work found before remains equal to the same stored real internal energy found before. Hence

$$\tfrac{1}{2}P\Delta_P = \tfrac{1}{2}\int_{\substack{\text{whole}\\\text{body}}} \sigma_R \varepsilon_R \, dV$$

But with the unit virtual force present when P is applied, additional internal energy and external work arise, for as P is applied this time the *unit force* undergoes real displacement Δ and the *internal elements* stressed with σ_u undergo real displacements $\varepsilon_R \, dL$. The external work and the stored internal energy have been shown to be equal for each step of the loading so far. Thus to satisfy conservation of energy for the combined loading of P and the unit virtual force these additional external work and internal energy terms must also be equal. Thus

$$(1)\Delta = \int_{\substack{\text{whole}\\\text{body}}} \sigma_u \, dA \, \varepsilon_R \, dL \tag{6.5a}$$

$$(1)\Delta = \int_{\substack{\text{whole}\\\text{body}}} \sigma_u \varepsilon_R \, dV \tag{6.5b}$$

which are expressions for the principle of virtual work. By superposition we may extend this formulation so that it is valid for more than a single external force P. The real strain ε_R may be developed by superposition of the strains related to individual forces. Hence ε_R may be treated as the function related to any general loading of the body, and Δ the desired displacement caused by that general loading.

Since the unit force appears on the left-hand side as a factor of 1, it is apparent that the right-hand side is directly equal to the displacement Δ.

We will use Eq. (6-5) frequently, but in several forms that are easier to apply. They will be derived for the specific situations of bending and axial force. As we develop them, it is important to remember the definitions of σ_u and ε_r:

σ_u is the expression for the stress throughout the body, caused by the *unit force* applied at the point and in the direction of its displacement to be calculated.

ε_r is the expression for the strain throughout the body caused by the *real* force system.

6.2 THE METHOD OF VIRTUAL WORK (UNIT LOAD METHOD)   121

Our first application of Eq. (6.5) will be to calculate beam deflections. Assume that we wish to calculate the vertical deflection of point A on the neutral axis of the beam loaded as shown.

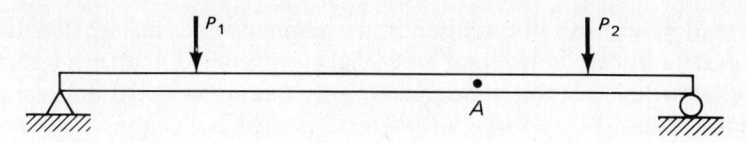

Figure 6.27

First apply a vertical unit virtual force at A.

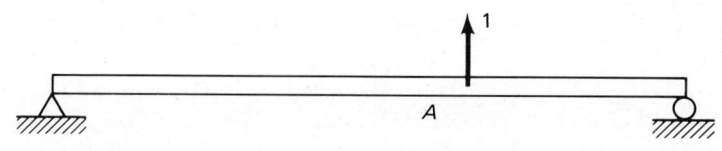

Figure 6.28

Stresses σ_u arising from the virtual force are

$$\sigma_u = -\frac{my}{I}$$

where $m = m(x)$ is the internal bending moment throughout the beam arising from the virtual force.

Next, for the real loading P_1 and P_2 the strains ε_r are

$$\varepsilon_r = -\frac{My}{EI}$$

where $M = M(x)$ is the internal bending moment throughout the beam caused by the real loading. Thus

$$(1)\Delta = \iint \sigma_u \, dA \, \varepsilon_r \, dL$$

$$= \iint \left(-\frac{my}{I}\right) dA \left(-\frac{My}{EI}\right) dx$$

Replacement of dL by dx is done, since for the beam all the stresses are along the beam axis, that is, the x direction. Rearranging,

$$\Delta = \int_{\text{length}} \frac{mM}{EI^2} \, dx \int_{\substack{\text{cross} \\ \text{section}}} y^2 \, dA$$

Separation of the integrals into a part related to the position along the beam, x, and a part related to the cross section is permissible since they are independent. Noting that

$$\int_{\substack{\text{cross} \\ \text{section}}} y^2 \, dA = I$$

the equation becomes

$$\Delta = \int_{\text{length}} \frac{mM}{EI} dx \tag{6.6}$$

To use Eq. (6.6) you need only write the expressions for m and M, substitute them in the integral, and solve it. If E or I are variable with position, their algebraic expressions must be included in the integrand. Usually E and I are constant and are taken out of the integral.

EXAMPLE 6.2.1

Find the deflection at the free end of the cantilevered beam.

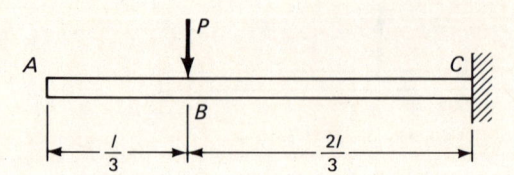

Figure 6.29

First expressions for M are determined. For region AB,

$$+\circlearrowleft \sum M_{\text{cut}} = M = 0$$

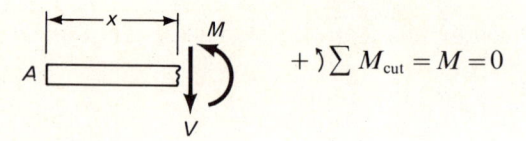

Figure 6.30

For region BC,

$$+\circlearrowleft \sum M_{\text{cut}} = M + Px = 0$$
$$M = -Px$$

Figure 6.31

Next the expressions for m are found. In doing this *it is necessary to use the same definitions of x and sign conventions that were used to find M.*
For region AB,

$$+\circlearrowleft \sum M_{\text{cut}} = m + 1x = 0$$
$$m = -x$$

Figure 6.32

For region BC,

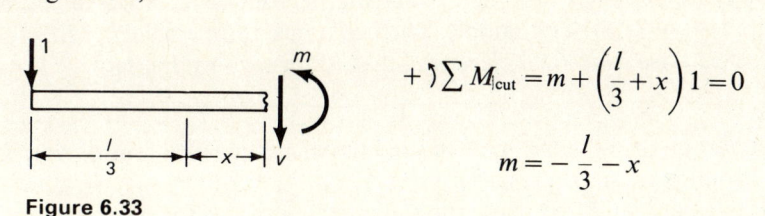

Figure 6.33

$$+\circlearrowleft\sum M_{cut} = m + \left(\frac{l}{3} + x\right)1 = 0$$

$$m = -\frac{l}{3} - x$$

These expressions are applied to the virtual work expression, Eq. (6.6).

$$\Delta = \int \frac{mM}{EI} dx$$

$$\Delta_A = \int_0^{l/3} \frac{-x(0)}{EI} dx + \int_0^{2l/3} \frac{(-l/3 - x)(-Px)}{EI} dx$$

$$= \frac{14}{81} \frac{Pl^3}{EI}$$

If a slope rather than a deflection is to be found, a unit couple is applied at the point where the slope is to be found, and the resulting m is used in Eq. (6.6) as before. Thus to find the slope at A in Example 6.2.1, the unit couple is applied at A.

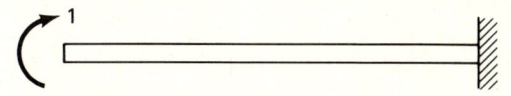

Figure 6.34

For region AB,

Figure 6.35

$$+\circlearrowleft\sum M_{cut} = -1 + m = 0$$

$$m = 1$$

For region BC

Figure 6.36

$$+\circlearrowleft\sum M_{cut} = -1 + m = 0$$

$$m = 1$$

$$\theta_A = \int_0^{l/3} \frac{1(0)}{EI} dx + \int_0^{2l/3} \frac{1(-Px)}{EI} dx$$

$$= -\frac{2Pl^2}{9EI}$$

The minus sign indicates that the beam has rotated at A in a direction opposite to that of the unit couple. The beam has a slope of $2Pl^2/9EI$ counterclockwise from the horizontal, that is, upward to the right.

The method of virtual work may be used to calculate deflections of rigid frames, trusses, and composite structures, too.

Rigid frames are analyzed using the same techniques as for beams. A unit force or a unit couple is applied at the point on the structure whose deflection or rotation is to be determined, and the resulting internal bending moment m is found. These are multiplied by the functions for M, the internal bending moments caused by the real loads, divided by EI, and integrated over the whole structure. Usually several expressions for m and M are necessary to account for all parts of the frame. Thus several integrals are usually required for the solution. As in beams, a positive sign for the solution indicates a deflection or rotation in the same direction as the unit force or couple.

EXAMPLE 6.2.2

Calculate the vertical deflection and rotation at point C if the material is steel, and $I = 500$ in^4 for all members.

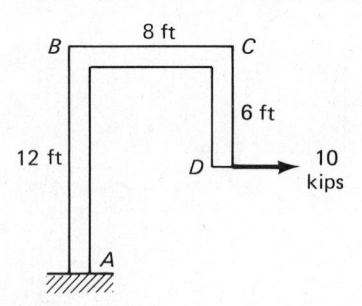

Figure 6.37

The bending moments M for this structure and the applied force were found in Example 4.3.3. We must find the internal moments m for the two deflections to be calculated, that is, due to a unit vertical force at C and due to a unit couple at C.

For the vertical deflection, the unit force is applied as shown.

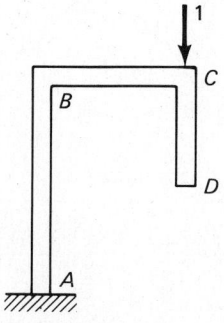

Figure 6.38

6.2 THE METHOD OF VIRTUAL WORK (UNIT LOAD METHOD)

For CD,

$$+\circlearrowleft \sum M_{cut} = m = 0$$

Figure 6.39

For BC,

$$+\circlearrowleft \sum M_{cut} = -m - 1(x) = 0$$
$$m = -x \quad \text{ft}$$

Figure 6.40

for AB,

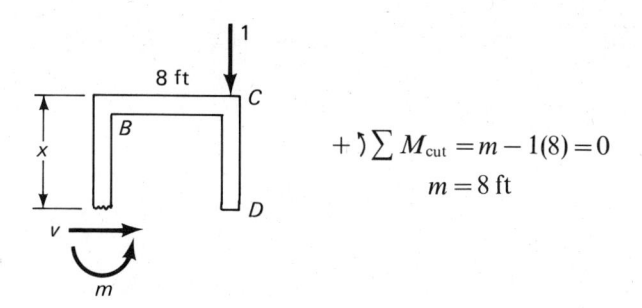

$$+\circlearrowleft \sum M_{cut} = m - 1(8) = 0$$
$$m = 8 \text{ ft}$$

Figure 6.41

The results are tabulated in Table 6.2.

Table 6.2

Member	Origin	Range of x (ft)	M (kip-ft)	m (ft)	EI (kip-in²)
AB	B	0→12	$10x - 60$	8	15×10^6
BC	C	0→ 8	60	$-x$	15×10^6
CD	D	0→ 6	$10x$	0	15×10^6

Now substituting these values in Eq. (6.7) to get $\Delta_{c\text{-vert}}$,

$$\Delta_{c\text{-vert}} = \int_0^{12\,\text{ft}} \frac{(10x - 60 \text{ kip-ft})(8 \text{ ft})(144 \text{ in}^2/\text{ft}^2)}{15 \times 10^6 \text{ kip-in}^2} \, dx$$

$$+ \int_0^{8\,\text{ft}} \frac{(60 \text{ kip-ft})(-x \text{ ft})(144 \text{ in}^2/\text{ft}^2)}{15 \times 10^6 \text{ kip-in}^2} \, dx$$

$$= -0.01843 \text{ ft}$$

The minus sign shows that the deflection is upward (opposite the downward unit force).

Next, to get the rotation at point C, we apply a unit couple there.

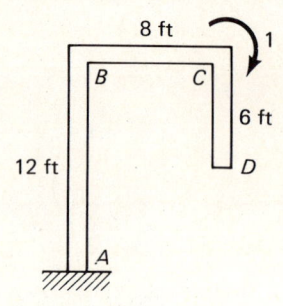

Figure 6.42

The resulting internal bending moments m can be shown to be

$m = 1$ for AB
$m = -1$ for BC
$m = 0$ for CD

(You are urged to verify these.) These values replace those for m in Table 6.2, so it now appears as Table 6.3.

Table 6.3

Member	Origin	Range of x (ft)	M (kip-ft)	m (ft)	EI (kip-in^2)
AB	B	$0 \to 12$	$10x - 60$	1	15×10^6
BC	C	$0 \to 8$	60	-1	15×10^6
CD	D	$0 \to 6$	$10x$	0	15×10^6

Substituting in Eq. (6.7) as before,

$$\theta_c = \int_0^{12\,\text{ft}} \frac{(10x - 60 \text{ kip-ft})(1)(144 \text{ in}^2/\text{ft}^2)}{15 \times 10^6 \text{ kip-in}^2} \, dx + \int_0^{8\,\text{ft}} \frac{60(-1)(144)}{15 \times 10^6} \, dx$$

$$= \left[\frac{(5x^2 - 60x)(144)}{15 \times 10^6} \right]_0^{12} - \left[\frac{60(x)(144)}{(15 \times 10^6)} \right]_0^{8}$$

$$= -0.00461 \text{ rad}$$

The reader should notice the care in the use of units, the consistency in defining and using the variable x in m and M, and the consistency of sign convention for m and M. It may have seemed puzzling at first that the unit force has no units. If you recall Eqs. (6.5) and (6.6), the unit force appears on the left-hand side of the equation as a multiplier of the displacement Δ. The unit force's magnitude being 1 causes no change when it is used as a multiplier of the displacement. Its being dimensionless also causes no change in the units of the displacement. It is for these conveniences that a dimensionless unit force is used.

Calculation of truss deflections requires a different formulation of σ_u and ε_r in Eq. (6.5). Truss members are axial force members, so that σ_u and ε_r would be associated with that loading. Further, the members are loaded at joints only, so there is no variation in axial force over the length of any member. For any member

$$\sigma_u = \frac{u}{A}$$

where

u = internal force in the member caused by the unit virtual force applied at the location and in the direction of the deflection to be calculated
A = cross-sectional area of member

For the same member

$$\varepsilon_r = \frac{N}{AE}$$

where

N = internal force in the member caused by real loading of the truss
E = modulus of elasticity of member

Substitution into Eq. (6.5) gives

$$\Delta = \sum \frac{u}{A} \times \frac{N}{AE} \times A \times L = \sum \frac{uNL}{AE} \tag{6.7}$$

where A has replaced dA and L has replaced dL. (The need to integrate is eliminated since the stress is uniform across the cross section and is constant over the length of each member.) The remaining summation implied by the integration is accomplished when the products uNL/AE for each member are summed, so that all the energy stored in all members is included.

Application of Eq. (6.7) requires determining the internal forces u and N by analyzing the truss with the unit virtual force and the real loading applied, respectively, calculating NuL/AE for each member and adding. In this the usual notation, + for tension and − for compression for N and u, will apply. The resulting sign will indicate the same sense (+) or opposite sense (−) for the deflection compared to the direction of the unit force.

EXAMPLE 6.2.3

Find the vertical deflection of point F of the truss.

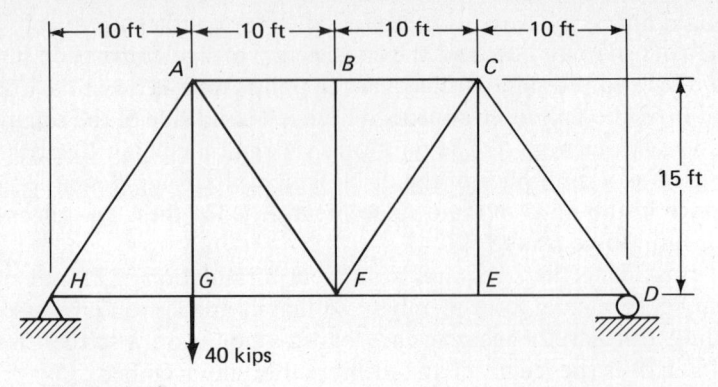

128 ELASTIC DEFLECTIONS OF PLANE STATICALLY DETERMINATE STRUCTURES

Figure 6.43

All the diagonal members AH, AF, CF, and CD have cross-sectional areas of 6 in^2, the verticals AG, BF, and CE have areas of 4 in^2, and the horizontals AB, BC, HG, GF, FE, and ED have areas of 3 in^2. The truss is steel with $E = 3 \times 10^4$ kip/in^2.

Examples 4.1.1 and 4.1.2 analyzed the forces carried in 9 of the 13 members. These, along with the forces in the members not previously analyzed, are listed in Table 6.4 as N.

In order to calculate the vertical deflection of joint F, we calculate the internal forces u caused by a unit virtual force applied vertically at F. These may be found using the method of joints or the method of sections:

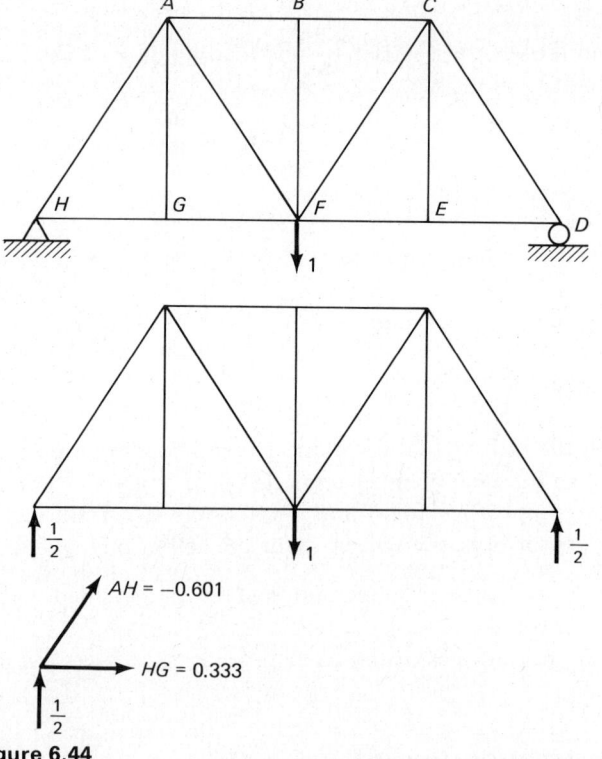

$AH = -0.601$

$HG = 0.333$

Figure 6.44

The structure with its unit force downward at F, the free-body diagram, and the free-body diagram for the joint at H are shown, giving $AH = -0.601$ and $HG = 0.333$. The internal forces for the other members are found using other joints, and are listed in the column headed u in Table 6.4.

Table 6.4

Member	L (ft)	A (in²)	N (kips)	u	NLu/A (kip-ft/in²)
AB	10	3	−13.3	−0.667	+29.57
BC	10	3	−13.3	−0.667	+29.57
AH	18	6	−36.1	−0.601	+65.09
AG	15	4	+40	0	0
AF	18	6	−12	+0.601	−21.64
BF	15	4	0	0	0
CF	18	6	+12	+0.601	+21.64
CE	15	4	0	0	0
CD	18	6	−12	−0.601	+21.64
HG	10	3	+20	+0.333	+22.20
GF	10	3	+20	+0.333	+22.20
FE	10	3	+6.67	+0.333	+7.40
ED	10	3	+6.67	+0.333	+7.40
					205.07

N, L, u, and A are combined according to NLu/A and are entered in a column so named. Assuming E to be the same for each member, the desired deflection $\sum(NLu/AE)$ is found conveniently by adding the column to get $\sum(NLu/A)$ and then dividing by E. Thus:

$$\Delta_F = \frac{205.07 \text{ kip-ft/in}^2}{3 \times 10^4 \text{ kips/in}^2} = 0.00684 \text{ ft} = 0.0820 \text{ in}$$

Let us return to the virtual work expression, Eq. (6.7), and look at it from a new viewpoint. The factor NL/AE appearing in each term is the change of length of the individual member arising because of the loading. Suppose that these changes in length were caused by something other than the loading. For example, suppose that each member were heated or cooled in such a way that each underwent a change in length exactly equal to its own NL/AE. The structure would be deformed exactly as if the loading had been applied; the Δ would be the same. Or suppose that each member were *fabricated* to be longer or shorter than its nominal length by its NL/AE. Upon assembly the unloaded structure would be "deformed" exactly as if each member had been made to its "correct" length and then the load applied. Again the Δ would be the same. Equation (6.7) can be altered slightly to provide a method for calculating displacements that are the result of member length changes due to any cause

$$\Delta = \sum u \Delta L \tag{6.8}$$

where

u = internal member force caused by the virtual unit force applied at the joint in the direction of the displacement component Δ

ΔL = change of length of individual member

EXAMPLE 6.2.4

Determine the vertical displacement of joint F of the truss of Example 6.2.3 if members AF and CF are both manufactured 0.333 in too short and the other member lengths are unchanged.

Since only two of the 13 members have changes of length, substitution in Eq. (6.8) involves only two terms, and there is no need for the tabular format used in Example 6.2.3. Recall that $u = +0.601$ for both AF and CF, so

$$\Delta_F = +0.601(-0.333) + 0.601(-0.333)$$
$$= -0.400 \text{ in}$$

So joint F would be displaced upward 0.400 in as a result of the two changes of length.

Occasionally the deflection of some point of a structure depends on bending, axial deformation, and torsion of various members. The unit load method is useful in these cases, with the deflection written using

$$\Delta = \int_{\substack{\text{whole} \\ \text{structure}}} \frac{Mm\,dx}{EI} + \sum_{\substack{\text{whole} \\ \text{structure}}} \frac{uNL}{AE} + \sum_{\substack{\text{whole} \\ \text{structure}}} \frac{TtL}{J_T G} \qquad (6.9)$$

You should recognize the first two terms as those related to bending and axial deformation. The third term, $\sum (TtL/J_T G)$, is similar to the others, except that it relates to torsional deformation. Its development, not given here, is exactly parallel to that of the other terms. The following defines its variables:

T = torque carried by the member, which is caused by the real loading of the structure

t = torque carried by the member that is caused by the unit force acting at and in the direction of the displacement to be determined

L = length of member

G = shear modulus of elasticity

J_T = torsional constant for cross section[2]

---

[2] The torsional constant for circular sections is the polar moment of inertia. For sections composed of several slender rectangular pieces, J_T may be taken as $\frac{1}{3}\sum ab^3$, where a is the long dimension of a rectangular piece and b is its thin dimension.

EXAMPLE 6.2.5

Determine the downward displacement of end D of the cable.

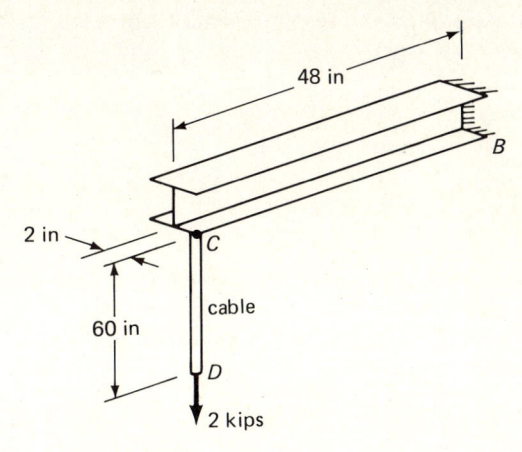

Figure 6.45

You may assume that the connection for the cable at C includes bracing at C such that there is no local distortion of the beam cross section.

For this loading the beam bends, the beam twists in torsion, and the cable stretches. For the loading,

$M = 2x$ kip-in

$T = 2(2) = 4$ kip-in

$N = 2$ kips

For the unit force at D,

$m = x$ in

$t = 2$

$u = 1$

Applying Eq. (6.9),

$$\Delta_D = \int_0^{48\,\text{in}} \frac{2x(x)\,dx}{3 \times 10^4 (11.3)} + \frac{1(2)(60)}{0.160(3 \times 10^4)} + \frac{4(2)(48)}{0.15(12 \times 10^3)}$$

$= 0.217$ in $+ 0.025$ in $+ 0.213$ in

$= 0.455$ in

6.3 CASTIGLIANO'S THEOREM

Castigliano's theorem[3] is another important energy method for determining displacement. It has many similarities to the method of virtual work.

Before we introduce the theorem itself, let us develop the idea of strain energy. A force F displaced dx along its direction does work

$$dW = F\,dx$$

For a finite displacement from x_1 to x_2 the work done is

$$W = \int_{x_1}^{x_2} F\,dx \tag{6.10}$$

Consider a force F applied to a structure and gradually increased from 0 to P as its point of application displaces Δ in the direction of F. If the force–displacement relation is linear and the structure ideal, that is, no energy losses due to joint friction and no backlash in the joints, the work done is

$$W = \int_0^{\Delta} F\,dx = \int_0^{\Delta} P\left(\frac{x}{\Delta}\right)dx = \frac{P\Delta}{2} \tag{6.11}$$

According to the law of conservation of energy, the work done is stored in the structure as an elastic potential energy called strain energy. For a truss the total strain energy would be

$$W = \sum_{\substack{\text{all}\\ \text{members}}} \frac{N\Delta L}{2}$$

where

N = axial force carried by an individual member
ΔL = change in length due to N in the individual member

Since $\Delta L = NL/AE$ for each member, the total strain energy for a truss becomes

$$W = \sum_{\substack{\text{all}\\ \text{members}}} \frac{N^2 L}{2AE} \tag{6.12}$$

Analogous to the preceding, a couple C applied to an ideal structure does work as it is gradually increased from 0 to M while its points of application rotates through θ. The work done and stored as strain energy is

$$W = \frac{M\theta}{2}$$

For a rigid frame or even for a beam it is necessary to account for the variation of moment with position. Thus

$$W = \tfrac{1}{2} \int_{\substack{\text{whole}\\ \text{structure}}} M\,d\theta$$

---

[3]There are two Castigliano's theorems. Often his first theorem is ignored in basic structural analysis books because it has lesser application, being primarily related to equilibrium and stability analysis. It is omitted in this book, too. Castigliano's second theorem is the more formal name for the very important theorem, which is called Castigliano's theorem in this book.

is used where M is considered constant over the differential element of length dx. Since

$$d\theta = \frac{M}{EI} dx$$

the total strain energy in the beam or frame is

$$W = \int_{\substack{\text{whole} \\ \text{structure}}} \frac{M^2}{2EI} dx \tag{6.13}$$

In cases where there are both internal axial forces and internal bending moments, the total strain energy is the sum of the strain energy for each,

$$W = \sum_{\substack{\text{all} \\ \text{members}}} \frac{N^2 L}{2AE} + \int_{\substack{\text{whole} \\ \text{structure}}} \frac{M^2}{2EI} dx \tag{6.14}$$

The summation and integration imply that all parts of the structure are included.

Castigliano's theorem: The first partial derivative of the total strain energy of a structure, taken with respect to an applied force (or couple), gives the displacement (or rotation) of the point of application and in the direction of the force (or couple).

Thus

$$\Delta = \frac{\partial W}{\partial P} = \frac{\partial}{\partial P} \left\{ \sum_{\substack{\text{all} \\ \text{members}}} \frac{N^2 L}{2AE} + \int_{\substack{\text{whole} \\ \text{structure}}} \frac{M^2}{2EI} dx \right\} \tag{6.15}$$

It is usually easier to take the derivative before integrating, so Eq. (6.15) becomes

$$\Delta = \frac{\partial W}{\partial P} = \sum_{\substack{\text{all} \\ \text{members}}} \frac{N(\partial N/\partial P) L}{AE} + \int_{\substack{\text{whole} \\ \text{structure}}} \frac{M(\partial M/\partial P)}{EI} dx \tag{6.16}$$

It should be obvious that appropriate terms will be zero for trusses and rigid frames. We will use the latter case to prove Castigliano's theorem. In Section 6.2 we defined

$m =$ internal bending moment due to a unit force (or couple) applied at the place and in the direction of the displacement

If, in addition, m_1 and m_2 are internal bending moments associated with unit forces at points 1 and 2, then the total bending moment M is

$$M = P_1 m_1 + P_2 m_2 + Pm$$

where P_1, P_2, and P are forces applied at points 1, 2, and the point in question. Taking the partial derivative

$$\frac{\partial M}{\partial P} = m$$

and substituting in Eq. (6.16),

$$\Delta = \int_{\substack{\text{whole} \\ \text{structure}}} \frac{Mm}{EI} dx$$

which is the virtual work expression for the deflection of the point in question due to the total loading P_1, P_2, and P.

A similar proof of Castigliano's theorem using the axial forces in a truss is left to the reader.

EXAMPLE 6.3.1

Use Castigliano's theorem to calculate the horizontal, vertical, and rotational displacements of point D. The material is steel, and $I = 500$ in^4.

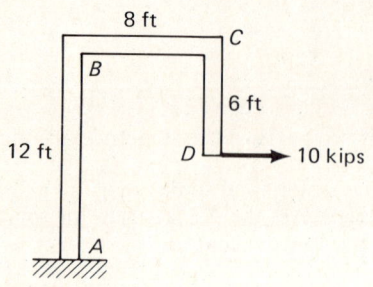

Figure 6.46

Castigliano's theorem requires that partial derivatives be taken with respect to the forces and couples acting at the point and in the direction of the displacement to be determined. The structural loading must include these forces and couples. So we will work with the structural loads shown here.

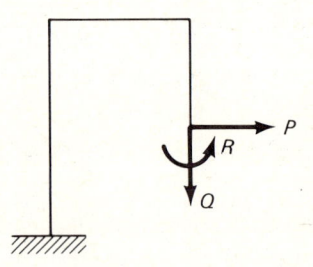

Figure 6.47

This illustrates two important features in the use of Castigliano's theorem. First, to calculate the vertical deflection and rotation of D, a vertical force and a couple, both at D, were necessary parts of the loading. Since neither are part of the real loading, they were added. But care must be taken that their addition does not change the real loading. Thus each will be given a value of 0 later in the calculation, Second, the horizontal force at D, 10 kips to the right in the real loading, is treated at the start with the algebraic symbol P. P will be given the value of 10 kips later in the calculation. All three, P, Q, and R, are kept as algebraic symbols, regardless of their actual values, so that partial derivatives may be taken with respect to them rather than to take a meaningless derivative with respect to a number.

Expressions for the internal bending moment[4] as a function of position and the loads P, Q, and R are found, and the partial derivatives are taken and tabulated in Table 6.5.

Table 6.5

Member	Range of x (ft)	Origin for x	M	$\dfrac{\partial M}{\partial P}$	$\dfrac{\partial M}{\partial Q}$	$\dfrac{\partial M}{\partial R}$
AB	0–12	B	$-R+8Q+P(x-6)$	$x-6$	8	-1
BC	0–8	C	$6P+R-Qx$	6	$-x$	1
CD	0–6	D	$Px+R$	x	0	1

In substituting these into Eq. (6.16), 10 kips is used for P and 0 for Q and R.

$$\Delta_{\text{horiz}} = \int_0^{12\text{ ft}} \frac{10(x-6\text{ kip-ft})(x-6\text{ ft})}{EI}\,dx$$

$$+ \int_0^{8\text{ ft}} \frac{6(10\text{ kip-ft})(6\text{ ft})}{EI}\,dx + \int_0^{6\text{ ft}} \frac{(10x\text{ kip-ft})(x\text{ ft})}{EI}\,dx$$

$$= \frac{1}{EI}\left\{\left[\frac{10}{3}x^3 - \frac{12x^2}{2} + 36x\right]_0^{12\text{ ft}} + \left[360x\right]_0^{8\text{ ft}} + \left[\frac{10}{3}x^3\right]_0^{6\text{ ft}}\right\}$$

$$= \frac{1}{EI}\{5328 + 2880 + 720\} = \frac{8929\text{ kip-ft}^3}{EI}$$

For the given E and I,

$$\Delta_{\text{horiz}} = \frac{(8929\text{ kip-ft}^3)(1728\text{ in}^3/\text{ft}^3)}{(3\times 10^4\text{ kip/in}^2)(500\text{ in}^4)} = 1.030\text{ in}$$

$$\Delta_{\text{vert}} = \int_0^{12} \frac{10(x-6)(8)}{EI}\,dx + \int_0^8 \frac{60(-x)}{EI} + \int_0^6 \frac{10x(0)}{EI}\,dx$$

$$= -\frac{1920\text{ kip-ft}^3}{EI} = -0.22\text{ in}$$

$$\theta = \int_0^{12} \frac{10(x-6)(-1)}{EI}\,dx + \int_0^8 \frac{60(1)}{EI}\,dx + \int_0^6 \frac{10x}{EI}\,dx$$

$$= \frac{660\text{ kip-ft}^2}{EI} = \frac{(660\text{ kip-ft}^2)(144\text{ in}^2/\text{ft}^2)}{(3\times 10^4\text{ kip/in}^2)(500\text{ in}^4)} = 0.00634\text{ rad}$$

---

[4]These expressions were found using the techniques treated in Section 4.3.

136   ELASTIC DEFLECTIONS OF PLANE STATICALLY DETERMINATE STRUCTURES

In statically determinate rigid frames

$$\sum \frac{(\partial N/\partial P)L}{AE}$$

is often ignored.

In the previous example, only BC carries an axial force, with $N = P$ and $\partial N/\partial P = 1$. For a cross-sectional area of 20 in² (consistent with $I = 500$ in⁴ for a W shape) the added term in Δ_{vert} would be

$$\frac{N(\partial N/\partial P)L}{AE} = \frac{(10 \text{ kips})(1)(8 \text{ ft})(12 \text{ in/ft})}{(20 \text{ in}^2)(3 \times 10^4 \text{ kips/in}^2)}$$

$$= 0.00160 \text{ in}$$

Clearly this is negligible in comparison to the 1.030-in horizontal displacement due to bending, which was calculated before.

The next example illustrates a case where both the bending of beams and the changes of the length of members are significant.

EXAMPLE 6.3.2

Use Castigliano's theorem to calculate the vertical displacement of C.

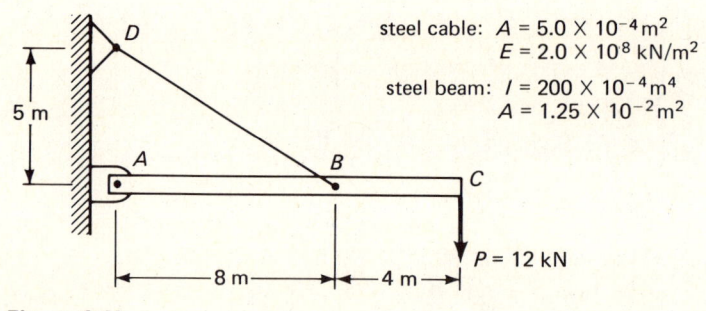

steel cable: $A = 5.0 \times 10^{-4} \text{ m}^2$
$E = 2.0 \times 10^8 \text{ kN/m}^2$

steel beam: $I = 200 \times 10^{-4} \text{ m}^4$
$A = 1.25 \times 10^{-2} \text{ m}^2$

$P = 12$ kN

Figure 6.48

The free-body diagram of the structure is shown with the support reactions.

2.83 P

2.4 P

0.5 P

P

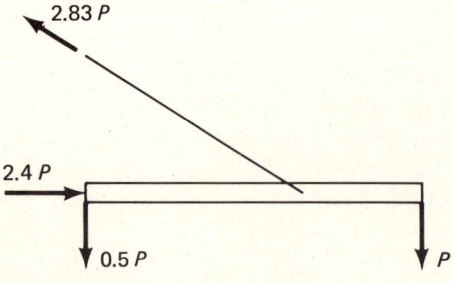

Figure 6.49

The internal forces and moments are given in Table 6.6.

Table 6.6

Member	Origin	Range of x (m)	M	$\dfrac{\partial M}{\partial P}$	N	$\dfrac{\partial N}{\partial P}$	L
AB	A	0–8	$-0.5Px$	$-0.5x$	$-2.4P$	-2.4	8
BC	C	0–4	$-Px$	$-x$	0	—	—
BD	—	—	—	—	$+2.83P$	2.83	9.434

Substituting in Eq. (6.16),

$$\Delta_{vert} = \int_0^{8m} \frac{(-0.5Px \text{ kN}\cdot\text{m})(-0.5x \text{ m})(dx \text{ m})}{EI}$$

$$+ \int_0^4 \frac{-Px \text{ kN}\cdot\text{m})(-x \text{ m})(dx \text{ m})}{EI}$$

$$+ \frac{(-2.4P \text{ kN})(-2.4)(8 \text{ m})}{A_{beam} E} + \frac{2.83P \text{ kN}(2.83)(9.434 \text{ m})}{A_{cable} E}$$

Integrating and substituting $P = 12$ kN, these become

$$\Delta_{vert} = \frac{512 \text{ kN}\cdot\text{m}^3}{EI} + \frac{256 \text{ kN}\cdot\text{m}^3}{EI} + \frac{552.96 \text{ kN}\cdot\text{m}}{A_{beam} E} + \frac{906.67 \text{ kN}\cdot\text{m}}{A_{cable} E}$$

Substituting for E, I, A_{beam}, and A_{cable},

$$\Delta_{vert} = 0.0128 + 0.0064 + 0.00022 + 0.00907 \text{ m}$$
$$= 0.0285 \text{ m}$$

6.4 FLEXIBILITY COEFFICIENTS

Flexibility coefficients will be used in the method of consistent deformations, our first venture into statically indeterminate structures. Since the flexibility coefficients in that application are displacements of statically determinate structures, it is appropriate to introduce them here.

The flexibility coefficient δ_{ij} is defined as the displacement (linear or rotational) of location i of a structure caused by a unit load (force or couple) acting at location j. Several examples follow in illustration. For the beam below consider 1 to refer to a downward linear displacement of B, 2 to refer to a downward linear displacement of C, and 3 to refer to a counterclockwise rotation of end D.

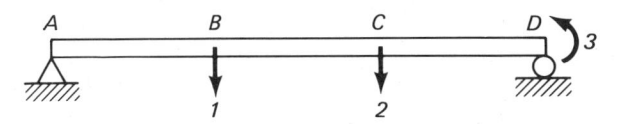

Figure 6.50

138 ELASTIC DEFLECTIONS OF PLANE STATICALLY DETERMINATE STRUCTURES

Consider a unit downward force in location 1 and the resulting deformation. The displacements δ_{11}, δ_{21}, and δ_{31} portray the flexibility coefficients.

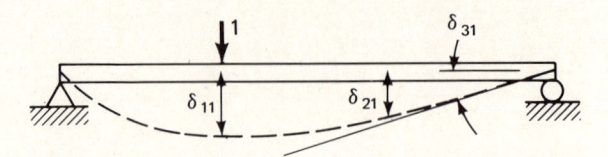

Figure 6.51

Now consider a unit counterclockwise couple acting in location 3 and its resulting deformation. The displacements indicated portray the flexibility coefficients.

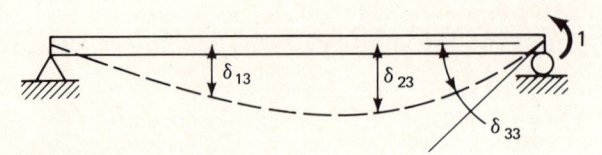

Figure 6.52

Recall Eq. (6.6),

$$\Delta = \int_{\substack{\text{whole}\\\text{structure}}} \frac{mM}{EI} dx \tag{6.6}$$

where

m = bending moment caused by a unit virtual load applied at and in the direction of Δ
M = bending moment caused by real loads

The Δ we calculate for a flexibility coefficient is δ_{ij}. Use of m_i for m adapts Eq. (6.6) to the proper place and direction of displacement. Use of m_j in place of M establishes the unit loading applied at j as the cause of the displacement. Thus Eq. (6.6) becomes

$$\delta_{ij} = \int_{\substack{\text{whole}\\\text{structure}}} \frac{m_i m_j}{EI} dx \tag{6.17}$$

EXAMPLE 6.4.1

Calculate the flexibility coefficients for the vertical and rotational displacements of end A.

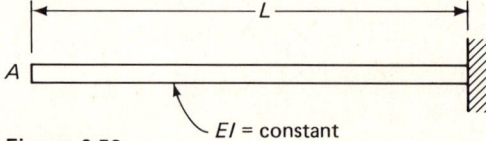

Figure 6.53

EI = constant

6.4 FLEXIBILITY COEFFICIENTS

First define 1 as a downward displacement of A and 2 as a counterclockwise rotation of A.

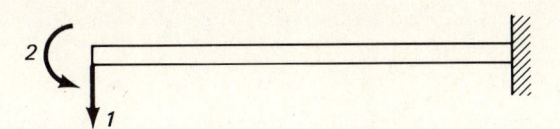

Figure 6.54

Determine m_1.

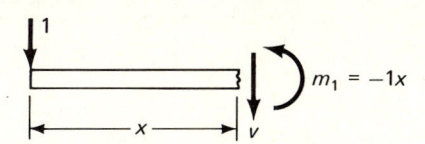

$m_1 = -1x$

Figure 6.55

Then determine m_2.

$m_2 = -1$

Figure 6.56

Substituting these in various combinations of $i = 1, 2$ and $j = 1, 2$,

$$\delta_{11} = \int_{\text{whole structure}} \frac{m_1 m_1}{EI} dx = \int_0^L \frac{-x(-x)}{EI} dx = \frac{L^3}{3EI}$$

$$\delta_{12} = \delta_{12} = \int_{\text{whole structure}} \frac{m_1 m_2}{EI} dx = \int_0^L \frac{-x(-1)}{EI} dx = \frac{L^2}{2EI}$$

$$\delta_{22} = \int_{\text{whole structure}} \frac{m_2 m_2}{EI} = \int_0^L \frac{-1(-1)}{EI} dx = \frac{L}{EI}$$

Dimensionally, δ_{11} is length/force, and δ_{12} and δ_{21} are radian/force and length/moment, respectively. You are urged to think about this a little to resolve this apparent paradox. Finally, δ_{22} is radian/moment.

In Section 6.3 an equivalence of $\partial M/\partial R$ and m was shown. It should be easy for you to extend this equivalence to show that an alternate, and very useful, equation for flexibility coefficients based on Castigliano's theorem is

$$\delta_{ij} = \int_{\text{whole structure}} \frac{(\partial M/\partial R_i)(\partial M/\partial R_j)}{EI} dx \tag{6.18}$$

It will be necessary to calculate flexibility coefficients for trusses, too. The unit load method for trusses gave displacements using

$$\Delta = \sum_{\text{whole structure}} \frac{uNL}{AE} \qquad (6.7)$$

This may be converted to an equation for flexibility coefficients in a way completely analogous to that just used for beams and frames. The set of member forces N is replaced by u_j, which is the set of member forces for a unit load at j. Then u becomes u_i, the set of member forces for the unit load at point i. Thus Eq. (6.7) becomes

$$\delta_{ij} = \sum_{\text{whole structure}} \frac{u_i u_j L}{AE} \qquad (6.19)$$

Notice that $\delta_{ij} = \delta_{ji}$ for any structure. This is a valuable characteristic of all structures: a computation for δ_{ij} provides δ_{ji} directly, reducing computational effort. This is a well-known phenomenon known as Maxwell's reciprocal theorem.

PROBLEMS

Sequenced Problem 1, parts (g), (h), (i), (j), (k), (l), and (m).
Sequenced Problem 2, parts (f), (g), (h), and (i).
Sequenced Problem 3, parts (c), (e), (f), (g), (h), (i), and (j).
Sequenced Problem 4, parts (c), (e), (f), (g), (h), (i), and (j).

6.1

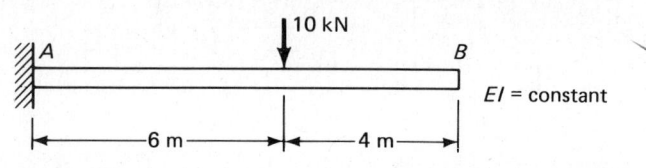

Figure P6.1

(a) Use the method of conjugate beams to determine the slope and deflection at B.
(b) Use moment-area methods to determine the slope and deflection at B.
(c) Compare the results and the expressions written in obtaining them in parts (a) and (b).

6.2

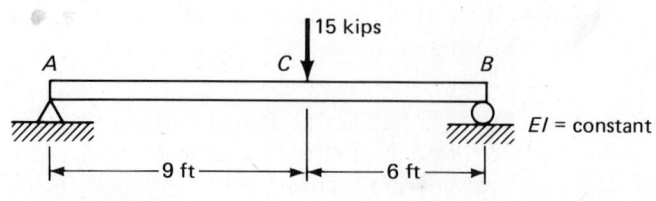

Figure P6.2

(a) Use the method of conjugate beams to determine the slope and deflection at C.

(b) Use the method of conjugate beams to determine the location and magnitude of the maximum deflection.
(c) Use moment-area methods to determine the slope at A, and the location and magnitude of the maximum deflection.
(d) Use Castigliano's theorem to determine the vertical deflection at C and the slope at C.
(e) Use Castigliano's theorem to determine the slopes at A and B.

6.3

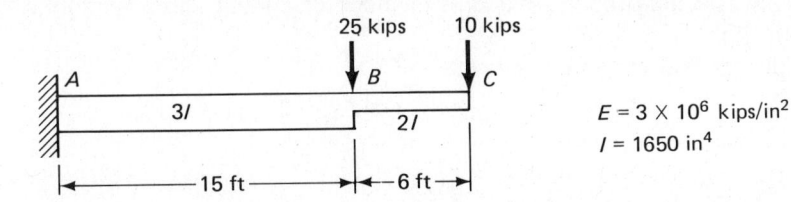

Figure P6.3

(a) Use conjugate beams to determine the slope and deflection at B.
(b) Use conjugate beams to determine the slope and deflection at C.
(c) Use moment-area methods to determine the slope and deflection at B.
(d) Use moment-area methods to determine the slope and deflection at C.
(e) Use the method of virtual work (unit load method) to determine the slope and deflection at B.
(f) Use the method of virtual work (unit load method) to determine the slope and deflection at C.
(g) Use Castigliano's theorem to determine the slope and deflection at B.
(h) Use Castigliano's theorem to determine the slope and deflection at C.

6.4

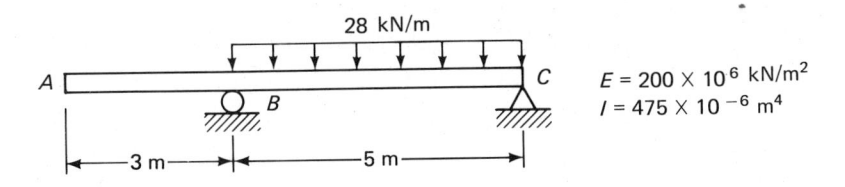

Figure P6.4

(a) Use the method of conjugate beams to determine the slope and deflection at A.
(b) Use moment-area methods to determine the slope and deflection at A.
(c) Use the method of virtual work (unit load method) to determine the slope and deflection at A.

6.5

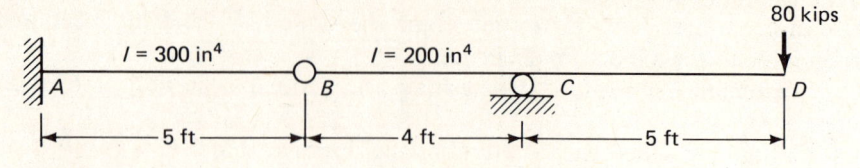

Figure P6.5

Determine the slopes at B (there are two slopes to be determined, one at end B of member AB and the other at end B of member BCD) and the deflection at B. $E = 3 \times 10^4$ kips/in^2.
(a) Use the method of conjugate beams.
(b) Use the method of virtual work (unit load method).

6.6

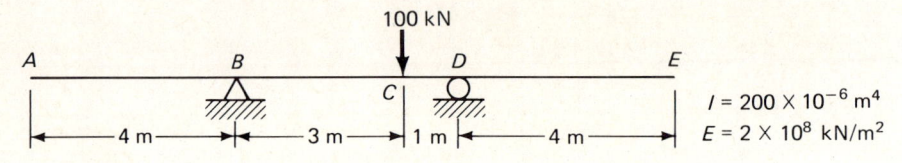

Figure P6.6

(a) Use the method of conjugate beams to determine the deflections at A and E.
(b) Use moment-area methods to determine the deflections at A and E.
(c) Use the method of virtual work (unit load method) to determine the deflections at A and E.

6.7

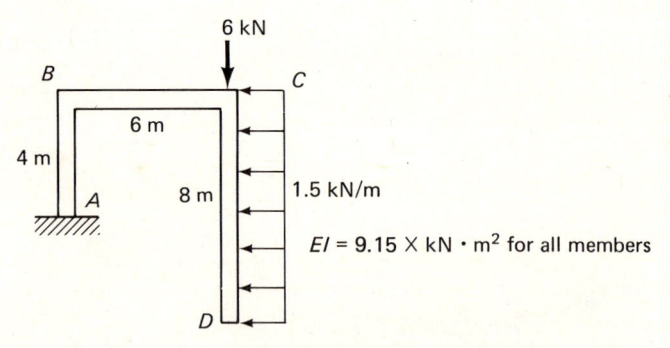

Figure P6.7

The structure and loads from Problem 4.4 are repeated here. Use the method of virtual work.
(a) Determine the vertical deflection at D.
(b) Determine the horizontal deflection at D.
(c) Determine the rotation at D.
(d) Determine the horizontal displacement and rotation at B.

6.8

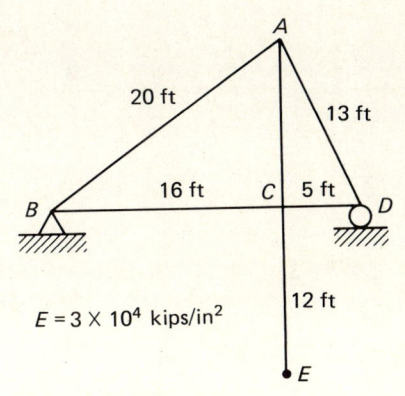

Figure P6.8

(Diagram: Triangular truss with joints A (top), B (left support, pinned), C (middle), D (right support, roller), and E (bottom). AB = 20 ft, AC = 16 ft, AD = 13 ft, CD = 5 ft, CE = 12 ft. $E = 3 \times 10^4$ kips/in²)

Cross-sectional areas are

$AB = 6 \text{ in}^2 \quad\quad CE = 4.0 \text{ in}^2$
$AD = 4.5 \text{ in}^2 \quad\quad BC = 2.5 \text{ in}^2$
$AC = 4.0 \text{ in}^2 \quad\quad CD = 2.5 \text{ in}^2$

Determine the vertical displacement of joint E if:
(a) 100 kips acts downward at A.
(b) A unit force acts downward at E.

6.9

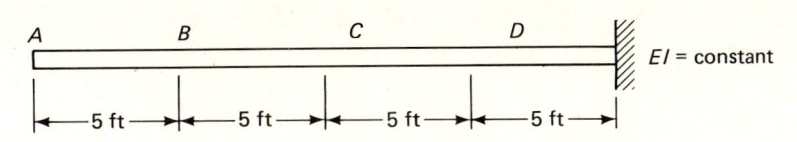

Figure P6.9

(Diagram: Cantilever beam fixed at right end with points A, B, C, D along the beam. Spacing: A to B = 5 ft, B to C = 5 ft, C to D = 5 ft, D to fixed end = 5 ft. EI = constant)

Assign subscripts 1, 2, 3, and 4 related to downward forces and displacements at A, B, C, and D, respectively. Determine all the flexibility coefficients.

6.10

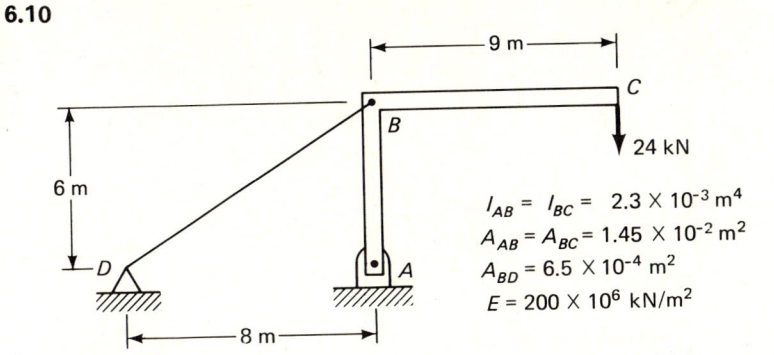

Figure P6.10

(Diagram: Frame with horizontal member BC of 9 m, vertical member from B going down 6 m to A (pinned), diagonal member from B to D (pinned support), horizontal distance D to A = 8 m. Load 24 kN downward at C.)

$I_{AB} = I_{BC} = 2.3 \times 10^{-3} \text{ m}^4$
$A_{AB} = A_{BC} = 1.45 \times 10^{-2} \text{ m}^2$
$A_{BD} = 6.5 \times 10^{-4} \text{ m}^2$
$E = 200 \times 10^6 \text{ kN/m}^2$

Use Castigliano's theorem to calculate the vertical displacement of C.

6.11

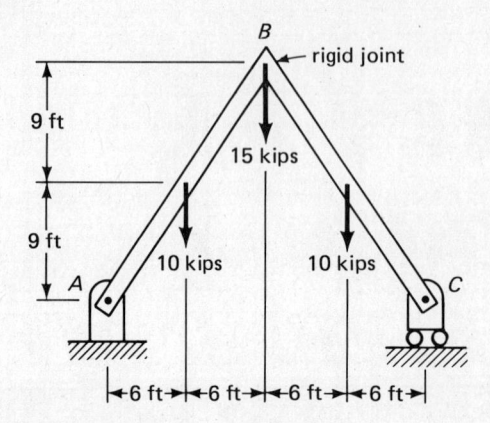

Figure P6.11

Determine the horizontal displacement of C if $EI = 3.15 \times 10^7$ kip-in^2 and $AE = 9.25 \times 10^5$ kips throughout.

Part II
STATICALLY INDETERMINATE STRUCTURAL ANALYSIS: FORCE/FLEXIBILITY METHODS

Two distinct methods of analysis are in use for analyzing statically indeterminate structures. We examine the first of these, known as both force methods and flexibility methods, in Chapters 7 through 9. Several variations on these methods will be examined, including consistent deformations, least work, and conjugate beams. There are others that are not included in this text.

Each approach to the force/flexibility methods supplements the equations of statics with flexibility equations in which the extra unknowns are forces.

7
Statically Indeterminate Analysis: Method of Consistent Deformations

7.0 INTRODUCTION

The track crews, having noticed a problem, are discussing who will pick up a few hundred feet of track and re-lay it with better alignment.

We begin our study of statically indeterminate structures with a method somewhat akin to the railroad track since it introduces misalignments of the structure and then corrects them. Fortunately our misalignments and corrections are done on paper and rarely are accompanied by a brawl. The method, known as consistent deformations, may be familiar to you already. Most courses in strength of materials introduce the method in a simplified form.

EXAMPLE 7.0.1

Determine the support reactions of the structure loaded and supported as shown.

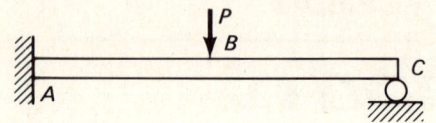

Figure 7.1

In strength of materials the support at C would be removed and the downward displacement Δ'_C of point C calculated.

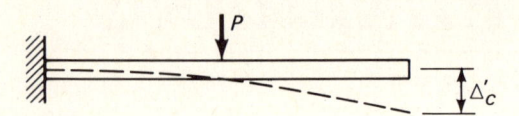

Figure 7.2

Then the value would be calculated for force R_C acting upward at C (it replaces the support), which would cause an equal upward displacement, Δ'_C, of point C.

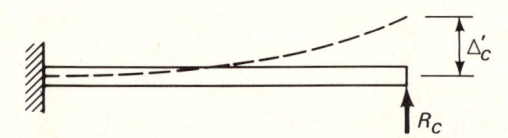

Figure 7.3

The superposition of these two cases gives a zero displacement of point C, a deformation consistent with the requirement of the support there. Since R_C has replaced the support at C and does exactly what support C would do, it must be the support reaction. With R_C known, the structure has become statically determinate, and the remaining support reactions, all at A, may be calculated using the three equilibrium equations.

7.1 BEAM AND FRAME ANALYSIS BY CONSISTENT DEFORMATIONS

The simplified form of consistent deformations used in Example 7.0.1 is packed with potential sign errors if used to analyze beams and frames several degrees indeterminate. A more disciplined method, which obviates these problems, is illustrated in the following example.

EXAMPLE 7.1.1

Determine the support reactions for the two-degree statically indeterminate beam loaded and supported as shown.

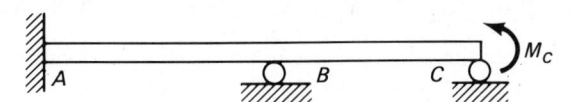

Figure 7.4

We remove structural constraints equal in number to the degree of static indeterminacy and replace them with the unknown forces or couples they provided. In our example we will remove the rotational support at A and the vertical support at B.[1]

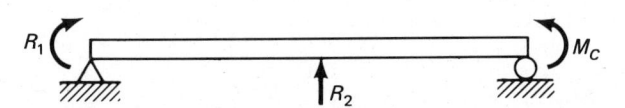

Figure 7.5

The structure with its remaining supports is called the primary structure. The forces or couples replacing these supports are called redundant forces and redundant couples; often they are just called *redundants*. Our two redundants, a couple at A and a vertical force at B, are designated R_1 and R_2. A single letter R is used for both, even though one is a force and the other a couple. The numerical subscripts that distinguish between them are chosen arbitrarily without any implication of a relation between them and the letter names for their point of application. Notice that the primary structure is stable and statically determinate.

More will be said in Section 7.2 on the selection of the primary structure. Suffice it to say here that the choice made can have a marked effect on the ease or difficulty of solution. In statically indeterminate structures of a high degree, for which a computer would be required for the solution, the choice of redundants can have a marked effect on the

---

[1] It will be shown in Section 7.2, that a better choice could have been made for the set of structural constraints to be removed. The two used here were selected because they are similar to typical choices made in strength of materials courses and, therefore, should be easily understood. Choice of this set permits introduction of the more disciplined notation without the distraction of other new concepts.

150 METHOD OF CONSISTENT DEFORMATIONS

accuracy of the solution, too. That may be a moot point since the method of consistent deformations is rarely used to analyze high degree statically indeterminate structures.

For our example, superposition of the three situations shown here is recommended.

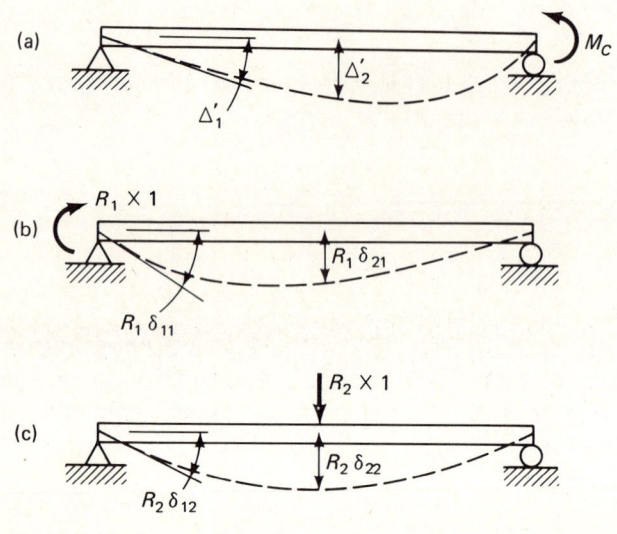

Figure 7.6

Figure 7.6(a) shows the primary structure with the original load M_C applied. The broken line represents its elastic curve with Δ'_1 and Δ'_2 as the resulting displacements of redundant points 1 and 2. Analogous to the use of the symbol R, Δ' is used for both linear and angular displacements. Again subscripts are used to distinguish one from the other, and it is essential that they match subscript for subscript with those used for the redundant force and couple.

Figure 7.6(b) is the primary structure with redundant R_1 applied. Its elastic curve is shown; the displacements of *both* redundant locations are indicated. For convenience in calculations, these displacements are treated as the products $R_1\delta_{11}$ and $R_1\delta_{21}$, where δ_{11} and δ_{21} are the flexibility coefficients. Since δ_{11} and δ_{21} represent displacements of redundant points 1 and 2 caused by a *unit* couple at 1, the products $R_1\delta_{11}$ and $R_1\delta_{21}$ represent the redundant point displacements caused by R_1. Figure 7.6(c) should require no explanation; its analogy to Figure 7.6(b) should be evident.

Values of $\Delta'_1, \Delta'_2, \delta_{11}, \delta_{12}, \delta_{21},$ and δ_{22} can be determined by the application of techniques studied in Chapter 6. Once they are known, the requirement that redundant points 1 and 2 have zero displacement, consistent with their original rigid supports, is satisfied by superposing the displacements at each for the three cases. Thus

7.1 BEAM AND FRAME ANALYSIS BY CONSISTENT DEFORMATIONS

$$\Delta'_1 + R_1\delta_{11} + R_2\delta_{12} = 0$$
$$\Delta'_2 + R_1\delta_{21} + R_2\delta_{22} = 0$$
(7.1)

These equations may be solved simultaneously for the redundants R_1 and R_2. Then the three remaining support reactions can be found using three equilibrium equations.

A structure statically indeterminate to the nth degree requires n redundants and n^2 flexibility coefficients which would be used in n equations:

$$\Delta'_1 + R_1\delta_{11} + R_2\delta_{12} + \cdots + R_n\delta_{1n} = 0$$
$$\Delta'_2 + R_1\delta_{21} + R_2\delta_{22} + \cdots + R_n\delta_{2n} = 0$$
$$\vdots$$
$$\Delta'_n + R_1\delta_{n1} + R_2\delta_{n2} + \cdots + R_n\delta_{nn} = 0$$
(7.2)

Readers acquainted with matrix algebra should recognize these as

$$\begin{Bmatrix} \Delta'_1 \\ \Delta'_2 \\ \vdots \\ \Delta'_n \end{Bmatrix} + \begin{bmatrix} \delta_{11} & \delta_{12} & \cdots & \delta_{1n} \\ \delta_{21} & \delta_{22} & \cdots & \delta_{2n} \\ & & & \\ \delta_{n1} & \delta_{n2} & \cdots & \delta_{nn} \end{bmatrix} \begin{Bmatrix} R_1 \\ R_2 \\ \vdots \\ R_n \end{Bmatrix} = 0$$
(7.3)

or as

$$\{\Delta'\} + [\delta]\{R\} = 0$$
(7.4)

EXAMPLE 7.1.2

Determine the support reactions for the rigid frame loaded and supported as shown.

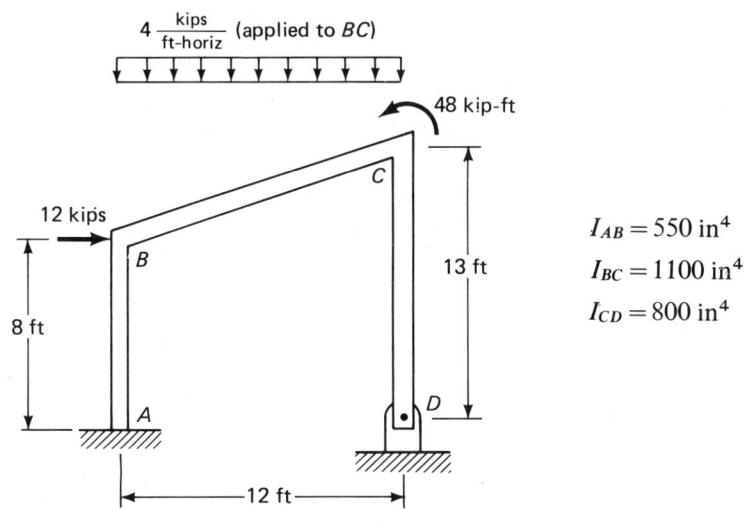

$I_{AB} = 550$ in^4
$I_{BC} = 1100$ in^4
$I_{CD} = 800$ in^4

Figure 7.7

A statically determinate structure with the same loads and dimensions was considered in Example 4.2.4, and the bending moments from the real loading were calculated. We will take advantage of this previous work and use the structure of Example 4.2.4 as the primary structure. Then we need only to determine the terms for bending moment related to the redundants, which we will add to the moment expressions calculated in Example 4.2.4.

The primary structure with its redundants R_1 and R_2 is shown below.

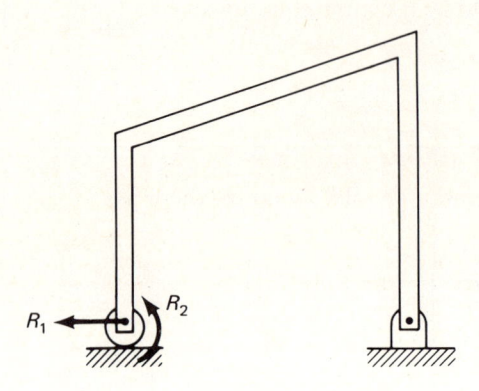

Figure 7.8

To find the additional terms in the moment expressions related to the redundants, it is first necessary to calculate the primary structure's support reactions in terms of R_1 and R_2. These are found through the equilibrium relations and are shown here with the redundants themselves.

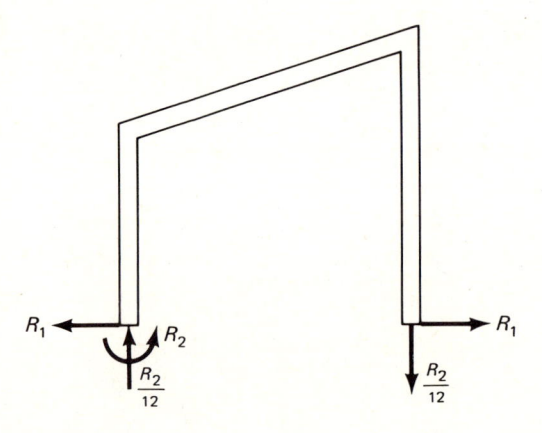

Figure 7.9

The members are sectioned and the internal moments found. For AB,

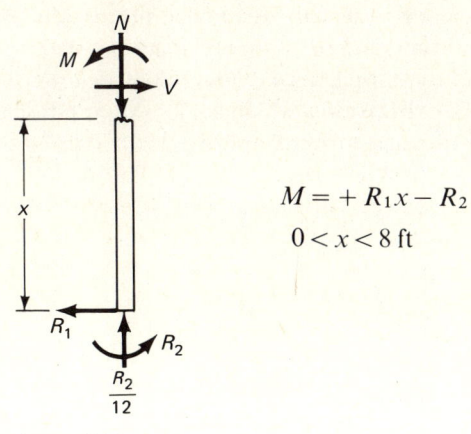

Figure 7.10

$$M = +R_1 x - R_2$$
$$0 < x < 8 \text{ ft}$$

For *BC*,

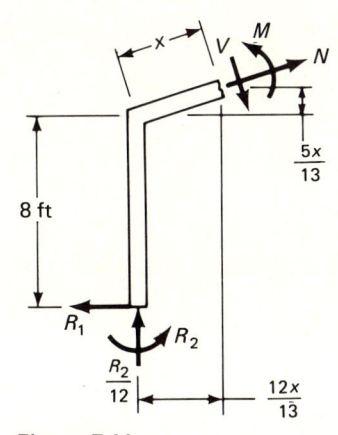

Figure 7.11

$$M = R_1\left(8 + \frac{5}{13}x\right) + R_2\left(\frac{x}{13}\right)$$
$$0 < x < 13 \text{ ft}$$

For *CD*,

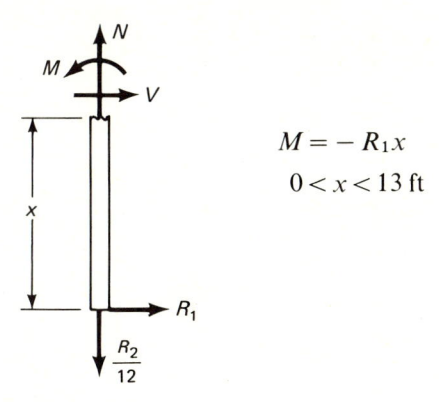

$$M = -R_1 x$$
$$0 < x < 13 \text{ ft}$$

Figure 7.12

METHOD OF CONSISTENT DEFORMATIONS

Although it may not have been obvious to you, the definitions of x and the sign conventions for the moment used in these added moment terms are identical to those used in Example 4.2.4. *This consistency is mandatory.* The complete moment expressions, which are the summations of those from Example 4.2.4 and those just found, are listed in Table 7.1.

Table 7.1

Member	Range of x (ft)	M (kip-ft)
AE	0–8	$R_1 x - R_2$
BC	0–13	$-1.70x^2 + 13.84x + R_1(8 + \frac{5}{13}) + R_2 \frac{x}{13}$
CD	0–13	$-12x - R_1 x$

Now Eqs. (6.16) and (6.18) are applied to get Δ'_i and δ_{ij}, respectively. Since Δ'_i represents deflections of the redundant points without the presence of the redundant forces, the values of R_i are taken as zero in Eq. (6.16), and the effects of axial forces, assumed small, are ignored.

$$\Delta'_1 = \int_{\text{whole structure}} \frac{M(\partial M/\partial R_1)}{EI} dx$$

$$= \int_0^{13\,\text{ft}} \frac{(-1.7x^2 + 13.84x \text{ kip-ft})(8 + \frac{5}{13}x \text{ ft})(dx \text{ ft})(144 \text{ in}^2/\text{ft}^2)}{(3 \times 10^4 \text{ kips/in}^2)(1100 \text{ in}^4)}$$

$$+ \int_0^{13\,\text{ft}} \frac{-12x(-x)(dx)(144)}{3 \times 10^4 (800)}$$

$$= 0.679 \text{ ft}$$

$$\Delta'_2 = \int_{\text{whole structure}} \frac{M(\partial M/\partial R_2)}{EI} dx$$

$$= \int_0^{13\,\text{ft}} \frac{(-1.7x^2 + 13.84x \text{ kip-ft})(x/13)(dx \text{ ft})(144 \text{ in}^2/\text{ft}^2)}{(3 \times 10^4 \text{ kips/in}^2)(1100 \text{ in}^4)}$$

$$= -6.723 \times 10^{-4} \text{ rad}$$

$$\delta_{11} = \int_{\text{whole structure}} \frac{(\partial M/\partial R_1)^2}{EI} dx$$

$$= \int_0^{8\,\text{ft}} \frac{(x^2 \text{ ft}^2)(dx \text{ ft})(144 \text{ in}^2/\text{ft}^2)}{(3 \times 10^4 \text{ kips/in}^2)(550 \text{ in}^4)} + \int_0^{13\,\text{ft}} \frac{(8 + \frac{5}{13}x)^2 (dx)(144)}{3 \times 10^4 (1100)}$$

$$+ \int_0^{13\,\text{ft}} \frac{(-x)^2 (dx)(144)}{3 \times 10^4 (800)}$$

$$= 1.225 \times 10^{-2} \text{ ft/kip}$$

7.1 BEAM AND FRAME ANALYSIS BY CONSISTENT DEFORMATIONS 155

$$\delta_{21} = \delta_{12} = \int_{\substack{\text{whole}\\\text{structure}}} \frac{(\partial M/\partial R_1)(\partial M/\partial R_2)}{EI} dx$$

$$= \int_0^{8\,\text{ft}} \frac{(x\,\text{ft})(-1)\,(dx\,\text{ft})(144\,\text{in}^2/\text{ft}^2)}{(3\times 10^4\,\text{kips/in}^2)(550\,\text{in}^4)} + \int_0^{13\,\text{ft}} \frac{(8+\tfrac{5}{13}x)(x/13)\,(dx)(144)}{3\times 10^4 (1100)}$$

$$= -4.215 \times 10^{-5}\,\text{rad/kip} \quad\text{or}\quad -4.215 \times 10^{-5}\,\text{ft/kip-ft}$$

$$\delta_{22} = \int_{\substack{\text{whole}\\\text{structure}}} \frac{(\partial M/\partial R_2)^2}{EI} dx$$

$$= \int_0^{8\,\text{ft}} \frac{-1(-1)\,(dx\,\text{ft})(144\,\text{in}^2/\text{ft}^2)}{(3\times 10^4\,\text{kips/in})(550\,\text{in}^4)} + \int_0^{13\,\text{ft}} \frac{(x/13)^2\,(dx)(144)}{3\times 10^4 (1100)}$$

$$= 8.873 \times 10^{-5}\,\text{rad/kip-ft}$$

Substituting in Eq. (7.2) gives

$$0.679\,\text{ft} + (R_1\,\text{kips})(1.225 \times 10^{-2}\,\text{ft/kip})$$
$$+ (R_2\,\text{ft-kip})(-4.215 \times 10^{-5}\,\text{ft/ft-kip}) = 0$$
$$-6.723 \times 10^{-4}\,\text{rad} + (R_1\,\text{kips})(-4.215 \times 10^{-5}\,\text{rad/kip})$$
$$+ (R_2\,\text{ft-kip})(8.873 \times 10^{-5}\,\text{rad/ft-kip}) = 0$$

and solving simultaneously gives

$$R_1 = -55.49\,\text{kips}$$
$$R_2 = -18.78\,\text{kip-ft}$$

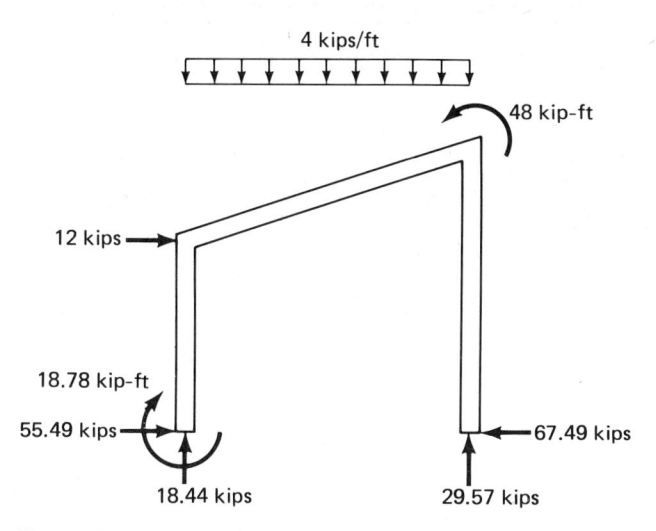

Figure 7.13

The remaining support reactions A_y, D_x, and D_y are found by using equilibrium and are shown on the free-body diagram.

7.2 SELECTION OF PRIMARY STRUCTURES

The selection of primary structures is a critical step. It is the dominant factor controlling the ease and accuracy of the solution.

Although we have not used them in examples before, there are two guidelines:

(1) Select a symmetric primary structure if the original structure is symmetric.
(2) If possible, select a primary structure that separates the original structure into several parts in such a way that if a single redundant force or couple were applied, only one or two of the parts of the structure would deform.

As noted in Section 7.1, primary structures must be statically determinate and stable.[2]

Consider the symmetric four-degree statically indeterminate beam.

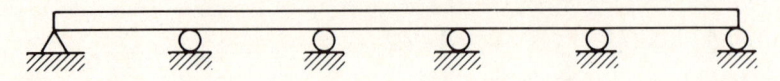

Figure 7.14

Several possible primary structures could be used. Three are illustrated.

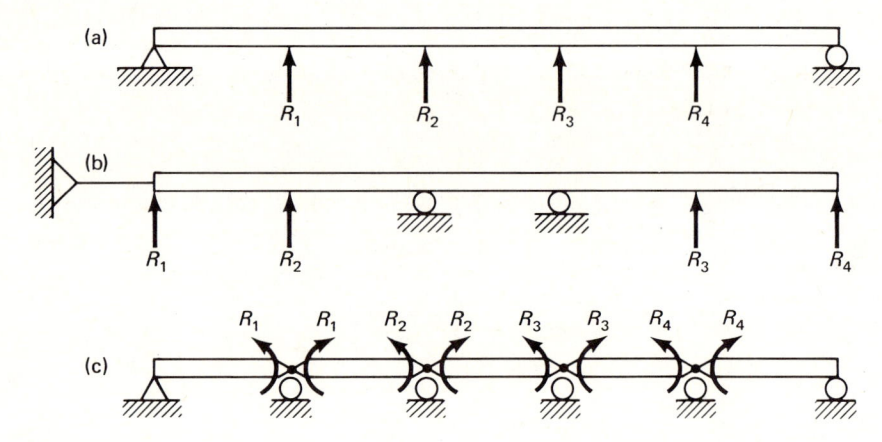

Figure 7.15

Redundant forces or couples act at the redundant points released from some of the constraints of the original structure. In Figure 7.15(a) and (b), redundant forces replace redundant vertical supports. These two primary structures are similar to those used in the earlier chapters of this book. Neither satisfies guideline (2), as will be shown. Figure 7.15(c), which satisfies both guidelines, releases the continuity of slope of the structure at each support by adding hinges that permit abrupt slope changes. The redundants are pairs of equal, opposite couples, ultimately found to have values that prevent any relative rotations across the hinges. These values, of course, will depend on the actual loading of the structure.

For each of the three primary structures a unit force (or couple) is shown applied

---

[2] A neutrally stable primary structure may be used as long as the original structure is stable, and if the application of the real loads would not require a nonzero value of one or more of the redundants in order to prevent rigid-body motion of any part of the primary structure.

at one of the redundant points. The resulting deformed shapes are illustrated by the broken lines.

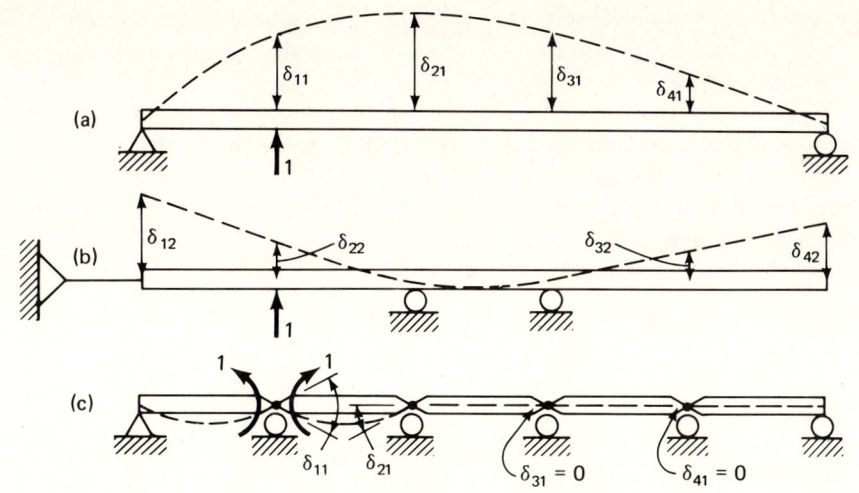

Figure 7.16

Displacements of redundant points for the applied unit forces and couples are labeled to represent the flexibility coefficients. In Figure 7.16(a) and (b) all parts of the structure deform, violating guideline (2). As a result, all of the δ_{ij} have nonzero values and must be calculated. Had the unit forces been applied at other redundant points in these two cases, we would find the same conclusion. For these two cases all of the δ_{ij} would have to be calculated.

That is not so for Figure 7.16(c). The flexibility coefficients, which represent relative rotations of the beam across the hinges, have some zero values. This is so since the unit couples at 1 cause only the left two spans to deform and leaves the other three undisturbed. It is safe to infer directly from such a sketch that $\delta_{31} = 0$ and $\delta_{41} = 0$. If the unit couples are applied at 2, 3, or 4, other $\delta_{ij} = 0$ are revealed. The savings in calculational effort should be obvious.[3]

Surprisingly, the reduced calculations for the δ_{ij} in Figure 7.15(c) may not be the most important feature. For this case, examination of the relative sizes of the δ_{ij} shows

$$|\delta_{11}| > |\delta_{21}| + |\delta_{31}| + |\delta_{41}|$$
$$|\delta_{22}| > |\delta_{12}| + |\delta_{32}| + |\delta_{42}|$$

and so on. The flexibility coefficient matrix is diagonally dominant.[4]

---

[3] One could philosophize that there is a virtue in laziness. Finding the primary structure, for which there are many directly noticeable $\delta_{ij} = 0$, may seem a lazy approach, but it also means fewer chances for calculational errors. If we reduce errors by working less strenuously, we are better analysts.

[4] A matrix

$$\begin{bmatrix} \delta_{11} & \delta_{12} & \delta_{13} & & \delta_{1n} \\ \delta_{21} & \delta_{22} & & & \delta_{2n} \\ \vdots & & & & \\ \delta_{n1} & \delta_{n2} & & & \delta_{nn} \end{bmatrix}$$

is diagonally dominant if, for every row, the absolute value of the diagonal element δ_{kk} of the row is larger than the sum of the absolute values of the remaining elements of the row.

It is known through numerical analysis theory that mathematical stability is certain for a system of simultaneous equations having a diagonally dominant coefficient matrix. So we conclude that the primary structure for Figure 7.15(c) will provide a well behaved set of equations that can be solved with ease and accuracy for the redundant couples.

You are urged to examine the δ_{ij} for Figure 7.15(a) and (b) and convince yourself that their flexibility coefficient matrices may not be diagonally dominant.

Occasionally it is difficult to sketch the deformed shape of a proposed primary structure. This may make it difficult to determine if there will be parts of it that will be undeformed by application of the various redundants. Equation (6.18) provides a convenient supplementary scheme:

$$\delta_{ij} = \int_{\substack{\text{whole} \\ \text{structure}}} \frac{(\partial M/\partial R_i)(\partial M/\partial R_j)}{EI} \tag{6.18}$$

δ_{ij} will be zero if either $(\partial M/\partial R_i)$ or $(\partial M/\partial R_j)$ is zero. If, for the whole structure, there are no parts having a bending moment that is a function of both R_i and R_j, then either $(\partial M/\partial R_i)$ or $(\partial M/\partial R_j)$ will be zero in each part, and that $\delta_{ij} = 0$. To test for $\delta_{ij} = 0$, therefore, write functional representations of the bending moment of each part and examine them for those redundant pairs that are not common to any member. Their $\delta_{ij} = 0$.

EXAMPLE 7.2.1

Consider the six-degree statically indeterminate rigid frame.

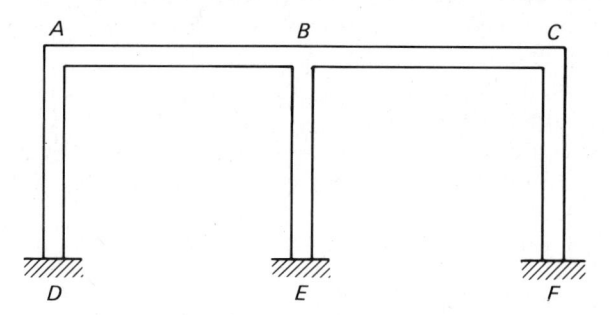

Figure 7.17

The proposed primary structure that follows is to be examined for zero-valued δ_{ij}.

$M_{AB} = f_1(R_1, R_3)$

$M_{BC} = f_2(R_4, R_6)$

$M_{AD} = f_3(R_1, R_2, R_3)$

$M_{BE} = f_4(R_2, R_3, R_5, R_6)$

$M_{CF} = f_5(R_4, R_5, R_6)$

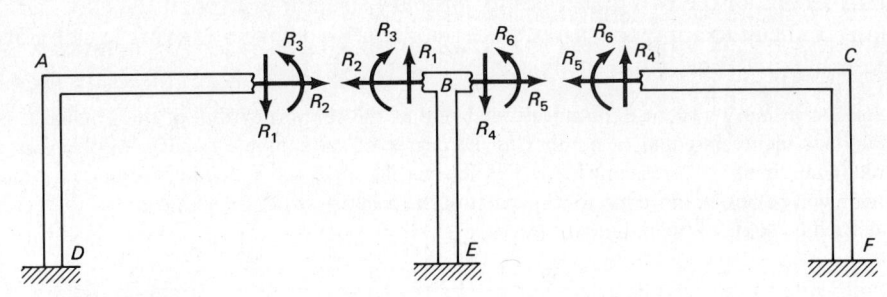

Figure 7.18

No member has a bending moment relating to both R_1 and R_4, so we conclude that

$$\delta_{14} = \delta_{41} = 0$$

Similarly, no members have R_1 and R_5 in common, R_1 and R_6 in common, R_2 and R_4 in common, or R_3 and R_4 in common. Therefore

$$\delta_{15} = \delta_{51} = 0$$
$$\delta_{16} = \delta_{61} = 0$$
$$\delta_{24} = \delta_{42} = 0$$
$$\delta_{34} = \delta_{43} = 0$$

Clearly this is a quick method to evaluate a candidate primary structure. The one just examined is evidently superior to both of the following.

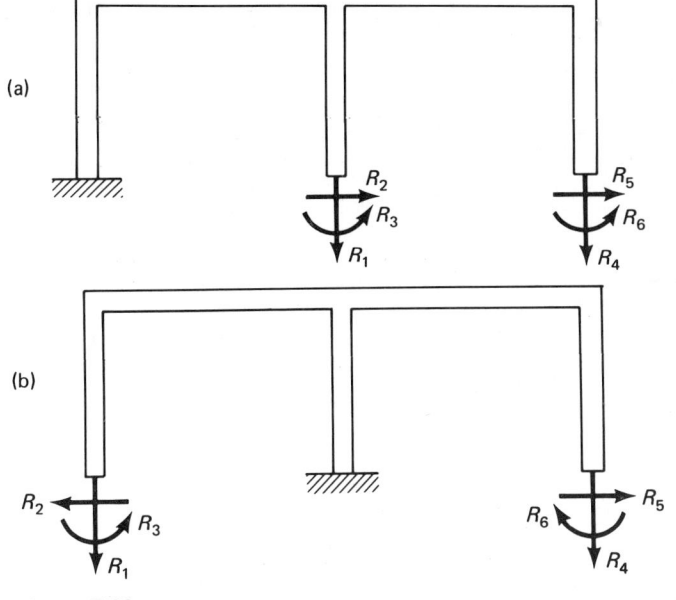

Figure 7.19

for which there are no zero-valued δ_{ij} at all.

The second of the two guidelines for primary structures given at the start of this section is difficult to apply to trusses, since their relative deformations are less obvious. A substitute guideline, more easily applied, is offered:

> Select a primary structure for a truss that retains the support system of the original truss and cuts members equal in number to the degree of static indeterminacy. Apply pairs of redundant member forces at each cut. Whenever possible, select members to be cut so that there will be one or more monolithic parts of the primary structure with member forces not affected by some of the redundant forces.

EXAMPLE 7.2.2

Select a primary structure for the three-degree statically indeterminate truss.

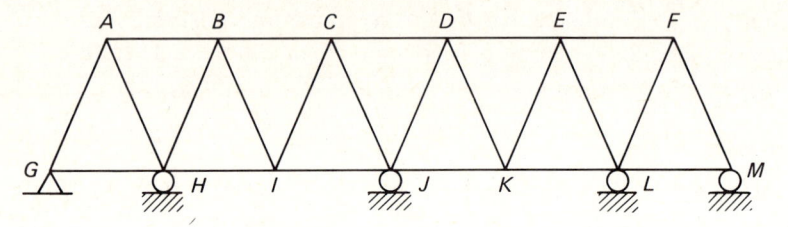

Figure 7.20

Cutting members AB, CD, and EF gives a symmetric primary structure satisfying the second guideline.

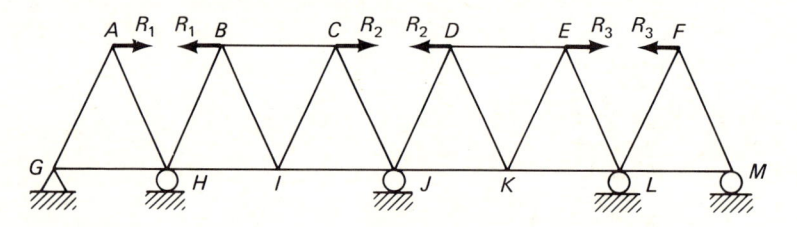

Figure 7.21

Parts $DEJKL$ and FLM are not affected by redundant R_1, parts AGH and FLM are not affected by R_2, while parts AGH and $BCHIJ$ are unaffected by R_3. No members share R_1 and R_3, permitting the immediate conclusion that $\delta_{13} = \delta_{31} = 0$.

7.3 TRUSS ANALYSIS BY CONSISTENT DEFORMATIONS

In principle the method of consistent deformations may be applied to a truss in exactly the same way as to beams and frames. Some changes in the details of application will be suggested.

Beams, frames, and trusses all begin with the selection of a primary structure and the development of its real load deflections and influence coefficients. These are used in Eq. (7.2) to evaluate the redundants. In trusses, when Eqs. (6.17) and (6.19) are applied using the tabular format of Example 6.2.3 (the tabular format used in Example 6.2.3 is that of Table 6.4), they produce the deflections and influence coefficients conveniently. With the redundants known, beam and frame analyses usually proceed with their simultaneous application to the primary structure along with the real loads. Routine calculations are then completed for all the required internal forces and moments.

In trusses that approach usually requires extra work. The tabular format lends itself well to the application of superposition for trusses and eliminates the added work. Extra columns, one for each redundant, are added to the table to list the redundant-caused member forces. They are the products of a particular redundant force and the previously tabulated member forces due to a unit force at that redundant point. The total member forces are then found by adding for each member the real-load-caused member forces and all of the redundant-caused member forces. Example 7.3.1 illustrates.

EXAMPLE 7.3.1

Determine all of the member forces for the truss of Example 7.2.2 if it is loaded with the two 10-kN forces shown. The members all have $A = 40 \text{ cm}^2$ and are steel.

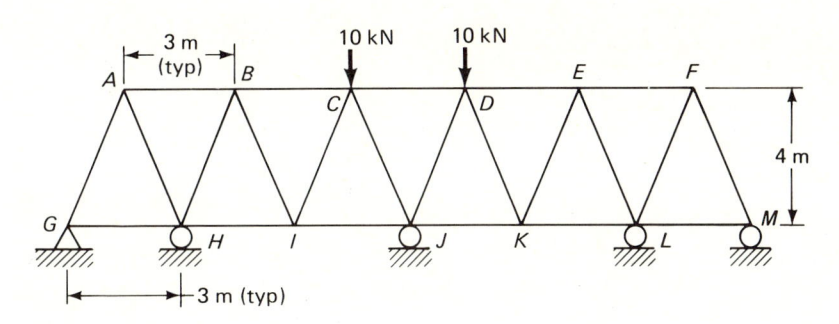

Figure 7.22

We will use the primary structure of Example 7.2.2. Separate analyses are used for the four sets of member forces: N', u_1, u_2, and u_3. N' is the set caused by real loads (two 10-kN forces), while u_1, u_2, and u_3 are sets for unit forces acting at R_1, R_2, and R_3, respectively. These analyses, using the techniques of Chapter 4, are not shown; but the member values are tabulated. You are encouraged to verify a few of them to be sure you understand their development. For convenience, Table 7.2 includes columns for the member lengths and areas.

Additional columns, giving appropriate multiplications/divisions of N', L, u_i, u_j, and A, are included, filled in, and totaled in Table 7.3.

Table 7.2

Member	L (m)	A (cm²)	N' (kN)	u_1	u_2	u_3
AB	3	40	0	1	0.	0
BC	3	40	−1.875	+0.5	+0.5	0
CD	3	40	0	0	+1	0
DE	3	40	−1.875	0	+0.5	0.5
EF	3	40	0	0	0	1
GH	3	40	0	+0.5	0	0
HI	3	40	+0.938	−0.75	−0.25	0
IJ	3	40	+2.813	−0.25	−0.75	0
JK	3	40	+2.813	0	−0.75	−0.25
KL	3	40	+0.938	0	−0.25	−0.75
LM	3	40	0	0	0	+0.5
AG	4.27	40	0	+1.424	0	0
AH	4.27	40	0	−1.424	0	0
BH	4.27	40	−2.67	−0.712	+0.712	0
BI	4.27	40	+2.67	+0.712	−0.712	0
CI	4.27	40	−2.67	−0.712	+0.712	0
CJ	4.27	40	−8.01	+0.712	−0.712	0
DJ	4.27	40	−8.01	0	−0.712	+0.712
DK	4.27	40	−2.67	0	+0.712	−0.712
EK	4.27	40	+2.67	0	−0.712	+0.712
EL	4.27	40	−2.67	0	+0.712	−0.712
FL	4.27	40	0	0	0	−1.424
FM	4.27	40	0	0	0	+1.424

Table 7.3

Member	$\frac{N'u_1 L}{A}$	$\frac{N'u_2 L}{A}$	$\frac{N'u_3 L}{A}$	$\frac{u_1^2 L}{A}$	$\frac{u_2^2 L}{A}$	$\frac{u_3^2 L}{A}$	$\frac{u_1 u_2 L}{A}$	$\frac{u_1 u_3 L}{A}$	$\frac{u_2 u_3 L}{A}$
AB	0	0	0	0.0750	0	0	0	0	0
BC	−0.07031	−0.07031	0	0.01875	0.01875	0	0.01875	0	0
CD	0	0	0	0	0.0750	0	0	0	0
DE	0	−0.07031	−0.07031	0	0.01875	0.01875	0	0	0.01875
EF	0	0	0	0	0	0.075	0	0	0
GH	0	0	0	0.01875	0	0	0	0	0
HI	−0.05276	−0.01759	0	0.04219	0.00469	0	0.01406	0	0
IJ	−0.05274	−0.15823	0	0.00469	0.00422	0	0.01406	0	0
JK	0	−0.15823	−0.05274	0	0.00422	0.00469	0	0	0.01406
KL	0	−0.01759	−0.05276	0	0.00469	0.04219	0	0	0.01406
LM	0	0	0	0	0	0.01875	0	0	0
AG	0	0	0	0.21647	0	0	0	0	0
AH	0	0	0	0.21647	0	0	0	0	0
BH	+0.20294	−0.20294	0	0.05412	0.05412	0	−0.05412	0	0
BI	+0.20294	−0.20294	0	0.05412	0.05412	0	−0.05412	0	0
CI	+0.20294	−0.20294	0	0.05412	0.05412	0	−0.05412	0	0
CJ	−0.60881	+0.60881	0	0.05412	0.05412	0	−0.05412	0	0
DJ	0	+0.60881	−0.60881	0	0.05412	0.05412	0	0	−0.05412
DK	0	−0.20294	+0.20294	0	0.05412	0.05412	0	0	−0.05412
EK	0	−0.20294	+0.20294	0	0.05412	0.05412	0	0	−0.05412
EL	0	−0.20294	+0.20294	0	0.05412	0.05412	0	0	−0.05412
FL	0	0	0	0	0	0.21647	0	0	0
FM	0	0	0	0	0	0.21647	0	0	0
	−0.1758	−0.4923	−0.1758	0.8088	0.5633	0.8088	−0.1696	0	−0.1696

7.3 TRUSS ANALYSIS BY CONSISTENT DEFORMATIONS

Now by dividing the totaled columns by $E = 2 \times 10^4$ kN/cm², a dimensionally convenient equivalent to $E = 200$ GPa, the various Δ'_i and δ_{ij} are found:

$$\Delta'_1 = -\frac{0.1758}{2 \times 10^4} = -8.79 \times 10^{-6} \text{ m}$$

$$\Delta'_2 = -\frac{0.4923}{2 \times 10^4} = -24.62 \times 10^{-6} \text{ m}$$

$$\Delta'_3 = -\frac{0.1758}{2 \times 10^4} = -8.79 \times 10^{-6} \text{ m}$$

$$\delta_{11} = \frac{0.8088}{2 \times 10^4} = +40.44 \times 10^{-6} \text{ m/kN}$$

$$\delta_{22} = \frac{0.5633}{2 \times 10^4} = +28.16 \times 10^{-6} \text{ m/kN}$$

$$\delta_{33} = \frac{0.8088}{2 \times 10^4} = +40.44 \times 10^{-6} \text{ m/kN}$$

$$\delta_{21} = \delta_{12} = -\frac{0.1696}{2 \times 10^4} = -8.48 \times 10^{-6} \text{ m/kN}$$

$$\delta_{31} = \delta_{13} = \frac{0}{2 \times 10^4} = 0$$

$$\delta_{32} = \delta_{23} = -\frac{0.1696}{2 \times 10^4} = -8.48 \times 10^{-6} \text{ m/kN}$$

Some economies of calculation that should have been taken were ignored to illustrate their validity; that $\delta_{13} = \delta_{31} = 0$ had been noted in Example 7.2.2. Its calculation was completed here, confirming the zero value. Symmetry should have indicated that $\Delta'_1 = \Delta'_2$ and $\delta_{11} = \delta_{33}$, so only one of each was necessary. Their equality is confirmed.

Once the Δ'_i and δ_{ij} are known, they are applied to Eq. (7.2) and the redundants calculated:

$$40.44 R_1 - 8.48 R_2 = 8.79$$
$$-8.48 R_1 + 28.16 R_2 - 8.48 R_3 = 24.62$$
$$-8.48 R_2 + 40.44 R_3 = 8.79$$

The 10^{-6} common to all terms was dropped.

$R_1 = 0.4587$ kN

$R_2 = 1.151$ kN

$R_3 = 0.4587$ kN

This is *not* the end of the solution for the truss. One more step remains: superposing the member forces for the situation where N', R_1, R_2, and R_3 all act simultaneously, for that is the real situation. For each member the calculation of

$$N = N' + R_1 u_1 + R_2 u_2 + R_3 u_3$$

is done in tabular form, where N is the total member force (Table 7.4).

Table 7.4

Member	$R_1 u_1$	$R_2 u_2$	$R_3 u_3$	N'	N
AB	0.4587	0	0	0	0.4587
BC	0.2294	0.5755	0	−1.875	−1.0700
CD	0	1.151	0	0	1.1510
DE	0	0.5755	0.2294	−1.875	−1.0700
EF	0	0	0.4587	0	0.4587
GH	0.2294	0	0	0	0.2294
HI	−0.3440	−0.2878	0	0.938	0.3062
IJ	−0.1147	−0.8633	0	2.813	1.8350
JK	0	−0.8633	−0.1147	2.813	1.8350
KL	0	−0.2878	−0.3440	0.938	0.3062
LM	0	0	0.2294	0	0.2294
AG	0.6532	0	0	0	0.6532
AH	−0.6532	0	0	0	−0.6532
BH	−0.3266	0.8195	0	−2.670	−2.1770
BI	0.3266	−0.8195	0	2.670	2.1770
CI	−0.3266	0.8195	0	−2.670	2.1770
CJ	0.3266	−0.8195	0	−8.010	−8.5030
DJ	0	−0.8195	0.3266	−8.010	−8.5030
DK	0	0.8195	−0.3266	−2.670	−2.1770
EK	0	−0.8195	0.3266	2.670	2.1770
EL	0	0.8195	−0.3266	−2.670	−2.1770
FL	0	0	−0.6532	0	−0.6532
FM	0	0	0.6532	0	0.6532

PROBLEMS

Sequenced Problem 1, parts (n), (o), and (p).
Sequenced Problem 2, parts (j) and (k).
Sequenced Problem 3, part (k).
Sequenced Problem 4, part (k).

7.1

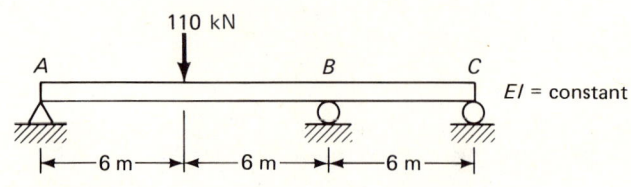

Figure P7.1

110 kN applied at 6 m from A on a beam ABC with supports at A, B, C; spans A–B = 6 m, with load at 6 m, B–C spans of 6 m and 6 m. EI = constant.

(a) Use the method of consistent deformations to determine the support reactions. Use a primary structure formed by releasing the vertical constraint at B.
(b) Use the method of consistent deformations to determine the support reactions. Use a primary structure formed by releasing the vertical constraint at C.
(c) Use the method of consistent deformations to determine the support reactions. Use a primary structure formed by adding a hinge at B.

7.2 Use the method of consistent deformations to evaluate all the member forces for the truss shown.

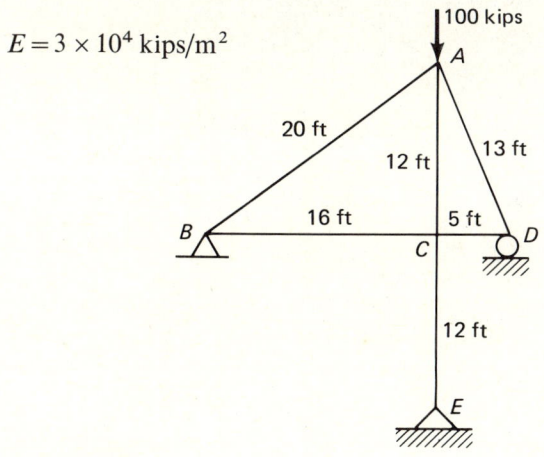

$E = 3 \times 10^4$ kips/m^2

Figure P7.2

Cross sectional areas are

$AB = 6$ in^2 $CE = 4.0$ in^2
$AD = 4.5$ in^2 $BC = 2.5$ in^2
$AC = 4.0$ in^2 $CD = 2.5$ in^2

Release the vertical support at E to form a primary structure.
Hint: Use displacements calculated in Problem 6.8.

7.3

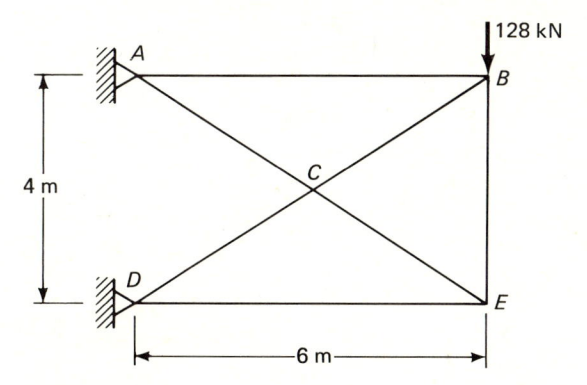

Figure P7.3

166 METHOD OF CONSISTENT DEFORMATIONS

If all members have $E = 2 \times 10^4$ kN/cm² and $A = 310$ cm², determine the internal force in each member using the method of consistent deformations.

7.4 Determine all the member forces if member AB of Problem 7.3 undergoes a temperature increase of 40°C due to sun exposure (the other members are shaded). The coefficient of thermal expansion is $\alpha = 1.2 \times 10^{-5}$ cm/cm · °C.

7.5

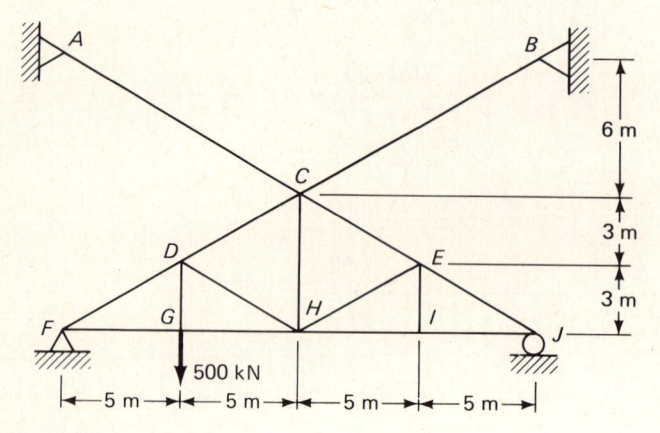

Figure P7.5

$E = 2 \times 10^4$ kN/cm²

Use the method of consistent deformations to determine the internal forces of each member. Form the primary structure by releasing at A and B. Member areas are

$AC = CB = 65.0$ cm²

$FD = DC = CE = EJ = DH = EH = 78.0$ cm²

$FG = GH = HI = IJ = 55.5$ cm²

$DG = CH = EI = 51.0$ cm²

7.6

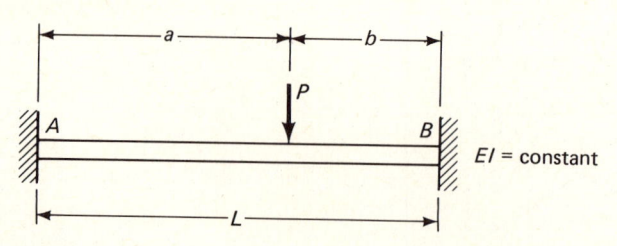

Figure P7.6

EI = constant

Use the method of consistent deformations to determine the reactions. Release all the supports at A to form the primary structure.

7.7

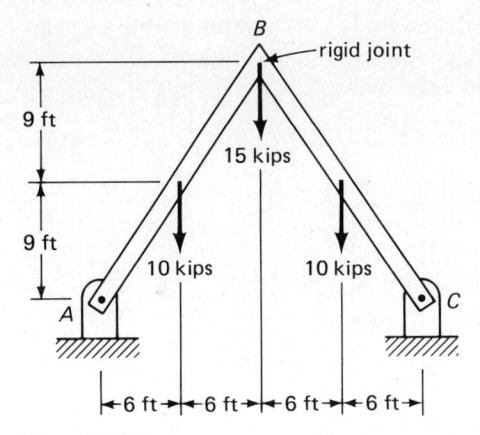

Figure P7.7

Use the method of consistent deformations to calculate the reactions at A and C if $EI = 3.15 \times 10^7$ kip-in^2 and $AE = 9.25 \times 10^5$ kips throughout.

7.8

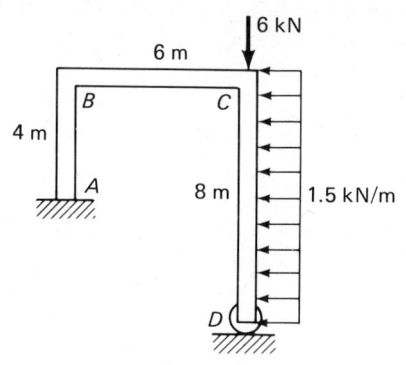

$EI = 9.15 \times 10^4$ kN·m^2 for all members

Figure P7.8

Using the structure of Problem 6.7 as the primary structure, determine the support reactions at A and D by the method of consistent deformations.

7.9

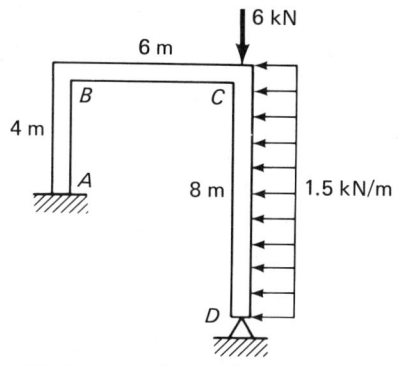

$EI = 9.15 \times 10^4$ kN·m^2 for all members

Figure P7.9

Using the structure of Problem 6.7 as the primary structure, determine the support reactions at A and D by the method of consistent deformations.

7.10 Recommend primary structures for the statically indeterminate structures shown. Label the redundants and list, for each, the flexibility coefficients δ_{ij} that would be zero.

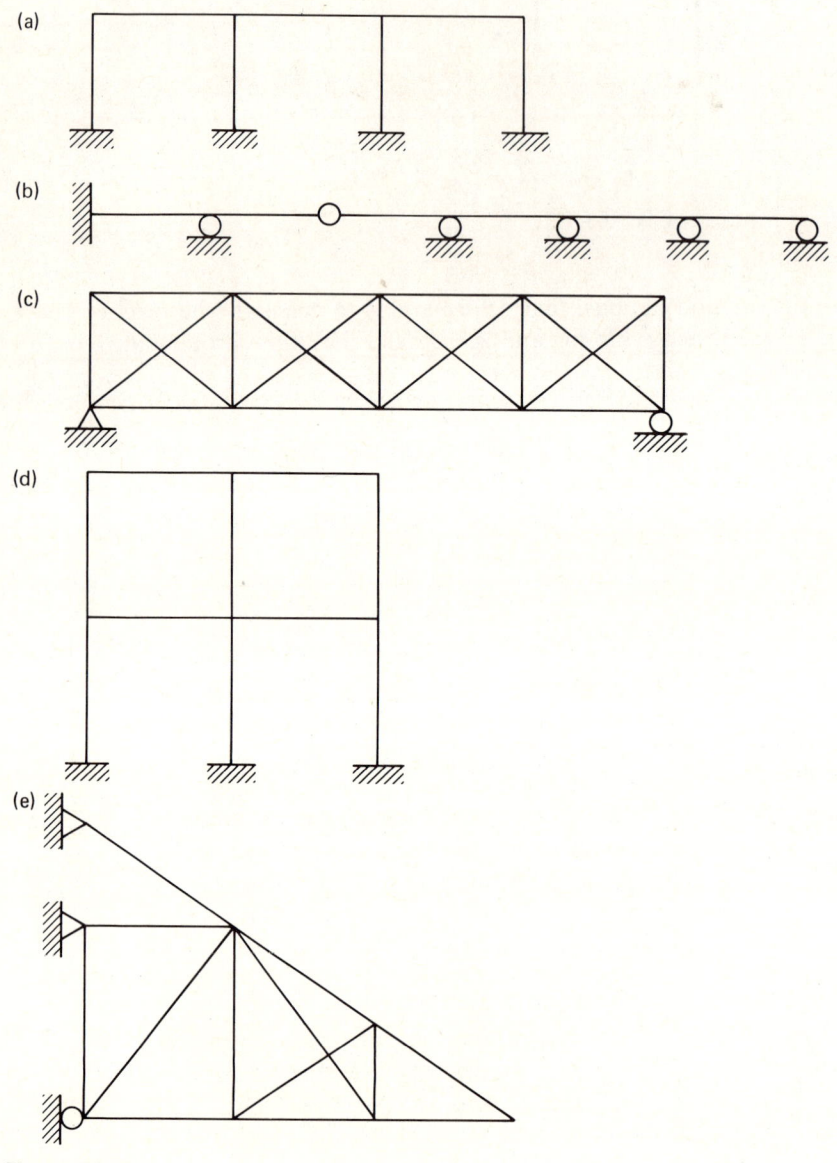

Figure P7.10

The Method of Least Work

He had answered an advertisement for a mystery adventure cruise, paid its fee, and was on the pier to embark.

A huge, terrifying man grabbed him, bound and gagged him, and threw him into a large boat with twenty-four oars. He was shackled to a bench at an oar and forced to row, accompanied by cracking whips and paced by the terrifying man beating cadence on an immense drum.

After two agonizing weeks the boat returned. The traveler was unshackled and told he could leave. Turning to his bench mate, he said, "I'm unfamiliar with the amenities for occasions such as this. Tell me, do we tip the drummer?"

8.0 INTRODUCTION

The vacationer must have been surprised at having to work so hard on his adventure cruise. Most students embark on the method of least work believing that its name implies that they won't work very hard and are surprised to discover that its name is not related to their efforts. Rather, it refers to the external work applied to the structure during loading and deforming, work that is stored internally as strain energy. A redundant elastic structure on unyielding supports takes a deformed shape that minimizes its potential (strain) energy.

Consider beam AB with one redundant (one degree statically indeterminate).

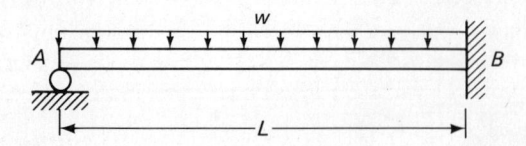

Figure 8.1

Using the support reaction at A as the redundant, the stable statically determinate cantilevered beam with load w and reaction R_A is the loaded primary structure.

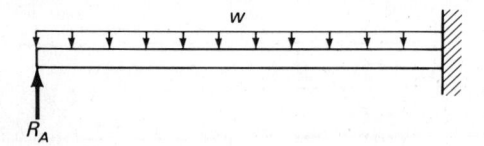

Figure 8.2

Sectioning, the bending moment is found.

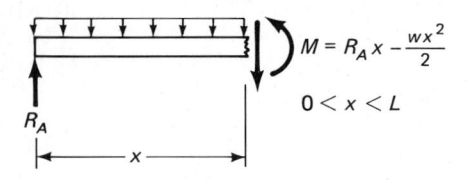

$$M = R_A x - \frac{wx^2}{2}$$
$$0 < x < L$$

Figure 8.3

Applying Eq. (6.13), the strain energy in the beam is

$$W = \int_{\substack{\text{whole}\\ \text{structure}}} \frac{M^2}{2EI} dx = \frac{1}{2EI} \int_0^L \left(R_A x - \frac{wx^2}{2} \right)^2 dx$$

The rigid support at A requires the deflection to be zero there. From Castigliano's theorem, therefore,

$$\Delta_A = \frac{\partial W}{\partial R_A} = \frac{1}{EI} \int_0^L \left(R_A x - \frac{wx^2}{2} \right) x \, dx = 0$$

That, evaluated, gives

$$R_A = \tfrac{3}{8}wL$$

Equating $\partial W/\partial R_A$ to zero is equivalent to requiring the relation between W and R_A to have a minimum, a maximum, or a stationary value of W at $R_A = \tfrac{3}{8}wL$. It is a minimum because $\partial^2 W/\partial R_A^2$ is positive, as may be shown easily.

8.1 STATICALLY INDETERMINATE BEAMS BY THE METHOD OF LEAST WORK

The procedure for analyzing statically indeterminate beams was illustrated in the preceding section. If the beam is several degrees indeterminate, a redundant must be established for each degree and included in the bending moment expressions. Castigliano's theorem is applied separately for each redundant, producing a set of equations to be solved simultaneously for the redundants.

EXAMPLE 8.1.1

Use the method of least work to analyze the two-degree statically indeterminate beam ABC.

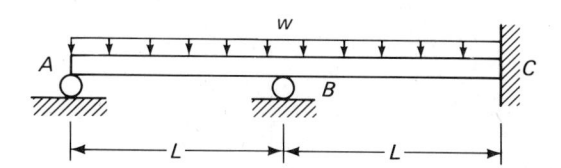

Figure 8.4

The guidelines for selecting primary structures found in Section 7.2 are applicable for the method of least work, too. Following them, we place hinges at B and C, making the internal moments there the redundants. The loaded primary structure is

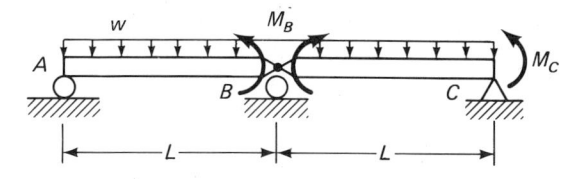

Figure 8.5

which can be treated conveniently as two separate statically determinate pieces, AB and BC. Their support reactions are written as functions of the loading and of the two redundants.

172 THE METHOD OF LEAST WORK

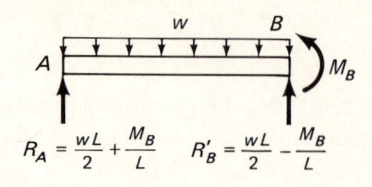

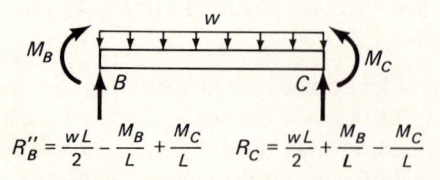

$R_A = \dfrac{wL}{2} + \dfrac{M_B}{L}$ $R'_B = \dfrac{wL}{2} - \dfrac{M_B}{L}$

$R''_B = \dfrac{wL}{2} - \dfrac{M_B}{L} + \dfrac{M_C}{L}$ $R_C = \dfrac{wL}{2} + \dfrac{M_B}{L} - \dfrac{M_C}{L}$

Figure 8.6

Each part is sectioned a distance x from its left end, the bending moments are found, and partial derivatives are taken.

For AB

$$M = \dfrac{wLx}{2} + \dfrac{M_B x}{L} - \dfrac{wx^2}{2} \quad \text{for } 0 \leq x \leq L$$

$$\dfrac{\partial M}{\partial M_B} = \dfrac{x}{L} \quad \text{and} \quad \dfrac{\partial M}{\partial M_C} = 0$$

For BC,

$$M = \dfrac{wLx}{2} + \dfrac{M_C x}{L} - \dfrac{M_B x}{L} - \dfrac{wx^2}{2} + M_B \quad \text{for } 0 \leq x \leq L$$

$$\dfrac{\partial M}{\partial M_B} = 1 - \dfrac{x}{L} \quad \text{and} \quad \dfrac{\partial M}{\partial M_C} = \dfrac{x}{L}$$

Castigliano's theorem applied for each of the two redundants gives

$$\dfrac{\partial W}{\partial M_B} = \dfrac{1}{EI}\int_0^L \left(\dfrac{wLx}{2} + \dfrac{M_B x}{L} - \dfrac{wx^2}{2}\right)\left(\dfrac{x}{L}\right)dx$$
$$+ \dfrac{1}{EI}\int_0^L \left(\dfrac{wLx}{2} + \dfrac{M_C x}{L} - \dfrac{M_B x}{L} - \dfrac{wx^2}{2} + M_B\right)\left(1 - \dfrac{x}{L}\right)dx = 0$$

and

$$\dfrac{\partial W}{\partial M_C} = \dfrac{1}{EI}\int_0^L \left(\dfrac{wLx}{2} + \dfrac{M_C x}{L} - \dfrac{M_B x}{L} - \dfrac{wx^2}{2} + M_B\right)\left(\dfrac{x}{L}\right)dx = 0$$

Evaluating and simplifying the simultaneous equations

$$4M_B + M_C = -\dfrac{wL^2}{2}$$

$$M_B + 2M_C = -\dfrac{wL^2}{4}$$

give

$$M_B = -\frac{3}{28}wL^2$$

$$M_C = -\frac{1}{14}wL^2$$

when solved.

The support reactions, already expressed as functions of the load and the redundants, are determined by substitution:

$$R_A = \frac{wL}{2} + \frac{M_B}{L} = \frac{wL}{2} + \left(-\frac{3}{28}wL^2\right)\left(\frac{1}{L}\right)$$

$$= \frac{11}{28}wL$$

$$R_B = R_B' + R_B'' = \left(\frac{wL}{2} - \frac{M_B}{L}\right) + \left(\frac{wL}{2} - \frac{M_B}{L} + \frac{M_C}{L}\right)$$

$$= wL - \frac{2M_B}{L} + \frac{M_C}{L}$$

$$R_B = wL - \frac{2}{L}\left(-\frac{3}{28}wL^2\right) + \frac{1}{L}\left(-\frac{1}{14}wL^2\right)$$

$$= \frac{8}{7}wL$$

$$R_C = \frac{wL}{2} + \frac{M_B}{L} - \frac{M_C}{L} = \frac{wL}{2} + \frac{1}{L}\left(-\frac{3}{28}wL^2\right) - \frac{1}{L}\left(-\frac{1}{14}wL^2\right)$$

$$= \frac{13}{28}wL$$

The complete set of supports is then

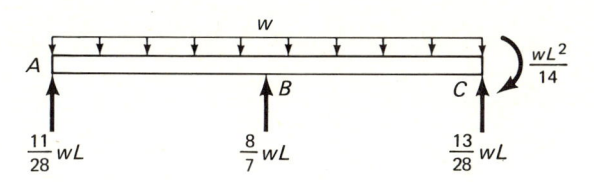

Figure 8.7

8.2 STATICALLY INDETERMINATE RIGID FRAMES BY THE METHOD OF LEAST WORK

The procedure for rigid frames parallels that used for beams.

EXAMPLE 8.2.1

Use the method of least work to evaluate the support reactions for frame ABC.

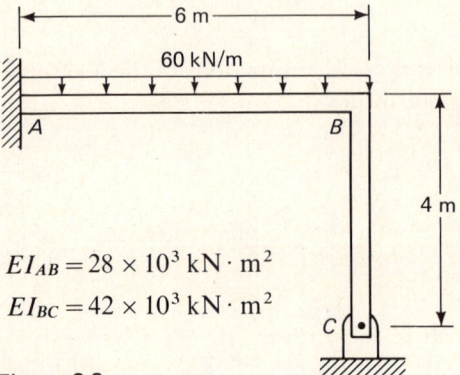

Figure 8.8

$EI_{AB} = 28 \times 10^3 \text{ kN} \cdot \text{m}^2$

$EI_{BC} = 42 \times 10^3 \text{ kN} \cdot \text{m}^2$

Selecting the horizontal and vertical support reactions at C as the redundants, the loaded primary structure is

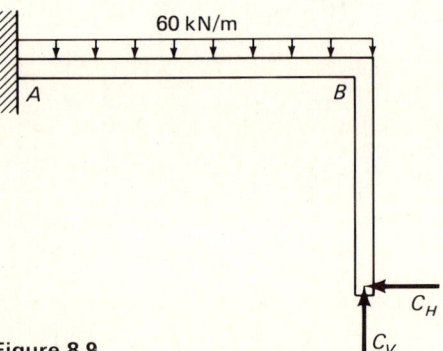

Figure 8.9

By sectioning and defining M and x as follows, we can delay calculating the support reactions at A and write the moment expressions directly.

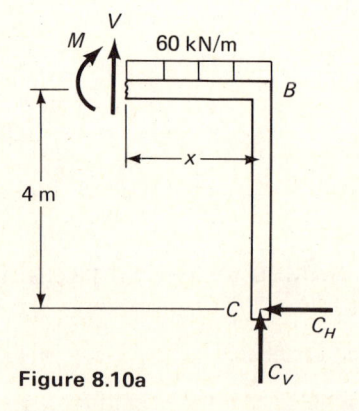

Figure 8.10a

STATICALLY INDETERMINATE RIGID FRAMES BY THE METHOD OF LEAST WORK

$$M = -\frac{60x^2}{2} - 4C_H + xC_V \quad \text{for } 0 \leq x \leq 6 \text{ m}$$

$$\frac{\partial M}{\partial C_H} = -4 \quad \frac{\partial M}{\partial C_V} = x$$

and

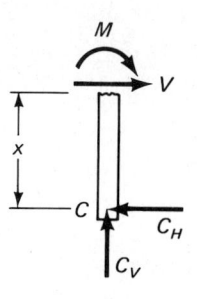

Figure 8.10b

$$M = -xC_H \quad \text{for } 0 \leq x \leq \text{m}$$

$$\frac{\partial M}{\partial C_H} = -x \quad \frac{\partial M}{\partial C_V} = 0$$

Applying Castigliano's theorem,

$$\frac{\partial W}{\partial C_H} = \frac{1}{28 \times 10^3} \int_0^{6\text{m}} \left(-\frac{60x^2}{2} - 4C_H + xC_V\right)(-4)\, dx$$

$$+ \frac{1}{42 \times 10^3} \int_0^{4\text{m}} \left(-xC_H\right)(-x)\, dx = 0$$

and

$$\frac{\partial W}{\partial C_V} = \frac{1}{28 \times 10^3} \int_0^6 \left(-\frac{60x^2}{2} - 4C_H + xC_V\right) x\, dx = 0$$

Integrating and simplifying gives

$$165.3 C_H - 108 C_V = -12960$$

and

$$-72 C_H + 72 C_V = 9720$$

which yield solutions

$$C_H = +28.3 \text{ kN}$$
$$C_V = +163.3 \text{ kN}$$

Equilibrium provides the reactions at A,

176 THE METHOD OF LEAST WORK

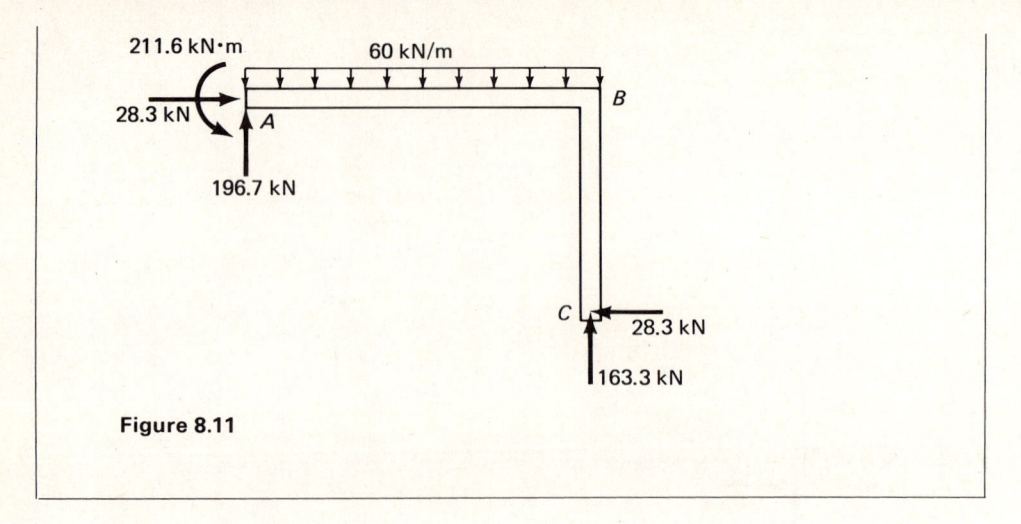

Figure 8.11

8.3 LEAST WORK ANALYSIS OF STATICALLY INDETERMINATE COMPOSITE STRUCTURES

Many statically indeterminate structures include several kinds of members, such as bending members, along with axial force members. They are called composite structures. Frequently a single member has significant loadings in several modes, such as in beam columns. The method of least work is particularly convenient for analyzing them.

EXAMPLE 8.3.1

Determine the tension in cable BC.

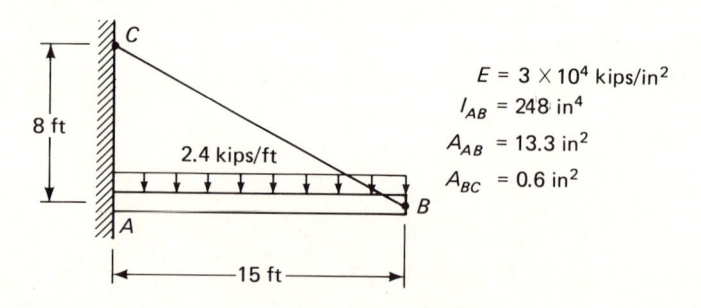

$E = 3 \times 10^4$ kips/in^2
$I_{AB} = 248$ in^4
$A_{AB} = 13.3$ in^2
$A_{BC} = 0.6$ in^2

Figure 8.12

We will consider cable BC the redundant. Disconnect it at B and require that there be zero relative displacement between its end and that of end B of the beam. The loaded primary structure is shown in two parts.

LEAST WORK ANALYSIS OF STATICALLY INDETERMINATE COMPOSITE STRUCTURES 177

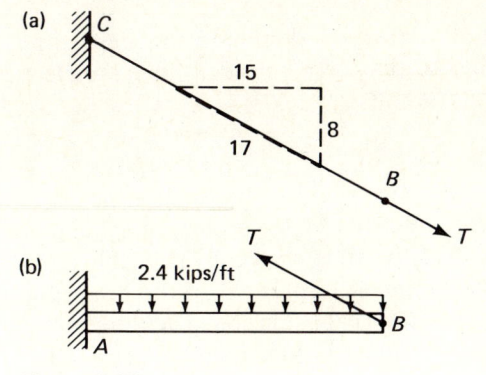

(a) [diagram: cable from C to B, with dimensions 15, 8, 17]

(b) [diagram: beam A to B with 2.4 kips/ft distributed load, tension T at angle]

Figure 8.13

Internal forces and moments in the beam are found from the free-body diagram.

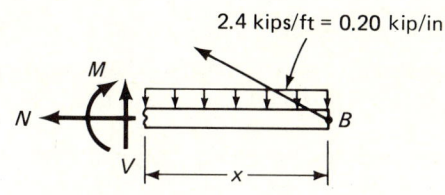

[free-body diagram showing N, V, M at cut section, 2.4 kips/ft = 0.20 kip/in, distance x to B]

Figure 8.14

$$N = -\frac{15}{17} T \quad \text{kip}$$

$$M = -0.2\frac{x^2}{2} + \frac{8}{17} Tx \quad \text{kip-in}$$

for $0 < x < 180$ in, and for the cable,

$$N = T \quad \text{kip}$$

After taking partial derivatives with respect to T, we apply Castigliano's theorem, using Eq. (6.16), a formulation that includes strain energy for both bending and axial deformations.

$$\frac{\partial W}{\partial T} = \frac{1}{EI} \int_{\substack{\text{whole} \\ \text{structure}}} M \frac{\partial M}{\partial T} dx + \sum_{\substack{\text{all} \\ \text{members}}} \frac{N(\partial N/\partial T)L}{AE} \quad (6.16)$$

$$= \frac{1}{3 \times 10^4 (248)} \int_0^{180} \left(-0.2\frac{x^2}{2} + \frac{8}{17} Tx\right)\left(\frac{8}{17}x\right) dx$$

$$+ \frac{-\frac{15}{17}T(-\frac{15}{17})(180)}{13.3(3 \times 10^4)} + \frac{T(1)(17 \times 12)}{0.6(3 \times 10^4)} = 0$$

Integrating and simplifying, this becomes

$$-1.661 + 5.791 \times 10^{-2}T + 3.51 \times 10^{-4}T + 1.133 \times 10^{-2}T = 0$$

$$T = 23.9 \text{ kips}$$

178 THE METHOD OF LEAST WORK

In most composite structures a few terms will dominate and, depending on the strength of their dominance, may be the only terms needed. In Example 8.3.1 the equation solved for the cable tension was

$$5.79 \times 10^{-2}T + 3.51 \times 10^{-4}T + 1.133 \times 10^{-2}T = 1.661$$

Each term expresses a displacement component along the cable, either of the cable end or of the beam end. The first term, $5.79 \times 10^{-2}T$, is the component of the vertical displacement of point B on the beam from bending caused by the cable tension. The second, $3.51 \times 10^{-4}T$, is the axial displacement of point B of the beam due to the beam's shortening from compression from the cable tension. The third is the stretch of the cable due to T. The right-hand side is the component of the beam deflection at B due to the distributed load.

Clearly $3.51 \times 10^{-4}T$ is several orders smaller than the other terms involving T. If it were ignored, the value of T would increase to $T = 24.0$ kips, a change of less than 0.5%. Ignoring the term implies that the beam does not shorten at all. Apparently that is a reasonable assumption.

On the other hand, if the stretch of the cable had also been ignored, T would increase to $T = 28.7$ kips. That is a change of 20%, a difference that would be unacceptable in most analyses. In ignoring both the beam compression and the cable stretch, you imply that the beam end does not move, that is, that the beam end at B is equivalent to a pin support. Its corresponding vertical reaction would be $\frac{8}{17} \times 28.7 = 13.5$ kips, exactly the value found from use of $\frac{3}{8}wL$, as found in Section 8.0.

8.4 THE METHOD OF LEAST WORK VIS-A-VIS THE METHOD OF CONSISTENT DEFORMATIONS

Although it is not obvious, the calculations for the methods of least work and consistent deformations are the same. In Section 8.0, for example, the least work equation was

$$\frac{\partial W}{\partial R_A} = \frac{1}{EI} \int_0^L \left(R_A x - \frac{wx^2}{2} \right) x \, dx = 0$$

for

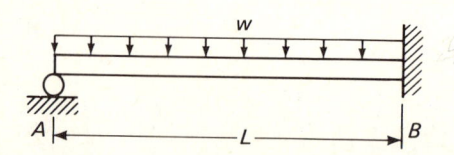

Figure 8.15

when R_A was taken as the redundant. The term

$$\frac{1}{EI} \int_0^L \left(-\frac{wx^2}{2} \right) x \, dx$$

is Δ'_A, the displacement of the redundant point A due to the loading w, as used in consistent deformations. It would be written easily using the unit load method or

Castigliano's theorem. Similarly,

$$\frac{1}{EI}\int_0^L (R_A x)x\, dx \quad \text{is} \quad R_A \delta_{AA}$$

the redundant point's displacement due to the redundant R_A. It is expressed as a force R_A times influence coefficients δ_{AA}. Evidently the least work formulation matches the consistent deformation formulation

$$\Delta'_A + R_A \delta_{AA} = 0$$

identically. The same correspondence can be shown for structures with higher degrees of static indeterminacy. Apparently the calculations used in least work are identical to those used in consistent deformations. In spite of this, there are reasons for choosing one method over the other in certain circumstances.

The method of least work is limited to energy changes arising from the application of loads only, it presumes no work is done by the support reactions, and it does not account for strain energy related to other phenomena such as thermal strain. Therefore, least work should not be used in analyses related to support settlements, thermal strains, construction misalignments, and so on. No such restriction exists for the method of consistent deformations.

In statically indeterminate beam analysis, consistent deformations may be the better choice, especially if influence coefficients and redundant point deflections can be written directly from available beam deflection formulas.

In statically indeterminate frames the equations may be written almost automatically using least work, while considerable extra care is often necessary when developing the same equations from consistent deformations.

Truss analysis requires that functions of the external loads and each redundant be written for each truss member in applying least work. The tabular solution form for trusses employed in consistent deformations, usually less cumbersome, is a preferable approach. For that reason least work analysis of trusses will not be considered.

Finally, the more automatic formulation of the least work equations makes that the preferred approach for composite structures.

PROBLEMS

Sequenced Problem 3, part (1).
Sequenced Problem 4, part (1).
8.1

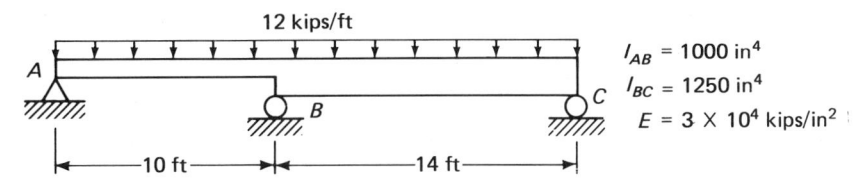

Figure P8.1

$I_{AB} = 1000\ in^4$
$I_{BC} = 1250\ in^4$
$E = 3 \times 10^4\ kips/in^2$

(Beam with 12 kips/ft distributed load; supports at A, B, C; spans 10 ft and 14 ft)

Use the method of least work to determine the reactions at A.

8.2

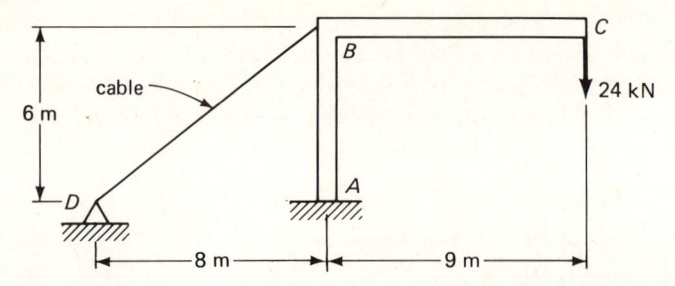

Figure P8.2

$$I_{AB} = I_{BC} = 2.3 \times 10^{-4} \text{ m}^4$$
$$A_{AB} = A_{BC} = 1.45 \times 10^{-2} \text{ m}^2$$
$$A_{BD} = 6.5 \times 10^{-4} \text{ m}^2$$
$$E = 200 \times 10^6 \text{ kN/m}^2$$

Use the method of least work to determine the cable tension.

8.3

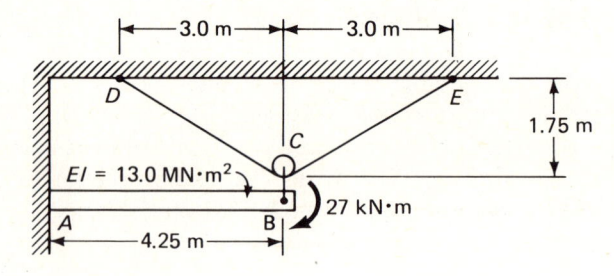

Figure P8.3

Cable *DCE* passes through pulley *C*. Use the method of least work to determine its tension by the 27-kN · m couple at *B*. The cable has a cross-sectional area of $8.1 \times 10^{-6} \text{ m}^2$ and is steel with $E = 200 \times 10^6 \text{ kN/m}^2$.

8.4 Solve Problem 7.7 by the method of least work.

8.5

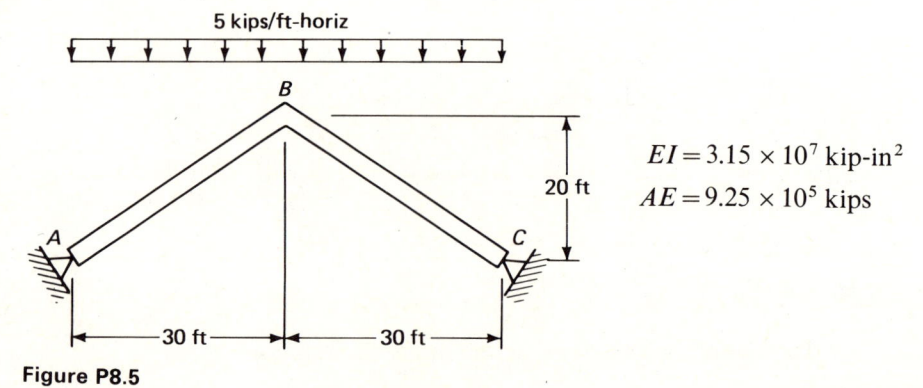

$EI = 3.15 \times 10^7$ kip-in^2
$AE = 9.25 \times 10^5$ kips

Figure P8.5

(a) Use the method of least work to determine the reactions.
(b) Use the method of least work to determine the reactions, but ignore the axial changes of the length of AB and BC.

8.6 In altering a structure to permit a load of 2.2 kips/ft on AB it is proposed that beam AB be supported at A by the arrangement shown. Use the method of least work to calculate the cable tension. Assume that C is a hinge connection.

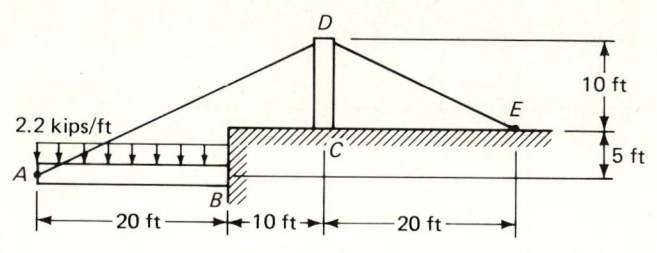

Figure P8.6

$A_{AB} = 32.7$ in^2 $I_{AB} = 1270$ in^4

$A_{DE} = A_{AD} = 0.938$ in^2 $E = 3 \times 10^4$ kips/in^2

$A_{CD} = 37.3$ in^2

Statically Indeterminate Structures: Conjugate Beam Analysis

The not-too-bright college sophomore
was seated next to a famous astronomer
at a dinner party and asked him what work
he was engaged in.
"I study astronomy," he replied.
"Oh really?" said the sophomore,
"I finished that last year."

9.0 INTRODUCTION

We take up conjugate beam analysis again. This time we will apply it in several different ways to statically indeterminate structures. You may have thought that you had finished conjugate beams, but you have not. This return may help overcome any difficulties with the topic that you had before and should bring out some of the subtleties of the method that may have escaped you the first time through. Applied to statically indeterminate beams, the conjugate beam method is particularly efficient for beams one or two degrees statically indeterminate.

9.1 STATICALLY INDETERMINATE BEAM ANALYSIS BY CONJUGATE BEAMS

Although the conjugate of a statically indeterminate beam is an unstable beam, its conjugate load must be such that it maintains equilibrium. From that we draw an efficient approach for the analysis of beams one or two degrees statically indeterminate.

EXAMPLE 9.1.1

Determine the support reaction at A of beam AB.

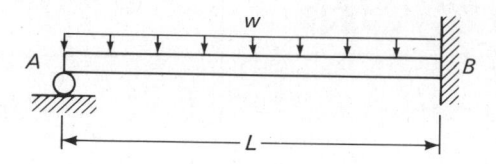

Figure 9.1

Since we want the reaction at A, we remove the support there, substituting R_A. We consider the remaining cantilevered beam as the primary structure, and for convenience consider its loading in superposition.

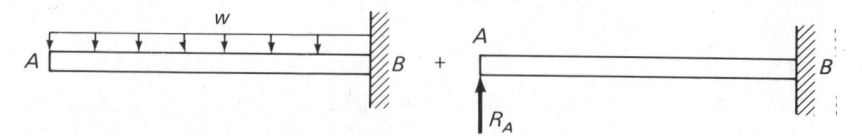

Figure 9.2

Moment diagrams for each are drawn.

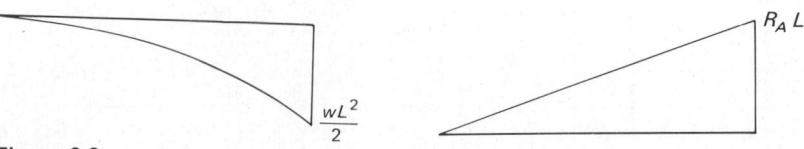

Figure 9.3

Converted to M/EI diagrams, these serve as the loads of the conjugate beam.

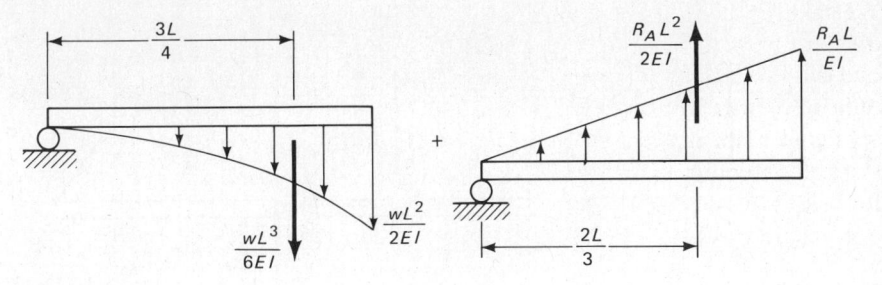

Figure 9.4

The two concentrated forces are the equivalents of the distributed forces (their areas) and are shown acting at the area centroids. Equilibrium is satisfied if $\sum \bar{M}_A = 0$. Therefore

$$+\circlearrowleft \sum \bar{M}_A = -\frac{wL^3}{6EI}\left(\frac{3L}{4}\right) + \frac{R_A L^2}{2EI}\left(\frac{2L}{3}\right) = 0$$

from which $R_A = \frac{3}{8}wL$.

EXAMPLE 9.1.2

Determine the moments at A and B of beam AB.

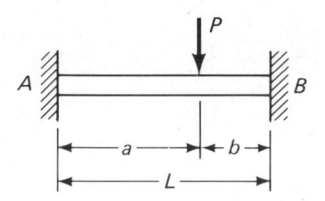

Figure 9.5

We select a simply supported beam as the primary structure and again use superposition.

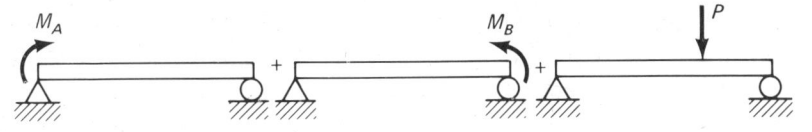

Figure 9.6

Moment diagrams are

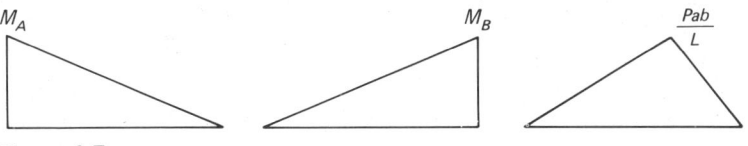

Figure 9.7

which are converted to M/EI diagrams for use as loads for the conjugate beam. In this case the conjugate beam is completely unsupported since the conjugate of each clamped end is a free end.

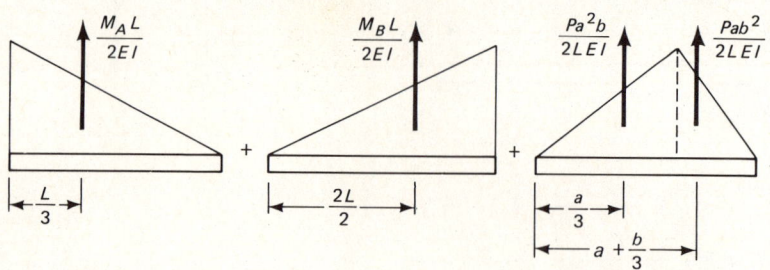

Figure 9.8

Satisfying equilibrium,

$$+\circlearrowright\sum \bar{M}_A = \frac{M_A L}{2EI}\left(\frac{L}{3}\right) + \frac{M_B L}{2EI}\left(\frac{2L}{3}\right) + \frac{Pa^2 b}{2LEI}\left(\frac{a}{3}\right) + \frac{Pab^2}{2LEI}\left(a + \frac{b}{3}\right) = 0$$

$$+\uparrow\sum \bar{F}_y = \frac{M_A L}{2EI} + \frac{M_B L}{2EI} + \frac{Pab}{LEI}\left(\frac{L}{2}\right) = 0$$

These reduce to

$$M_A + 2M_B = -\frac{Pab}{L^3}(2a^2 + 3ab + b^2)$$

$$M_A + M_B = -\frac{Pab}{L}$$

giving

$$M_A = -\frac{Pab^2}{L^2}$$

$$M_B = -\frac{Pa^2 b}{L^2}$$

This conjugate beam method is quick and direct for deriving stiffnesses, carryover factors, and fixed end moments, all significant parts of the slope deflection and moment distribution methods presented in the next four chapters.

9.2 INFLUENCE LINES FOR STATICALLY INDETERMINATE STRUCTURES: THE MÜLLER–BRESLAU PRINCIPLE

The Müller–Breslau principle was stated in Section 5.2.1:

> The influence line for any response function (support reaction, beam shear at a point, and so on) is identical to the geometry of the displaced structure when the restraint provided by the response function itself is removed and a unit displacement in the direction of the response function is imposed there.

INFLUENCE LINES FOR THE STATICALLY INDETERMINATE STRUCTURES

The Müller–Breslau principle is applicable to all stable, linearly elastic structures. It was proven for statically determinate ones in our earlier studies. We prove it now for statically indeterminate structures.

Consider a stable, linearly elastic, statically indeterminate beam, such as *ABCD*.

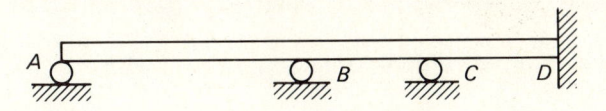

Figure 9.9

We wish to show that the influence line for a support reaction, such as at *A*, is identical to the elastic curve of the structure with a unit displacement imposed there.

First we remove support *A* and apply a unit force there. The deflected shape is

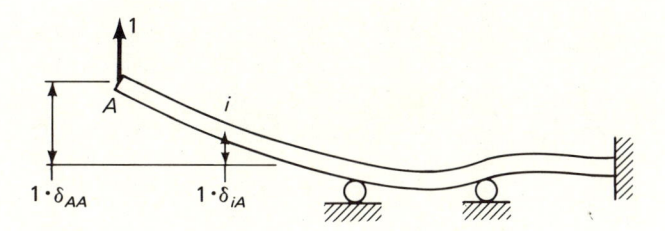

Figure 9.10

Shown are two deflections, one at *A*, the other at generic point *i* on the beam. We define two influence coefficients

δ_{AA} = deflection at *A* due to a unit force at *A*

δ_{iA} = deflection at *i* due to a unit force at *A*

We change the loading at *A* to $1/\delta_{AA}$ to change the displacement at *A* to a unit displacement, matching that of the proposed influence line. The new elastic curve is

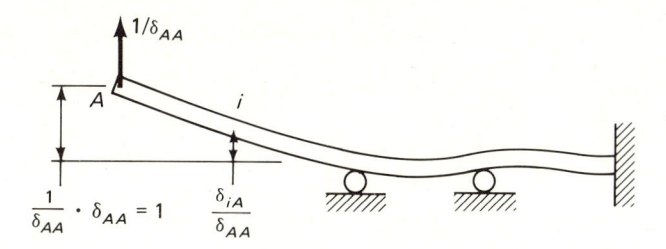

Figure 9.11

It is evident that when there is a unit displacement at *A*, there is a displacement δ_{iA}/δ_{AA} at generic point *i*.

Now let us return to structure *ABCD* and use the method of consistent deformations to calculate r_A, the support reaction at *A* from a unit force at a generic point *i*.

188 STATICALLY INDETERMINATE STRUCTURES: CONJUGATE BEAM ANALYSIS

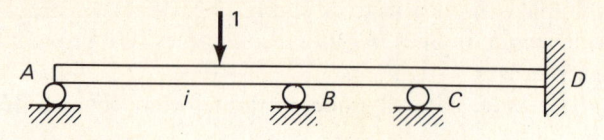

Figure 9.12

We use superposition, having removed support A and replaced it with r_A.

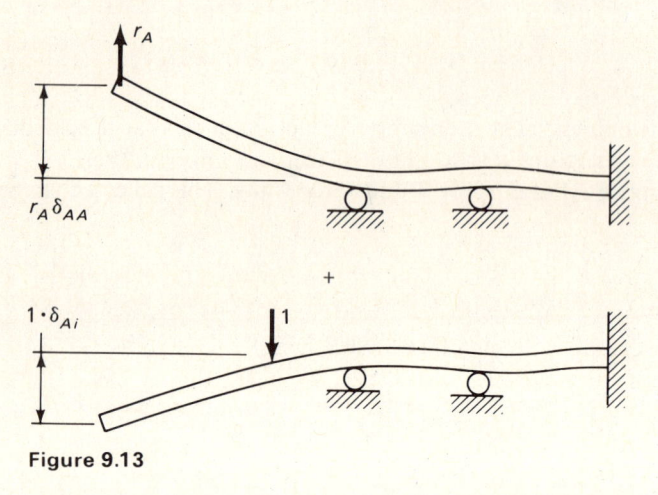

Figure 9.13

From consistent deformations

$$r_A \delta_{AA} = \delta_{Ai}$$

so that

$$r_A = \frac{\delta_{Ai}}{\delta_{AA}}$$

Applying Maxwell's relation $\delta_{Ai} = \delta_{iA}$,

$$r_A = \frac{\delta_{iA}}{\delta_{AA}}$$

and it is shown that r_A is identical in value to the displacement at i of the deflected structure when A is removed and a unit displacement imposed there. Thus the Müller–Breslau principle is proved for statically indeterminate structures.

The primary use of influence lines is to determine worst loading combinations for a member or support. Several examples for statically determinate beams were given in Section 5.2.3. Their use in statically indeterminate structures is the same. The influence line values must be known, and the Müller–Breslau principle combined with conjugate beam analysis provides a direct approach for evaluation.

EXAMPLE 9.2.1

Develop the influence line for the reaction at A.

INFLUENCE LINES FOR THE STATICALLY INDETERMINATE STRUCTURES **189**

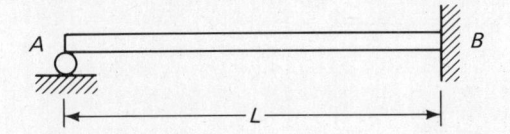

Figure 9.14

We remove the support at A and impose a unit displacement there, forming the influence line.

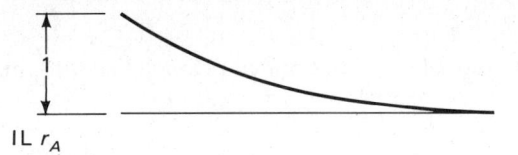

IL r_A
Figure 9.15

If such a displacement in a real beam is to occur, there must be a concentrated force of appropriate magnitude h applied at A.

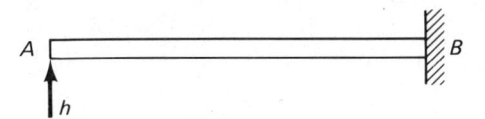

Figure 9.16

The M/EI diagram for that beam and load is applied as loading of the conjugate beam, along with a moment $\bar{M}_A = 1$ at A to account for the real beam unit deflection there.

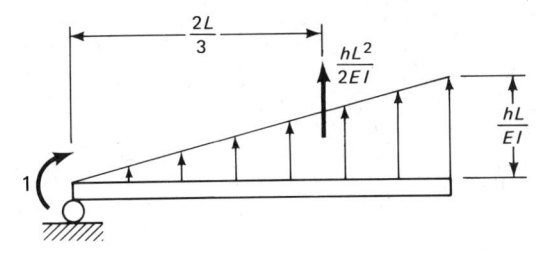

Figure 9.17

For equilibrium,

$$+\sum \bar{M}_A = -1 + \frac{hL^2}{2EI}\left(\frac{2L}{3}\right) = 0$$

$$h = \frac{3EI}{L^3}$$

The conjugate beam and its support then are

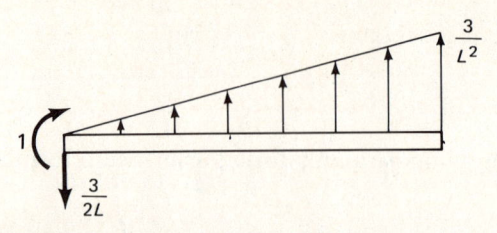

Figure 9.18

Now the conjugate bending moments, hence the real beam deflections, may be found by taking sections.

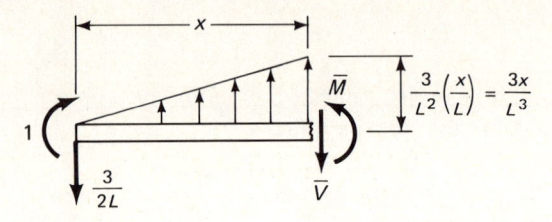

Figure 9.19

$$+\circlearrowleft\sum \bar{M}_{cut} = -1 + \frac{3x}{2L} - \frac{3x}{L^3}\left(\frac{x}{2}\right)\left(\frac{x}{3}\right) + \bar{M} = 0$$

$$\bar{M} = 1 - \frac{3x}{2L} + \frac{x^3}{2L^3} = r_A$$

With r_A known, equilibrium will give the remaining reactions.

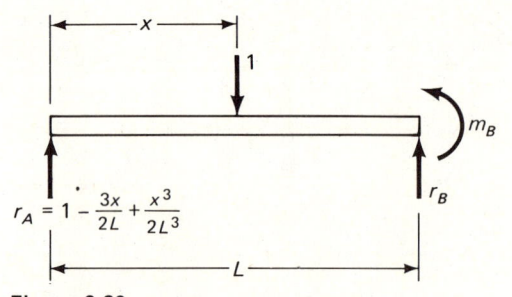

Figure 9.20

They are

$$r_B = \frac{3x}{2L} - \frac{x^3}{2L^3}$$

$$m_B = \frac{x^3}{2L^2} - \frac{x}{2}$$

INFLUENCE LINES FOR THE STATICALLY INDETERMINATE STRUCTURES

Plotted, the three influence lines are

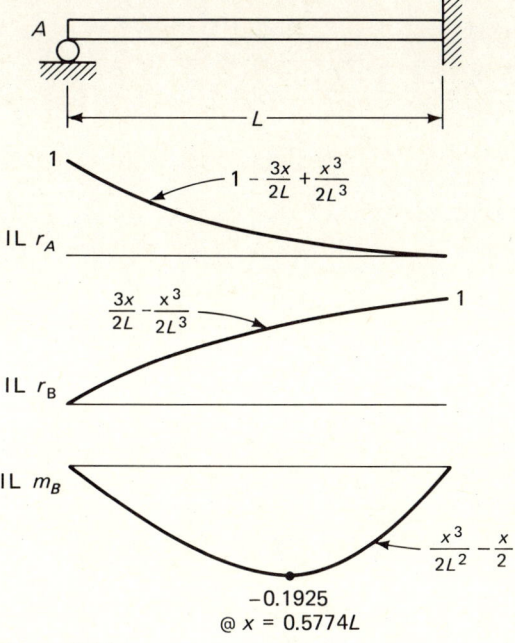

IL r_A: $1 - \dfrac{3x}{2L} + \dfrac{x^3}{2L^3}$

IL r_B: $\dfrac{3x}{2L} - \dfrac{x^3}{2L^3}$

IL m_B: $\dfrac{x^3}{2L^2} - \dfrac{x}{2}$

-0.1925 @ $x = 0.5774L$

Figure 9.21

EXAMPLE 9.2.2

Determine the influence line for the shear force and moment at point C located $L/3$ from A.

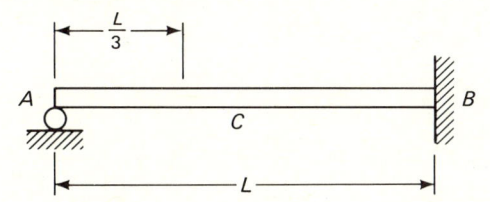

Figure 9.22

The position of the influence lines to the left of C is most easily found using the free-body diagram for the part to the right of C, part CB. This is comparable to the method used in Section 5.2.3.

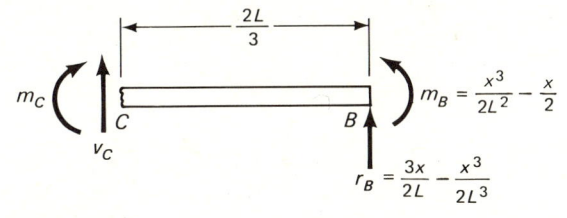

$m_B = \dfrac{x^3}{2L^2} - \dfrac{x}{2}$

$r_B = \dfrac{3x}{2L} - \dfrac{x^3}{2L^3}$

Figure 9.23

192 STATICALLY INDETERMINATE STRUCTURES: CONJUGATE BEAM ANALYSIS

From equilibrium,

$$v_C = -r_B = -\frac{3x}{2L} + \frac{x^3}{2L^3}$$

$$m_C = \frac{2L}{3}r_B + m_B = \frac{x}{2} + \frac{x^3}{6L^2}$$

for $0 < x < L/3$.

For the portion to the right of C, we use the free-body diagram to the left of C.

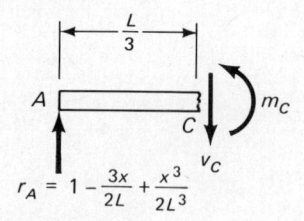

Figure 9.24

From equilibrium

$$v_C = r_A = 1 - \frac{3x}{2L} + \frac{x^3}{2L^3}$$

$$m_C = \frac{L}{3}r_A = \frac{L}{3} - \frac{x}{2} + \frac{x^3}{6L^2}$$

Plotted, the influence lines are

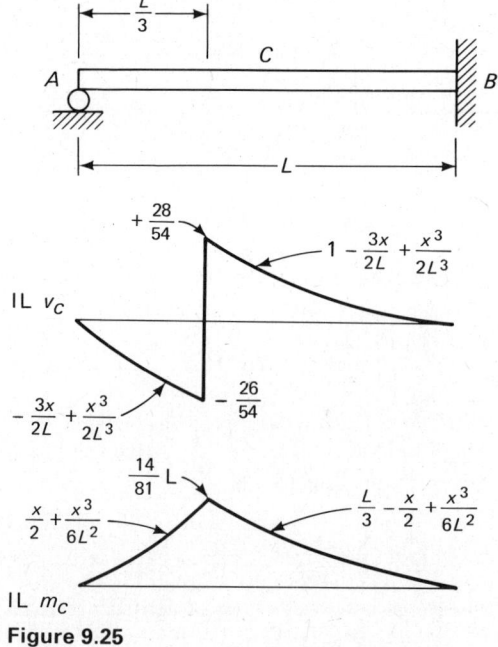

IL v_C

IL m_C

Figure 9.25

INFLUENCE LINES FOR THE STATICALLY INDETERMINATE STRUCTURES

Determining the exact influence line with ordinate values known may be accomplished easily for one or two statically indeterminate beams. But for high degrees the labor required increases markedly. The Müller–Breslau principle can be used easily to establish a qualitative influence line, that is, one with an established form but lacking ordinate values. These are used to determine probable worst case loading combinations.

EXAMPLE 9.2.3

Use the Müller–Breslau principle to sketch the influence line for the bending moment at the midpoint G of CD.

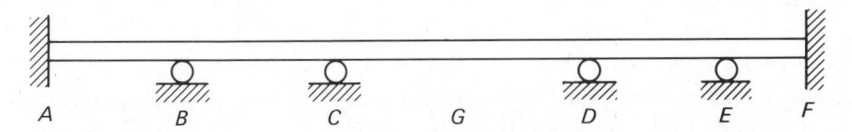

Figure 9.26

A hinge is presumed at G, a unit relative rotation imposed, and a reasonable elastic curve sketched.

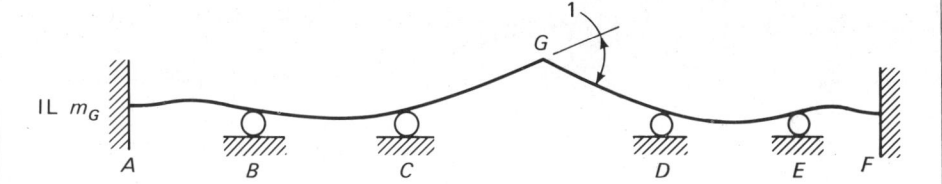

Figure 9.27

For maximum bending moment at G, spans AB, CD, and EF should be loaded, while BC and DE remain unloaded.

The same method may be used for rigid frames of very high static indeterminacy.

EXAMPLE 9.2.4

Sketch the influence line for the shear force near F of member FG in the rigid frame.

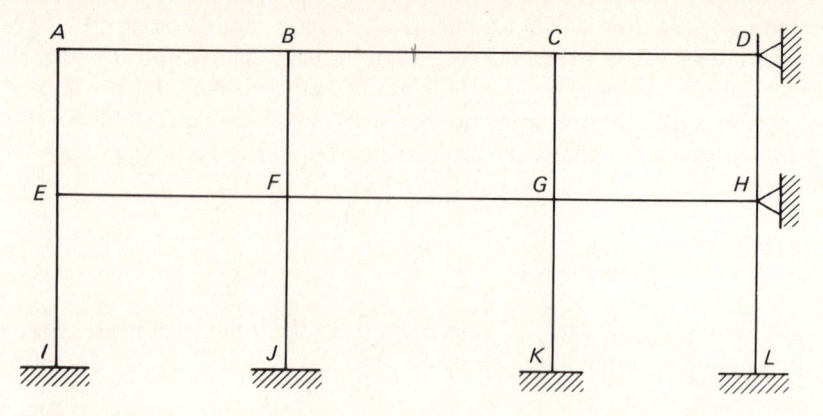

Figure 9.28

We sever *FG* near *F* and impose a unit shear displacement at the severed ends while maintaining the displaced ends parallel. The resulting deformed structure is sketched, allowing the natural joint rotations to occur. There may be some uncertainty about relative magnitudes of deflections, since they do depend on the relative stiffnesses of the members, but reasonable and usable sketches can be made. For our shear case,

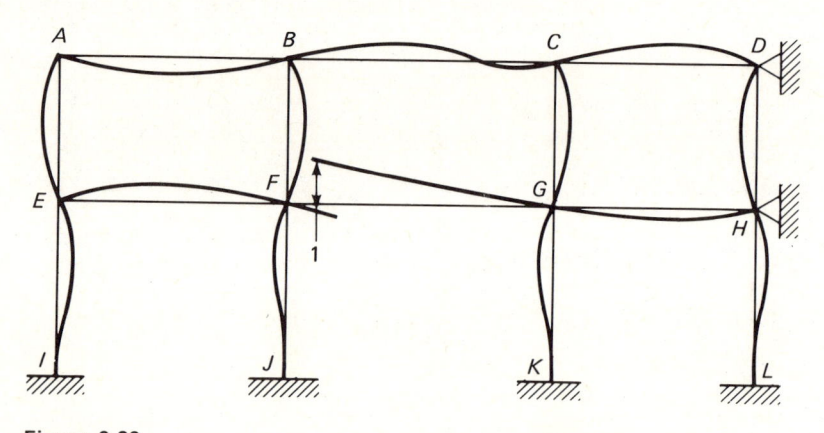

Figure 9.29

From it we conclude that a maximum shear near *F* on member *FG* will occur when gravity loads act simultaneously on *CD*, *EF*, *FG*, and the left-hand portion of *BC*, while the other horizontal members remain unloaded. But notice, too, that if horizontal forces to the right were applied to *AE*, *DH*, *FJ*, and *GK*, the shear force would increase. Also notice that it would further increase if, in addition to the gravity loads, upward forces were applied to *AB* and *GH*. It may not be physically possible for all these loads to exist simultaneously. But imagine that the floor is loaded from *E* to *G*, that snow has accumulated and drifted to load the roof from *B* to *D*, and that a high wind from left to right exerts pressure on *AE* and suction on *AB* and *DH*. For that not unlikely combination, the worst loading case previously described would exist almost exactly. If shear failure were imminent, this loading would have a high potential of causing it.

PROBLEMS

9.1 Use the method of conjugate beams to determine the support reactions for each of the statically indeterminate beams.

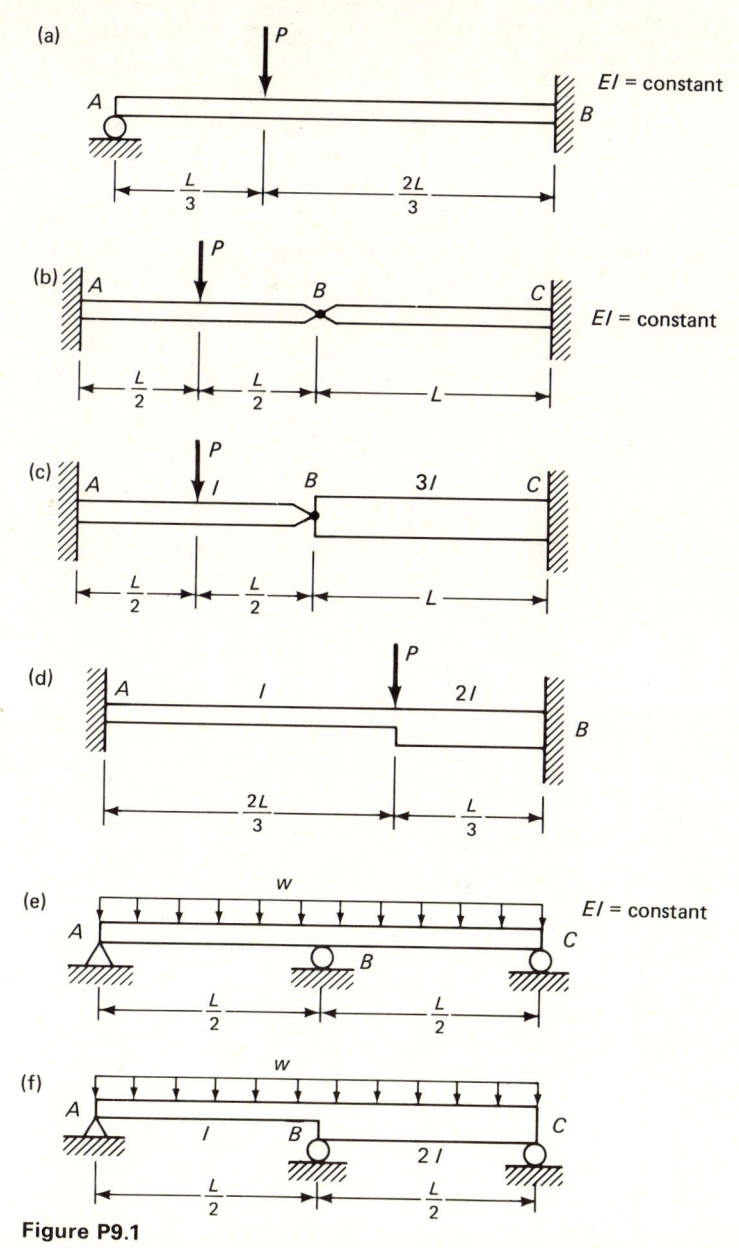

Figure P9.1

9.2 Develop algebraic expressions for the indicated influence lines and plot them in relation to the structure.

(a) Moment at A, vertical reaction at A, and shear at D.

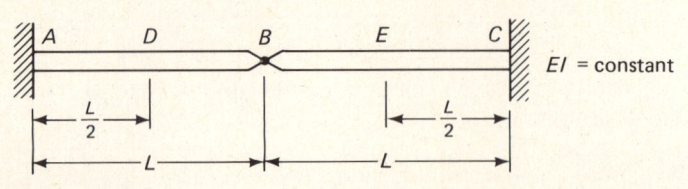

Figure P9.2A

(b) Moment at C, vertical reaction at C, and shear at E for the beam of part (a).
(c) Bending moment at D and shear at B for the beam of part (a).
(d) Support reaction at A, support reaction at B, and shear at D.

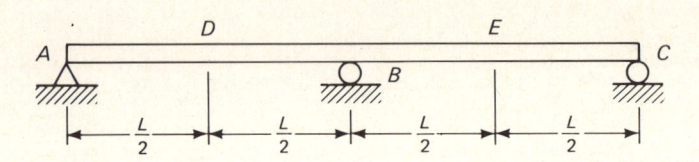

Figure P9.2D

(e) Support reactions at C and bending moment at E for the structure of part (d).
(f) Moment at D and vertical reaction at D.

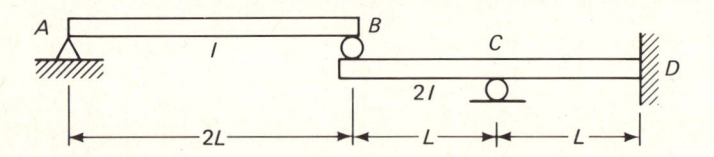

Figure P9.2F

(g) Bending moment at C and reaction at C for beam of part (f).

9.3 Use the Müller–Breslau principle to sketch the indicated influence lines. Interpret these for "worst loading" situations.

(a) Shear and bending moment at G (a point slightly to the right of B).

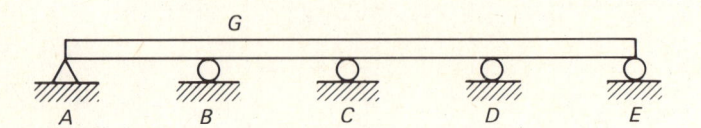

Figure P9.3A

(b) Shear and bending moment at J (a point slightly above joint D).

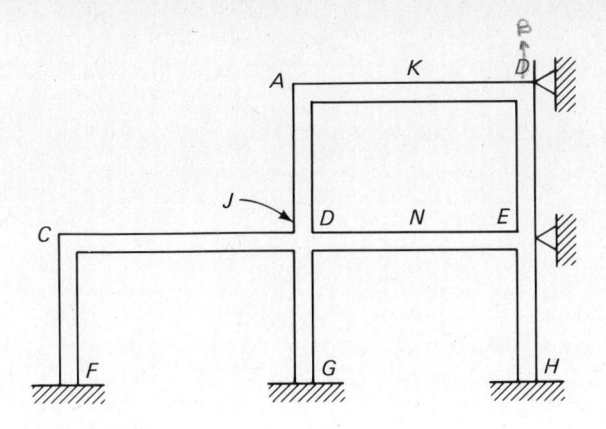

Figure P9-3B

(c) Shear and bending moment at K (midpoint of AB) in the structure for part (b).
(d) Shear and bending moment at N (midpoint of DE) in the structure for part (b).
(e) Bending moment at the midpoint of member BC in Example 9.2.4
(f) Bending moment at the midpoint of member AB in Example 9.2.5.

Part III
STATICALLY INDETERMINATE STRUCTURAL ANALYSIS: DISPLACEMENT/STIFFNESS METHODS

The second general method of analysis of statically indeterminate structures is known by two names: displacement methods and stiffness methods. We will examine this approach in Chapters 10 through 14, using several variations: slope deflection and moment distribution, which are both classical approaches, followed by a matrix formulation of the methods.

Each application of the displacement/stiffness methods illustrated uses displacements as the unknowns, later solving forces from these displacements. The basic equations in use are stiffness equations.

These methods constitute a marked departure from the previous approaches studied. An open mind, ready to look at new schemes, is an important beginning.

Statically Indeterminate Structures: The Slope Deflection Method

As teenage boys they had competed in everything, a competition that continued into their adult years and became increasingly spirited with time. After they had gone their separate ways, one as a priest and the other as a military officer, the competition continued by mail, but limited to written put-downs and insults.

Many years later they chanced upon each other at a railroad station. The priest by then was a very portly bishop and the officer a much decorated general. Approaching the general, the bishop asked, "Conductor, please, when is the next train for Albany?"

The general replied, "Please forgive my personal intrusion, Madam, but I suggest that in your condition you not travel at all. Expectant mothers should remain near the hospital, especially when delivery is imminent."

10.0 INTRODUCTION

Early in this century slope deflection was the most popular method in use for analyzing statically indeterminate frames. It was developed by Professor G. A. Maney and began its reign of popularity almost immediately after its publication in 1915.[1]

Fifteen years later Professor Hardy Cross introduced the moment distribution method and there began a period of spirited professional competition over the merits of the two methods, with moment distribution eventually emerging as the "winner," primarily because of its speed and simplicity.

But the competition has not ended. Today, although moment distribution continues as the more popular method, there remain many contemporary engineers who prefer slope deflection. They contend that in performing a slope deflection analysis the engineer can acquire a better "intuitive feel" for the structure than through the use of any other method.

More significant, though, slope deflection has gained renewed importance with the advent of the computer, serving as the central method used for structural analysis software. We will look at the method in its classical form in this chapter and again, in its software form, in Chapter 14.

Slope deflection differs markedly from the methods presented in early chapters. Its focus is on individual members, their loads, and certain conditions at their ends. In using this method, simultaneous equations are written and solved that have displacements, rather than forces or moments, as the unknowns. It employs a simple sign convention: all variables related to a *member* are *positive* if they are *clockwise*.

Consider member AB with moments and shears acting at its ends and deformed as illustrated:

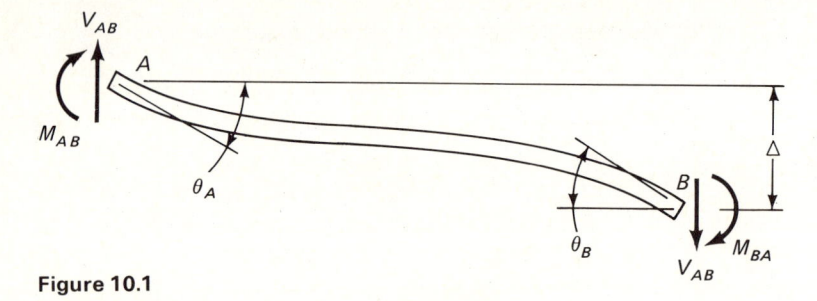

Figure 10.1

Each item shown is shown in its positive sense:

M_{AB} and M_{BA} are clockwise end moments.
V_{AB} and V_{BA} are end shear forces tending to cause a clockwise rotation of AB.
θ_A and θ_B are clockwise end rotations.
Δ is a relative linear displacement of ends A and B that matches a clockwise rotation of AB.

Slope deflection uses a double subscript notation for its end moments and shears. The first subscript is the end where the moment or shear acts, while the second is the other end of the member. Thus M_{AB} is a moment acting at end A of member AB.

A set of equations, called slope deflection equations, relates the moments at the

---
[1] Maney, G. A., *Studies in Engineering—No. 1*, University of Minnesota, 1915.

INTRODUCTION **203**

two ends of each member to the loads the member carries and to its end displacements. We will develop these equations term by term, considering first the term related to the member loads. To avoid including end displacements in this term, we *prevent* them by assuming that both ends are fixed. Appropriately, these terms are called *fixed end moments* and are written FEM_{AB} or FEM_{BA}.

Table 10.1

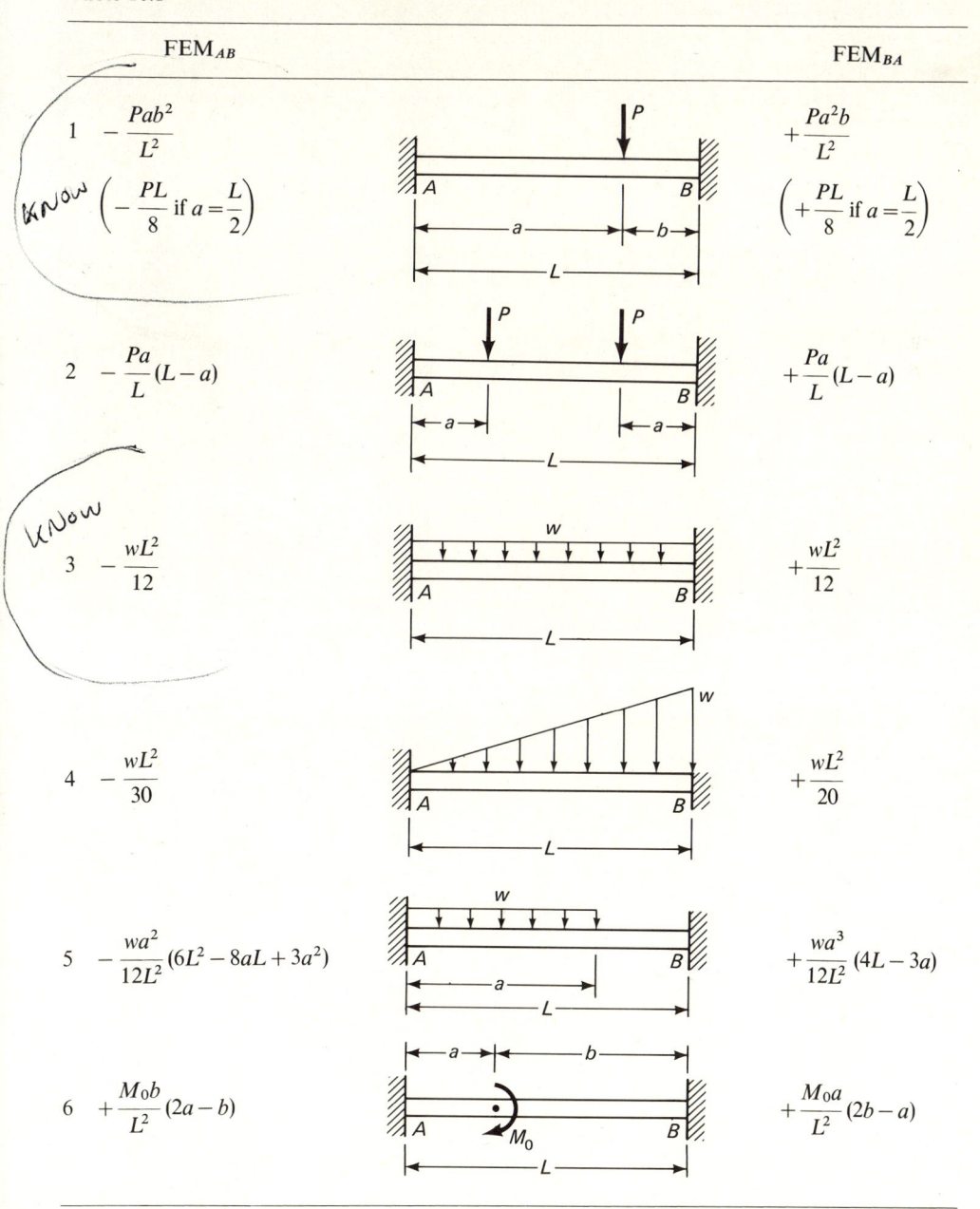

	FEM_{AB}		FEM_{BA}
1	$-\dfrac{Pab^2}{L^2}$ $\left(-\dfrac{PL}{8} \text{ if } a = \dfrac{L}{2}\right)$	Concentrated load P at distance a from A, b from B	$+\dfrac{Pa^2b}{L^2}$ $\left(+\dfrac{PL}{8} \text{ if } a = \dfrac{L}{2}\right)$
2	$-\dfrac{Pa}{L}(L-a)$	Two equal loads P at distance a from each end	$+\dfrac{Pa}{L}(L-a)$
3	$-\dfrac{wL^2}{12}$	Uniform load w over span L	$+\dfrac{wL^2}{12}$
4	$-\dfrac{wL^2}{30}$	Triangular load, max w at B	$+\dfrac{wL^2}{20}$
5	$-\dfrac{wa^2}{12L^2}(6L^2 - 8aL + 3a^2)$	Uniform load w over distance a from A	$+\dfrac{wa^3}{12L^2}(4L - 3a)$
6	$+\dfrac{M_0b}{L^2}(2a-b)$	Applied moment M_0 at distance a from A, b from B	$+\dfrac{M_0a}{L^2}(2b-a)$

204 STATICALLY INDETERMINATE STRUCTURES: THE SLOPE DEFLECTION METHOD

In Example 9.1.2 end moments for

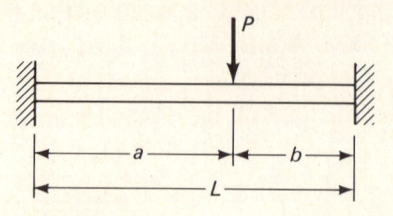

Figure 10.2

were found to be

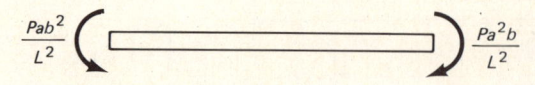

Figure 10.3

Therefore

$$\text{FEM}_{AB} = -\frac{Pab^2}{L^2}$$

$$\text{FEM}_{BA} = +\frac{Pa^2b}{L^2}$$

for that loading.

Calculations similar to those of Example 9.1.2 would give the fixed end moments for other load arrangements. Fixed end moments for common load arrangements are given in Table 10.1.

Although it is impossible to include formulas for every possible load combination in Table 10.1, the fixed end moments for nearly any loading can be established by superposing those for several of the loadings of Table 10.1.

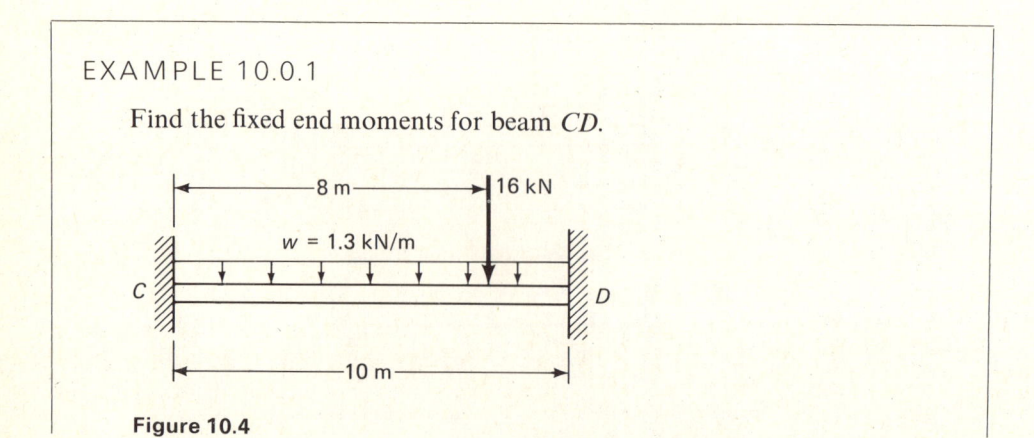

EXAMPLE 10.0.1

Find the fixed end moments for beam CD.

Figure 10.4

INTRODUCTION 205

Superposing cases 1 and 3 of Table 10.1,

$$FEB_{CD} = -\frac{(16\text{ kN})(8\text{ m})(2^2\text{ m}^2)}{10^2\text{ m}^2} - \frac{(1.3\text{ kN/m})(10^2\text{ m}^2)}{12}$$

$$= -15.95\text{ kN}\cdot\text{m}$$

$$FEM_{DC} = +\frac{16(8^2)(2)}{10^2} + \frac{1.3(10^2)}{12}$$

$$= +31.3\text{ kN}\cdot\text{m}$$

Next we will develop the terms in the slope deflection equations that relate to θ_A. As before, we constrain the beam ends so that other displacements cannot occur. Further, we use no external loads except those at the ends.

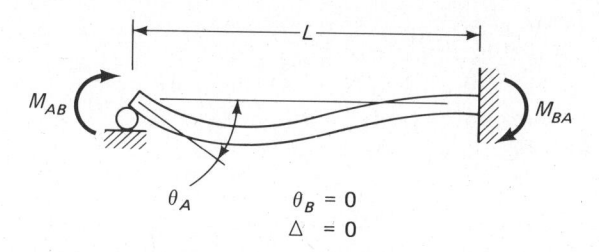

Figure 10.5

M_{BA} is the reaction of the clamped end at B, preventing rotation there as end A rotates θ_A, while M_{AB} is the moment at A that causes θ_A. It is convenient to construct the moment diagram in two parts.

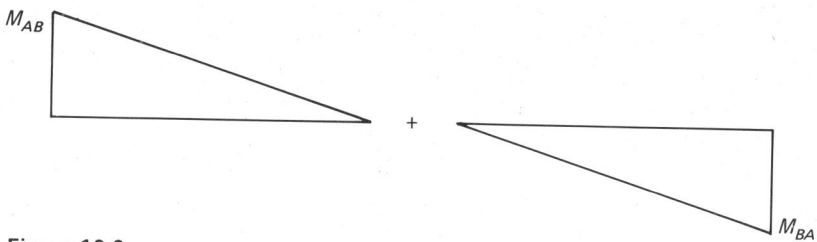

Figure 10.6

The moment diagram sign convention remains unaffected by the slope deflection sign convention: the moment is plotted on the compression side of the member.

The moment diagrams, converted to M/EI diagrams, are applied as loads to the conjugate beam. Thus

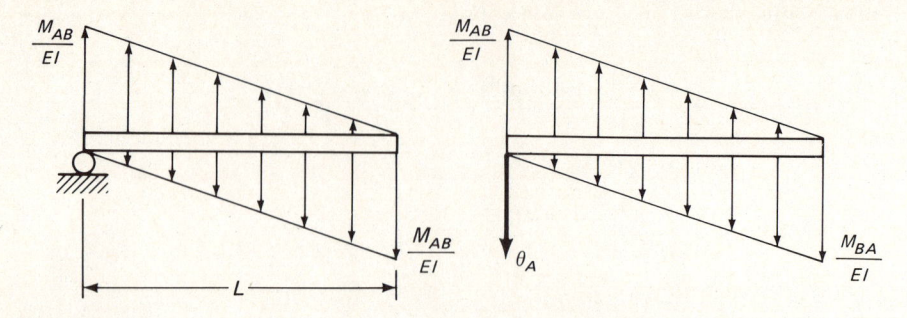

Figure 10.7

In conjugate beam conventions, real beam slopes that are downward to the right are negative and, therefore, have negative conjugate shears. Thus a downward reaction at A equal to θ_A is the conjugate support reaction at A. From equilibrium, the equations

$$+\circlearrowright \sum \bar{M}_A = \frac{M_{AB}L}{2EI}\left(\frac{L}{3}\right) - \frac{M_{BA}L}{2EI}\left(\frac{2L}{3}\right) = 0$$

$$+\uparrow \sum \bar{F}_y = \frac{M_{AB}L}{2EI} - \frac{M_{BA}L}{2EI} - \theta_A = 0$$

solved simultaneously, give

$$M_{AB} = \frac{4EI}{L}\theta_A$$

$$M_{BA} = \frac{2EI}{L}\theta_A$$

Generalized, the end moment at the rotated end is $(4EI/L)\theta$, while the end moment at the fixed end is $(2EI/L)\theta$. Thus for a rotation of end B while end A is fixed,

$$M_{AB} = \frac{2EI}{L}\theta_B$$

$$M_{BA} = \frac{4EI}{L}\theta_B$$

Next we clamp both ends to prevent their rotations, and then introduce downward displacement Δ of end B.

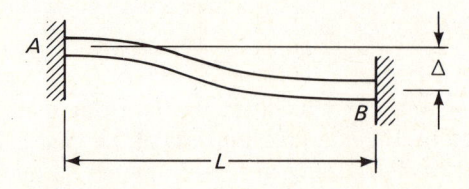

Figure 10.8

INTRODUCTION 207

Moments M_{AB} and M_{BA} arising from this displacement lead to the moment diagrams

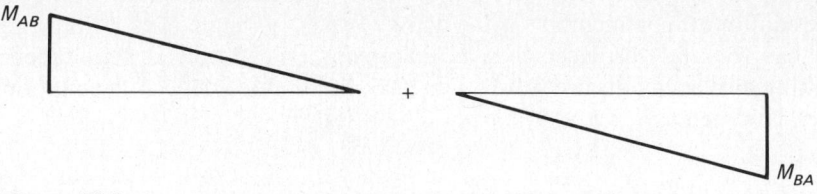

Figure 10.9

which can be converted to M/EI diagrams and applied as loads to the conjugate beam. It is completely unsupported in this case since free ends are the conjugates of the two real clamped ends.

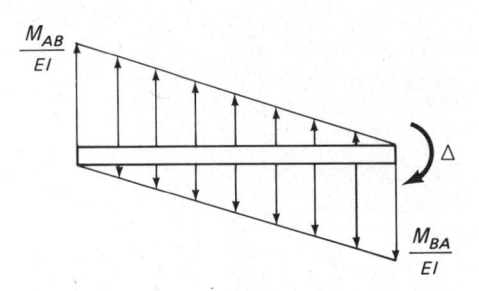

Figure 10.10

Couple Δ at B is the conjugate of the downward displacement of the real beam at B. For equilibrium,

$$+\uparrow \sum F_y = \frac{M_{AB}L}{2EI} - \frac{M_{BA}L}{2EI} = 0$$

$$+\sum \bar{M}_A = \frac{M_{AB}L}{2EI}\left(\frac{L}{3}\right) - \frac{M_{BA}L}{2EI}\left(\frac{2L}{3}\right) - \Delta = 0$$

from which

$$M_{AB} = M_{BA} = -\frac{6EI}{L^2}\Delta$$

The complete slope deflection equations for M_{AB} and M_{BA} are the superpositions of the four preceding parts: θ_A, θ_B, Δ, and loads. Thus

$$M_{AB} = \frac{4EI}{L}\theta_A + \frac{2EI}{L}\theta_B - \frac{6EI}{L^2}\Delta + \text{FEM}_{AB}$$

$$M_{BA} = \frac{2EI}{L}\theta_A + \frac{4EI}{L}\theta_B - \frac{6EI}{L^2}\Delta + \text{FEM}_{BA}$$

(10.1)

Analysis by slope deflection begins with use of Eqs. (10.1) to write separate expressions for the end moments at each end of each member. Equilibrium is then imposed, using moment equilibrium at joints rotated an unknown θ and transverse force equilibrium on members with an unknown Δ. A system of equations is produced that has the end displacements as unknowns. When solved simultaneously, the resulting displacements are substituted back in the slope deflection equations, giving the end moments.

10.1 STATICALLY INDETERMINATE BEAM ANALYSIS USING SLOPE DEFLECTION

We will illustrate the slope deflection method applied to a statically indeterminate beam with several examples.

EXAMPLE 10.1.1

Determine the end shears and moments and the support reactions for beam ABC. Construct its shear and moment diagrams.

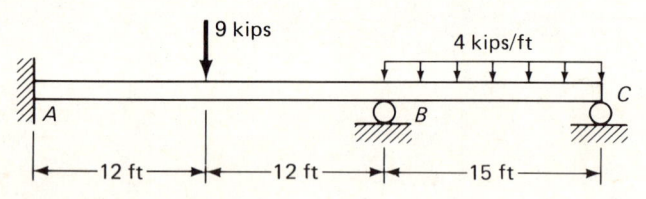

Figure 10.11

There are only two unknown end displacements, θ_B and θ_C. θ_A is known to be zero since it is clamped at A, and $\Delta_{AB} = \Delta_{BC} = 0$ since vertical displacements are prevented at the supports.

Using Table 10.1, the fixed end moments are

$$\text{FEM}_{AB} = -\frac{PL}{8} = -\frac{(9 \text{ kips})(24 \text{ ft})}{8} = -27 \text{ kip-ft}$$

$$\text{FEM}_{BA} = +\frac{PL}{8} = +27 \text{ kip-ft}$$

$$\text{FEM}_{BC} = -\frac{wL^2}{12} = -\frac{(4 \text{ kips/ft})(15^2 \text{ ft}^2)}{12} = -75 \text{ kip-ft}$$

$$\text{FEM}_{CB} = +\frac{wL^2}{12} = +75 \text{ kip-ft}$$

These, with the unknowns θ_B and θ_C, are substituted in the slope deflection equations,

$$M_{AB} = \frac{2EI}{24}\theta_B - 27$$

STATICALLY INDETERMINATE BEAM ANALYSIS USING SLOPE DEFLECTION 209

$$M_{BA} = \frac{4EI}{24}\theta_B + 27$$

$$M_{BC} = \frac{4EI}{15}\theta_B + \frac{2EI}{15}\theta_C - 75$$

$$M_{CB} = \frac{2EI}{15}\theta_B + \frac{4EI}{15}\theta_C + 75$$

There must be equal opposites of M_{BA} and M_{BC} acting on joint B.

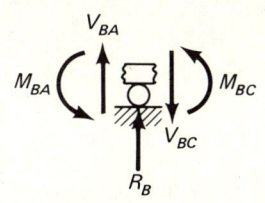

Figure 10.12

We apply moment equilibrium at the joint,

$$+\circlearrowleft\sum M_B = M_{BA} + M_{BC} = 0$$

so that

$$\frac{4EI}{24}\theta_B + 27 + \frac{4EI}{15}\theta_B + \frac{2EI}{15}\theta_C - 75 = 0$$

which simplifies to

$$\frac{13}{30}EI\theta_B + \frac{2}{15}EI\theta_C = 48$$

Similarly, summing moments at joint C,

$$+\circlearrowleft\sum M_C = M_{CB} = 0$$

so that

$$\frac{4EI}{15}\theta_C + \frac{2EI}{15}\theta_B + 75 = 0$$

Simultaneous solution gives

$$\theta_B = \frac{233.2}{EI} \quad \text{and} \quad \theta_C = -\frac{397.8}{EI}$$

which, upon substitution in the slope deflection equations, gives

$$M_{AB} = \frac{2EI}{24}\left(\frac{233.2}{EI}\right) - 27 = -7.57 \text{ kip-ft}$$

$$M_{BA} = \frac{4EI}{24}\left(\frac{233.2}{EI}\right) + 27 = +65.9 \text{ kip-ft}$$

$$M_{BC} = \frac{4EI}{15}\left(\frac{233.2}{EI}\right) + \frac{2EI}{15}\left(-\frac{397.8}{EI}\right) - 75 = -65.9 \text{ kip-ft}$$

$$M_{CB} = \frac{2EI}{15}\left(\frac{233.2}{EI}\right) + \frac{4EI}{15}\left(-\frac{397.8}{EI}\right) + 75 = 0$$

In this example the end moments are independent of EI. This will always be so if values of the end displacements are *not* specified and if EI is constant. With the end moments known, beam elements AB and BC have become statically determinate, and their end shears can be found from equilibrium.

For AB

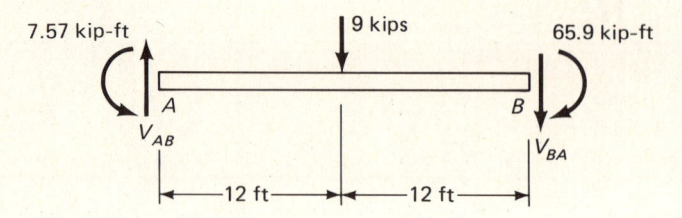

Figure 10.13

$$+\circlearrowright \sum M_A = -7.57 + 65.9 + 9(12) + 24V_{BA} = 0$$

$$V_{BA} = -6.93 \text{ kips}$$

$$+\uparrow \sum F_y = V_{AB} - 9 - V_{BA} = 0$$

$$V_{AB} = 9 + V_{BA} = 2.07 \text{ kips}$$

For BC,

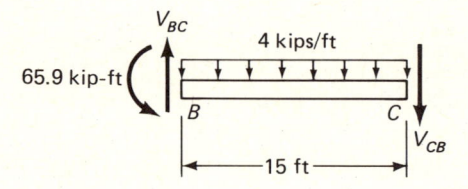

Figure 10.14

$$+\circlearrowright \sum M_B = -65.9 + 4(15)(7.5) + 15V_{CB} = 0$$

$$V_{CB} = -25.6 \text{ kips}$$

$$+\uparrow \sum F_y = V_{BC} - 4(15) - V_{CB} = 0$$

$$V_{BC} = 60 + (-25.6) = 34.4 \text{ kips}$$

The support reactions at A and C are the end shears V_{AB} and V_{CB}, while the support reaction at B is the change in the end shears at B, that is, the change from V_{BA} to V_{BC}. The beam, with loads and support reactions, is shown along with its shear and moment diagrams.

STATICALLY INDETERMINATE BEAM ANALYSIS USING SLOPE DEFLECTION **211**

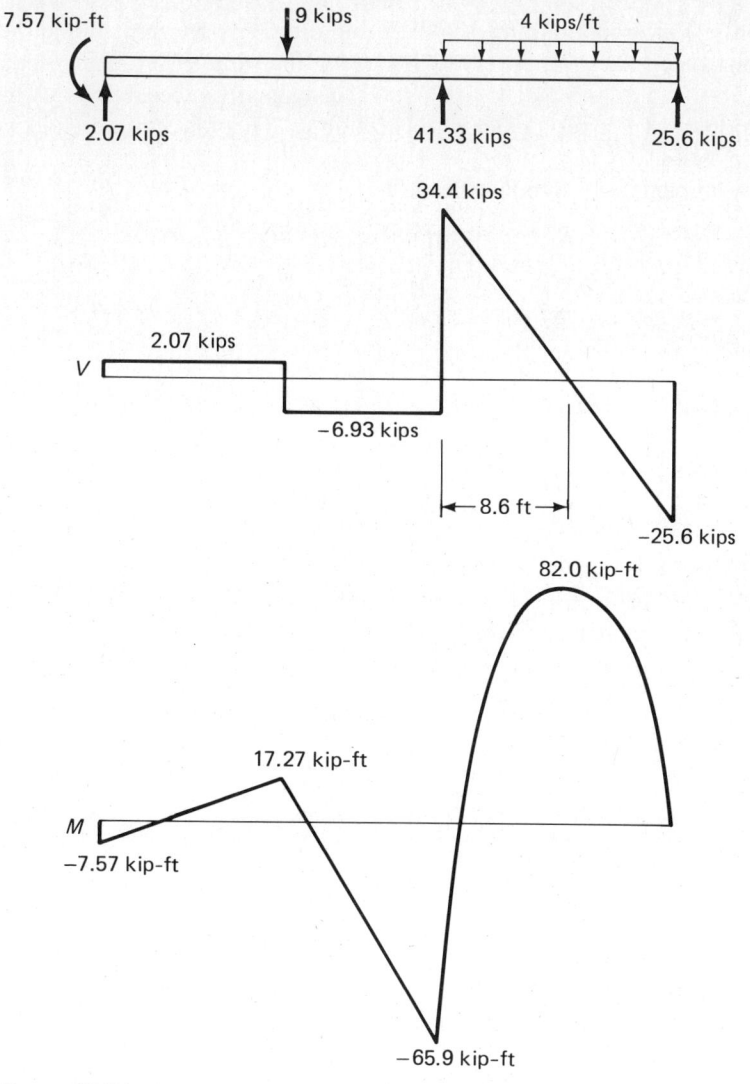

[Beam diagram: 7.57 kip-ft moment at left end, 9 kips point load, 4 kips/ft distributed load; reactions 2.07 kips, 41.33 kips, 25.6 kips]

[Shear (V) diagram: 2.07 kips, −6.93 kips, 34.4 kips, −25.6 kips; 8.6 ft marked]

[Moment (M) diagram: −7.57 kip-ft, 17.27 kip-ft, 82.0 kip-ft, −65.9 kip-ft]

Figure 10.15

EXAMPLE 10.1.2

Use slope deflection to determine the end shears and moments for beam ABC if support B settles downward $\frac{3}{4}$ in.

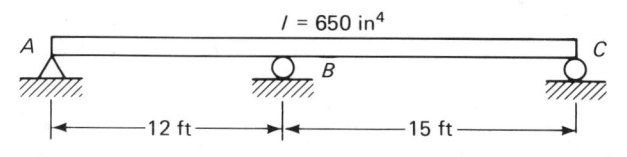

[Beam ABC: A (pin), B (roller), C (roller); $I = 650$ in^4; spans 12 ft and 15 ft]

Figure 10.16

With no external loading, the fixed end moments are zero. While the rotations at the three supports are unknown, the support settlement leads to $\Delta_{AB} = +0.75$ in and $\Delta_{BC} = -0.75$ in. The signs are so because a downward motion of B gives a clockwise rotation to AB and a counterclockwise rotation to BC.

The slope deflection equations are

$$M_{AB} = \frac{4EI}{12}\theta_A + \frac{2EI}{12}\theta_B - \frac{6EI}{12}\left(\frac{0.75}{144}\right)$$

$$M_{BA} = \frac{2EI}{12}\theta_A + \frac{4EI}{12}\theta_B - \frac{6EI}{12}\left(\frac{0.75}{144}\right)$$

$$M_{BC} = \frac{4EI}{15}\theta_B + \frac{2EI}{15}\theta_C - \frac{6EI}{15}\left(-\frac{0.75}{180}\right)$$

$$M_{CB} = \frac{2EI}{15}\theta_B + \frac{4EI}{15}\theta_C - \frac{6EI}{15}\left(-\frac{0.75}{180}\right)$$

Notice the dimensional consistency used term by term. Factors EI/L use L in feet throughout, while θ and Δ/L are expressed as radians and inches per inch, respectively.

Three equilibrium equations are written by taking moments at each of the joints that have unknown rotations.

$$\sum M_A = M_{AB} = 0$$

$$\frac{4EI}{12}\theta_A + \frac{2EI}{12}\theta_B - \frac{6EI}{12}\left(\frac{0.75}{144}\right) = 0$$

$$\sum M_B = M_{BA} + M_{BC} = 0$$

$$\frac{2EI}{12}\theta_A + \frac{4EI}{12}\theta_B - \frac{6EI}{12}\left(\frac{0.75}{144}\right) + \frac{4EI}{15}\theta_B + \frac{2EI}{15}\theta_C + \frac{6EI}{15}\left(\frac{0.75}{180}\right) = 0$$

which reduces to

$$30\frac{EI}{180}\theta_A + 108\frac{EI}{180}\theta_B + 24\frac{EI}{180}\theta_C = \frac{30.375}{180}\left(\frac{EI}{180}\right)$$

$$\sum M_C = M_{CB} = 0$$

$$\frac{2EI}{15}\theta_B + \frac{4EI}{15}\theta_C + \frac{6EI}{15}\left(\frac{0.75}{180}\right) = 0$$

EI can be factored out of each equation, showing that the values of θ_A, θ_B, and θ_C are independent of EI.

$\theta_A = 7.29 \times 10^{-3}$ rad

$\theta_B = 1.042 \times 10^{-3}$ rad

$\theta_C = -6.77 \times 10^{-3}$ rad

But when these values are substituted in the slope deflection equations, EI reappears. Therefore, the end moments are dependent on EI if a specific displacement is prescribed.

STATICALLY INDETERMINATE BEAM ANALYSIS USING SLOPE DEFLECTION 213

Only M_{BA} and M_{BC} need be evaluated, although M_{AB} and M_{CB} should be calculated as a check. They should be zero.

$$M_{BA} = \frac{2(3 \times 10^4 \text{ kips/in}^2)(650 \text{ in}^4)}{12 \times 12 \text{ in}} 7.29 \times 10^{-3}$$

$$+ \frac{4(3 \times 10^4)(650)}{12 \times 12} 1.042 \times 10^{-3}$$

$$- \frac{6(3 \times 10^4)(650)}{12 \times 12} \left(+ \frac{0.75}{144} \right)$$

$$= -1693 \text{ kip-in}$$

$$M_{BC} = \frac{4(3 \times 10^4)(650)}{15 \times 12} 1.042 \times 10^{-3}$$

$$+ \frac{2(3 \times 10^4)(650)}{15 \times 12} (-6.77 \times 10^{-3})$$

$$- \frac{6(3 \times 10^4)(650)}{15 \times 12} \left(- \frac{0.75}{180} \right)$$

$$= +1693 \text{ kip-in}$$

An important check, that M_{BA} and M_{BC} are equal and opposite in sign, is provided.

Shear forces are found next.

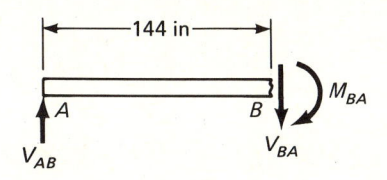

Figure 10.17

$$+\circlearrowleft \sum M_B = 144 V_{AB} + M_{BA} = 0$$

$$V_{AB} = -\frac{M_{BA}}{144} = -\frac{(-1693)}{144} = 11.76 \text{ kips}$$

$$+\uparrow \sum F_y = V_{AB} - V_{BA} = 0$$

$$V_{BA} = V_{AB} = +11.76 \text{ kips}$$

Similar calculations for *BC* give

$$V_{CB} = V_{BC} = -9.41 \text{ kips}$$

Thus

Figure 10.18

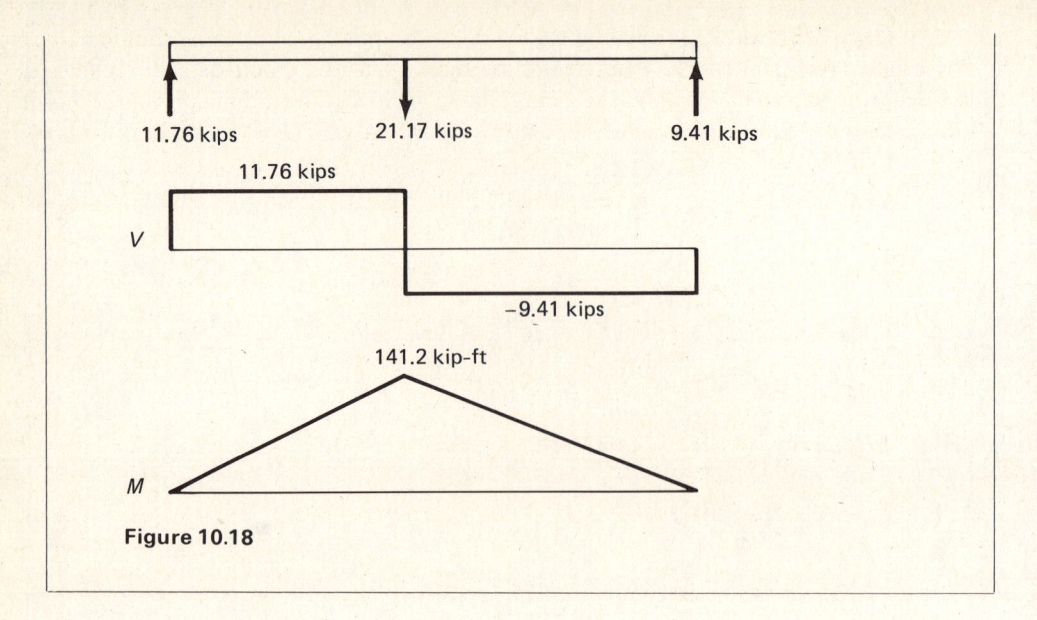

10.2 STATICALLY INDETERMINATE FRAMES USING SLOPE DEFLECTION

Slope deflection analysis of rigid frames is similar to that of beams, although there may be added considerations for sidesway.

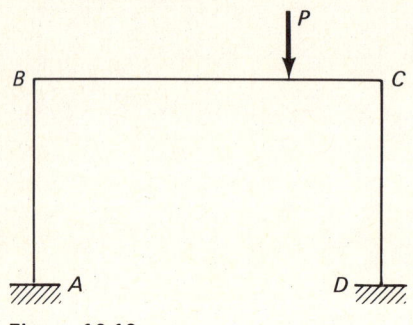

Figure 10.19

Frame *ABCD* is likely to have an elastic curve such as

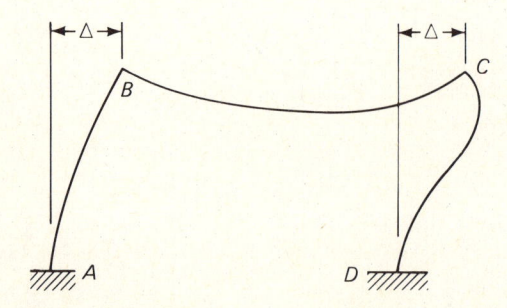

Figure 10.20

where B and C have displaced a distance Δ to the right. We presume that the axial deformations of the three members are negligible and that their departures from a straight line are small (despite the exaggerated drawing). We conclude that joints B and C cannot move vertically so that the Δ for BC is zero. B and C can move horizontally, however, but both must move the same direction and distance Δ. That Δ must be included in the slope deflection equations for AB and CD, along with the rotations at B and C.

EXAMPLE 10.2.1

Find the end moments and draw the shear and moment diagrams for frame $ABCDE$ subjected to a 625-kN/m load.

$EI_{AB} = 10^5 \text{ kN} \cdot \text{m}^2$
$EI_{BC} = EI_{CD} = 3 \times 10^5 \text{ kN} \cdot \text{m}^2$
$EI_{CE} = 1.4 \times 10^5 \text{ kN} \cdot \text{m}^2$

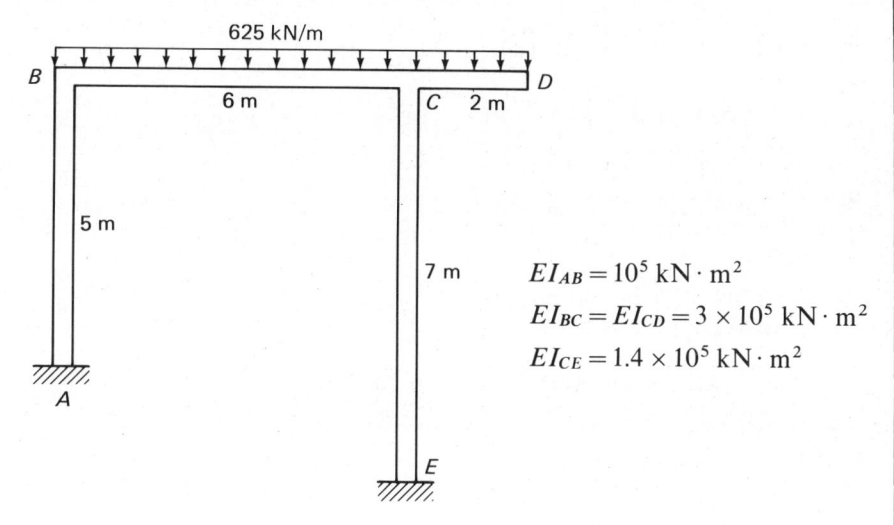

Figure 10.21

At first it may seem necessary to include θ_D and Δ_{CD} with other unknown end displacements, for certainly end D will deflect vertically and rotate. But member CD is statically determinate, so its end moments can be written using equilibrium alone. Thus

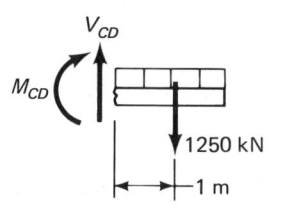

Figure 10.22

$+\circlearrowleft \sum M_C = M_{CD} + 1250(1) = 0$
$M_{CD} = -1250 \text{ kN} \cdot \text{m}$

For member BC the fixed end moments are

$$M_{BC} = -M_{CB} = -\frac{(625 \text{ kN/m})(6^2 \text{ m}^2)}{12} = -1875 \text{ kN} \cdot \text{m}$$

The slope deflection equations are

$$M_{AB} = \frac{2(10 \times 10^4 \text{ kN} \cdot \text{m}^2)}{5 \text{ m}} \theta_B - \frac{6(10 \times 10^4 \text{ kN} \cdot \text{m}^2)}{5^2 \text{ m}^2} \Delta_{AB}$$

$$= 4 \times 10^4 \theta_B - 2.4 \times 10^4 \Delta_{AB}$$

$$M_{BA} = \frac{4(10 \times 10^4)}{5} \theta_B - \frac{6(10 \times 10^4)}{5^2} \Delta_{AB}$$

$$= 8 \times 10^4 \theta_B - 2.4 \times 10^4 \Delta_{AB}$$

$$M_{BC} = \frac{4(30 \times 10^4)}{6} \theta_B + \frac{2(30 \times 10^4)}{6} \theta_C - 1875$$

$$= 20 \times 10^4 \theta_B + 10 \times 10^4 \theta_C - 1875$$

$$M_{CB} = 10 \times 10^4 \theta_B + 20 \times 10^4 \theta_C + 1875$$

$$M_{CE} = \frac{4(14 \times 10^4)}{7} \theta_C - \frac{6(14 \times 10^4)}{7^2} \Delta_{CE}$$

$$= 8 \times 10^4 \theta_C - 1.714 \times 10^4 \Delta_{CE}$$

$$M_{EC} = 4 \times 10^4 \theta_C - 1.714 \times 10^4 \Delta_{CE}$$

Two of the needed equilibrium equations can be established using moment equilibrium at joints B and C. They are

$$\sum M_B = M_{BA} + M_{BC} = 0$$
$$8 \times 10^4 \theta_B - 2.4 \times 10^4 \Delta_{AB} + 20 \times 10^4 \theta_B + 10 \times 10^4 \theta_C - 1875 = 0$$

which reduces to

$$28\theta_B + 10\theta_C - 2.4\Delta_{AB} = 0.1875$$

and

$$\sum M_C = M_{CB} + M_{CD} + M_{CE} = 0$$
$$10 \times 10^4 \theta_B + 20 \times 10^4 \theta_C + 1875 - 1250 + 8 \times 10^4 \theta_C - 1.714 \times 10^4 \Delta_{AB}$$
$$= 0$$

which reduces to

$$10\theta_B + 28\theta_C - 1.714\Delta_{AB} = -0.0625$$

In the second equation Δ_{AB} replaced its equal Δ_{CE}.

A third equation, needed because there are three unknowns, is found by writing $\sum F_x = 0$ for the whole structure, where x corresponds to the direction of displacement Δ_{AB}. Preliminary for this summation, we write expressions for the shear forces V_{AB} and V_{EC}.

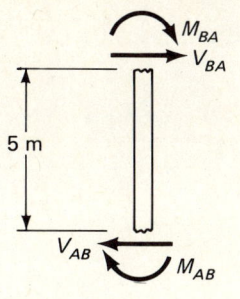

Figure 10.23

$$+\circlearrowleft \sum M_B = 5V_{AB} + M_{AB} + M_{BA} = 0$$

$$V_{AB} = -\frac{M_{AB} + M_{BA}}{5}$$

$$V_{AB} = -\frac{4 \times 10^4 \theta_B - 2.4 \times 10^4 \Delta_{AB} + 8 \times 10^4 \theta_B - 2.4 \times 10^4 \Delta_{AB}}{5}$$

$$= -2.4 \times 10^4 \theta_B + 0.96 \times 10^4 \Delta_{AB}$$

A similar process produces

$$V_{EC} = -1.714 \times 10^4 \theta_C + 0.499 \times 10^4 \Delta_{AB}$$

Considering the horizontal forces only on the structure,

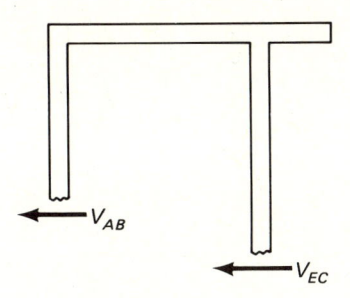

Figure 10.24

$$\pm \sum F_x = V_{AB} + V_{EC} = 0$$
$$-2.4 \times 10^4 \theta_B + 0.96 \times 10^4 \Delta_{AB} - 1.714 \times 10^4 \theta_C + 0.499 \times 10^4 \Delta_{AB} = 0$$

which reduces to

$$2.4\theta_B + 1.714\theta_C - 1.459\Delta_{AB} = 0$$

The three equilibrium equations, solved simultaneously, give

$$\theta_B = 9.28 \times 10^{-3} \text{ rad}$$
$$\theta_C = -4.97 \times 10^{-3} \text{ rad}$$
$$\Delta_{AB} = 9.43 \times 10^{-3} \text{ m}$$

Substituting these values in the slope deflection equations gives the end moments,

$$M_{AB} = 4 \times 10^4(9.28 \times 10^{-3}) - 2.4 \times 10^4(9.43 \times 10^{-3})$$
$$= 145 \text{ kN} \cdot \text{m}$$
$$M_{BA} = 8 \times 10^4(9.28 \times 10^{-3}) - 2.4 \times 10^4(9.43 \times 10^{-3})$$
$$= 516 \text{ kN} \cdot \text{m}$$
$$M_{BC} = 20 \times 10^4(9.28 \times 10^{-3}) + 10 \times 10^4(-4.97 \times 10^{-3}) - 1875$$
$$= -516 \text{ kN} \cdot \text{m}$$
$$M_{CB} = 10 \times 10^4(9.28 \times 10^{-3}) + 20 \times 10^4(-4.97 \times 10^{-3}) + 1875$$
$$= 1809 \text{ kN} \cdot \text{m}$$
$$M_{CE} = 8 \times 10^4(-4.97 \times 10^{-3}) - 1.714 \times 10^4(9.43 \times 10^{-3})$$
$$= -559 \text{ kN} \cdot \text{m}$$
$$M_{CD} = -1250 \text{ kN} \cdot \text{m}$$
$$M_{EC} = 4 \times 10^4(-4.97 \times 10^{-3}) - 1.714 \times 10^4(9.43 \times 10^{-3})$$
$$= -360 \text{ kN} \cdot \text{m}$$

Also, the shear forces V_{AB} and V_{EC} are

$$V_{AB} = -2.4 \times 10^4(9.28 \times 10^{-3}) + 0.96 \times 10^4(9.43 \times 10^{-3})$$
$$= -132 \text{ kN}$$
$$V_{EC} = -1.714 \times 10^4(-4.97 \times 10^{-3}) + 0.499 \times 10^4(9.4 \times 10^{-3})$$
$$= +132 \text{ kN}$$

Shear values for *BC*, not yet considered, are calculated now. They must include the distributed force carried by *BC*.

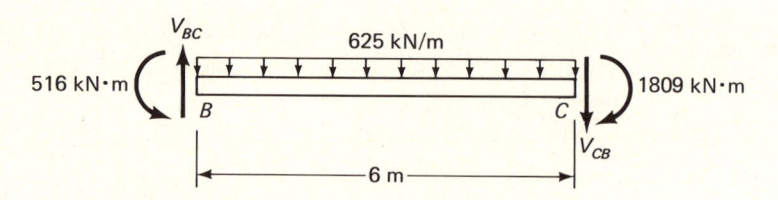

Figure 10.25

$$+\curvearrowleft \sum M_C = -516 + 1809 - 625(6)(3) + 6V_{BC} = 0$$
$$V_{BC} = 1660 \text{ kN}$$
$$+\uparrow \sum F_y = V_{BC} - 625(6) - V_{CB} = 0$$
$$V_{CB} = V_{BC} - 625(6)$$
$$= 1660 - 3750$$
$$= -2090 \text{ kN}$$

STATICALLY INDETERMINATE FRAMES USING SLOPE DEFLECTION   219

The free-body, shear, and moment diagrams are

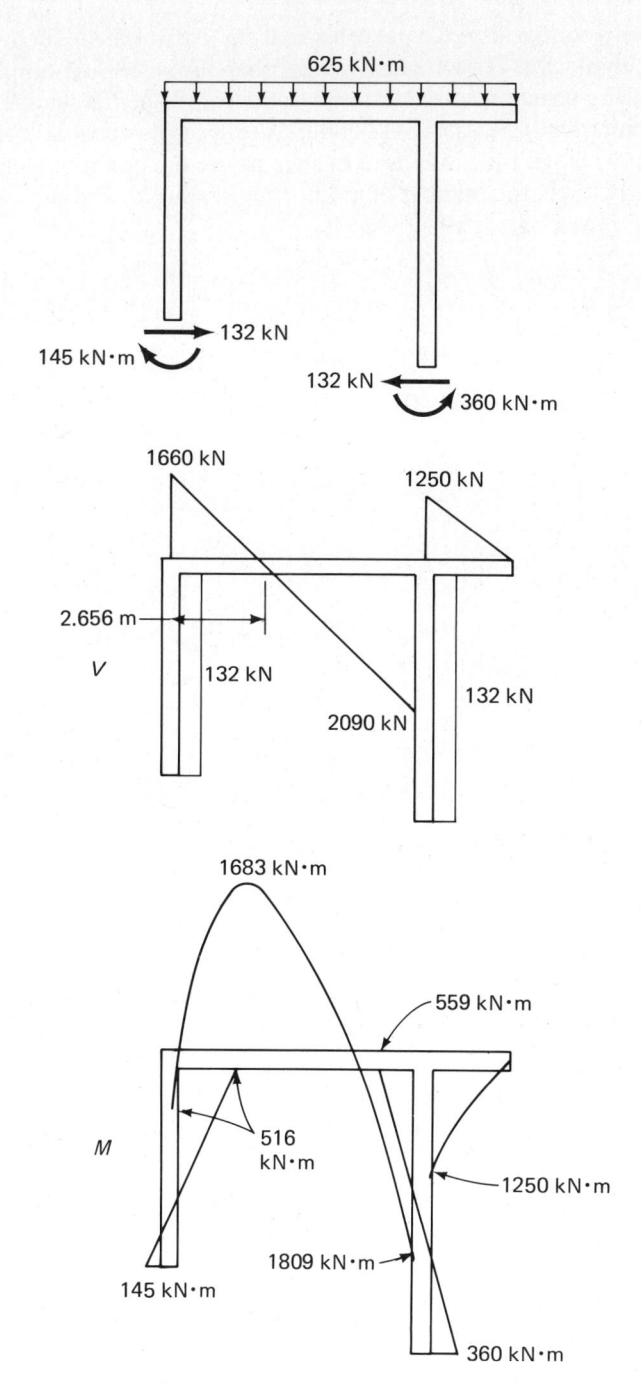

Figure 10.26

10.3 SYMMETRY AND ANTISYMMETRY

When half of a structure is a mirror image of its other half, so that there are identical members that match in orientation at each location and there are identical supports at corresponding points, the structure is said to be symmetric. When a symmetric structure is loaded symmetrically, it deforms symmetrically; rotations at corresponding joints are equal in magnitude, opposite in sign, as are the end moments at matching locations. Clearly, then, the number of unknowns is reduced, and the effort required in calculating end moments is also reduced.

EXAMPLE 10.3.1

Determine the end moments in $ABCD$.

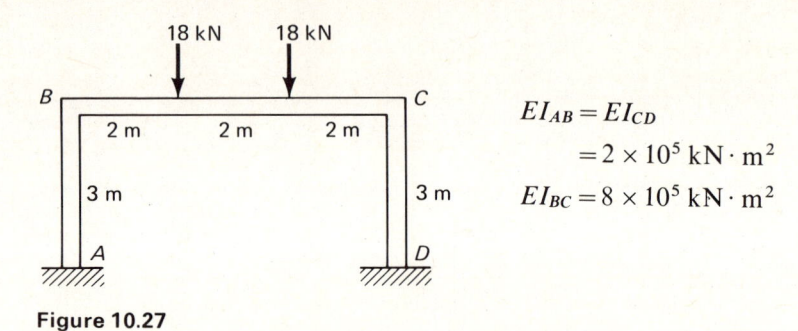

$EI_{AB} = EI_{CD}$
$= 2 \times 10^5$ kN·m²
$EI_{BC} = 8 \times 10^5$ kN·m²

Figure 10.27

The structure and its loads are symmetric with respect to a vertical line bisecting member BC. We assume, therefore, that $M_{AB} = -M_{DC}$, $M_{BA} = -M_{CD}$, and $M_{BC} = -M_{CB}$, that $\theta_B = -\theta_C$, and that there is no sidesway (joints B and C do not move to the left or right).

That there is no sidesway may be difficult to justify. It is easily demonstrated if a slope deflection analysis is employed in which symmetry is ignored. The value of Δ for AB and CD, associated with the sidesway, would be found to be zero.

The slope deflection equations are

$$M_{AB} = \frac{2(2 \times 10^5)}{3}\theta_B = 1.333 \times 10^5 \theta_B$$

$$M_{BA} = \frac{4(2 \times 10^5)}{3}\theta_B = 2.667 \times 10^5 \theta_B$$

$$M_{BC} = \frac{4(8 \times 10^5)}{6}\theta_B + \frac{2(8 \times 10^5)}{6}(-\theta_B) - \frac{18(2)(4)}{6}$$
$$= 2.667 \times 10^5 \theta_B - 24$$

In the equation for M_{BC}, θ_C has been written as $-\theta_B$ and the last term is the fixed-end moment for the two 18-kN forces, case 2 in Table 10.1.

Since there is only one unknown, only one equation is needed:

$$\sum M_B = M_{BA} + M_{BC} = 0$$
$$2.667 \times 10^5 \theta_B + 2.667 \times 10^5 \theta_B - 24 = 0$$

Solving, $\theta_B = 4.5 \times 10^{-5}$ rad. The end moments are then

$$M_{AB} = 1.333 \times 10^5 (4.5 \times 10^{-5}) = 6 \text{ kN} \cdot \text{m}$$
$$M_{BA} = 2.667 \times 10^5 (4.5 \times 10^{-5}) = 12 \text{ kN} \cdot \text{m}$$
$$M_{BC} = 2.667 \times 10^5 (4.5 \times 10^{-5}) - 24 = -12 \text{ kN} \cdot \text{m}$$

From our assumptions related to symmetry,

$$M_{DC} = -M_{AB} = -6 \text{ kN} \cdot \text{m}$$
$$M_{CD} = -M_{BA} = -12 \text{ kN} \cdot \text{m}$$
$$M_{CB} = -M_{BC} = +12 \text{ kN} \cdot \text{m}$$

There is another kind of loading, called *antisymmetric* loading, that is important when applied to a *symmetric structure*. It is loading that causes symmetrically located joints to rotate the same amount *and in the same direction*. For this loading, corresponding end moments are equal and have the *same* sign.

EXAMPLE 10.3.2

Determine the end moments for *ABCDEF*.

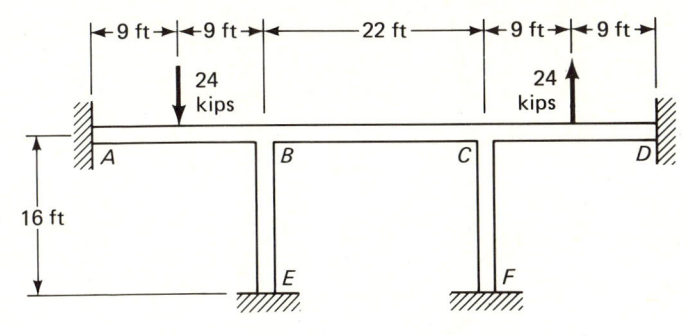

$EI_{AB} = EI_{BC} = EI_{CD} = 7.8 \times 10^7$ kip-in^2
$EI_{BE} = EI_{CF} = 6 \times 10^7$ kip-in^2

Figure 10.28

A probable elastic curve is shown,

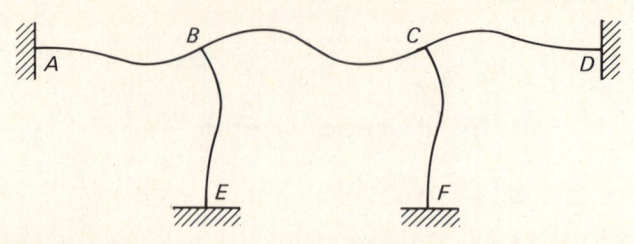

Figure 10.29

from which we conclude that

$$\theta_B = \theta_C$$

The slope deflection equations are

$$M_{AB} = \frac{2(7.8 \times 10^7 \text{ kip-in}^2)}{(18 \text{ ft})(144 \text{ in}^2/\text{ft}^2)} \theta_B - \frac{24(18)}{8}$$

$$= 6.02 \times 10^4 \theta_B - 54 \text{ kip-ft}$$

$$M_{BA} = \frac{4(7.8 \times 10^7)}{18(144)} \theta_B + 54 = 12.04 \times 10^4 \theta_B + 54$$

$$M_{BC} = \frac{4(7.8 \times 10^7)}{22(144)} \theta_B + \frac{2(7.8 \times 10^7)}{22(144)} \theta_B = 14.77 \times 10^4 \theta_B$$

$$M_{BE} = \frac{4(6 \times 10^7)}{16(144)} \theta_B = 10.42 \times 10^4 \theta_B$$

$$M_{EB} = \frac{2(6 \times 10^7)}{16(144)} \theta_B = 5.21 \times 10^4 \theta_B$$

Moment equilibrium at joint B gives

$$\sum M_B = M_{BA} + M_{BC} + M_{BE} = 0$$

$$12.04 \times 10^4 \theta_B + 54 + 14.77 \times 10^4 \theta_B + 10.42 \times 10^4 \theta_B = 0$$

$$\theta_B = -1.450 \times 10^{-4} \text{ rad}$$

Substituting θ_B into the slope deflection equations gives the end moments,

$$M_{AB} = 6.02 \times 10^4 (-1.45 \times 10^{-4}) = -8.73 \text{ kip-ft}$$
$$M_{BA} = 12.04 \times 10^4 (-1.45 \times 10^{-4}) + 54 = +36.53 \text{ kip-ft}$$
$$M_{BC} = 14.77 \times 10^4 (-1.45 \times 10^{-4}) = -21.42 \text{ kip-ft}$$
$$M_{BE} = 10.42 \times 10^4 (-1.45 \times 10^{-4}) = -15.17 \text{ kip-ft}$$
$$M_{EB} = 5.21 \times 10^4 (-1.45 \times 10^{-4}) = -7.55 \text{ kip-ft}$$

Since corresponding end moments in antisymmetry have the same sign and magnitude, the remaining end moments are

$M_{DC} = M_{AB} = -8.73$ kip-ft
$M_{CD} = M_{BA} = +36.53$ kip-ft
$M_{CB} = M_{BC} = -21.42$ kip-ft
$M_{CF} = M_{BE} = -15.11$ kip-ft
$M_{FC} = M_{EB} = -7.55$ kip-ft

Antisymmetric loading of a symmetric structure will produce sidesway unless the structure is braced to prevent it, as by the supports at A and D in Example 10.3.2. Reductions in calculation accrue nevertheless; but sidesway must be added as an unknown so that equilibrium by a force summation in its direction remains necessary.

EXAMPLE 10.3.3

The structure of Example 10.3.1 is loaded antisymmetrically, as shown. Determine the end moments.

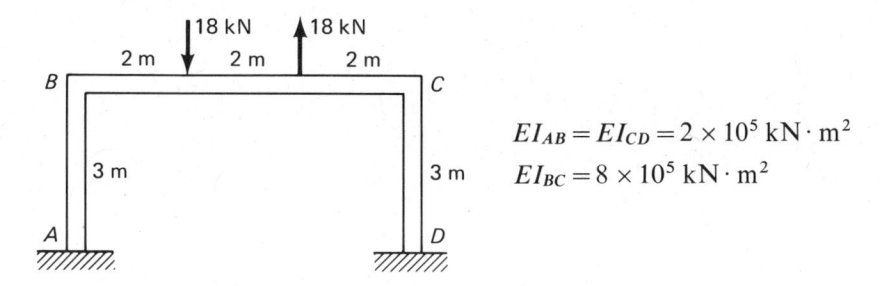

$EI_{AB} = EI_{CD} = 2 \times 10^5$ kN·m²
$EI_{BC} = 8 \times 10^5$ kN·m²

Figure 10.30

The antisymmetric loading can be expected to cause the deformed shape shown, where $\theta_b = \theta_c$.

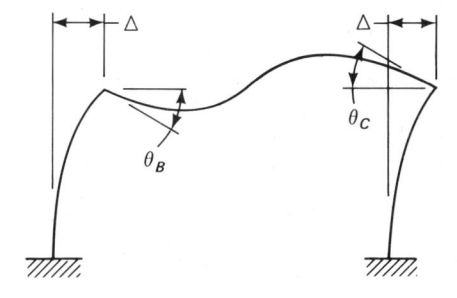

Figure 10.31

Slope deflection equations for the left half of the structure are

$$M_{AB} = \frac{2(2 \times 10^5)}{3}\theta_B - \frac{6(2 \times 10^5)}{3^2}\Delta$$

$$= 1.333 \times 10^5 \theta_B - 1.333 \times 10^5 \Delta$$

$$M_{BA} = \frac{4(2 \times 10^5)}{3}\theta_B - \frac{6(2 \times 10^5)}{32}\Delta$$

$$= 2.667 \times 10^5 \theta_B - 1.333 \times 10^5 \Delta$$

$$M_{BC} = \frac{4(8 \times 10^5)}{6}\theta_B + \frac{2(8 \times 10^5)}{6}\theta_B - \frac{18(2)(4)^2}{6^2} + \frac{18(4)(2)^2}{6^2}$$

$$= 8 \times 10^5 \theta_B - 8 \text{ kN} \cdot \text{m}$$

Preparatory to writing $\sum F_x = 0$, we calculate the shear at A, V_{AB},

$$V_{AB} = -\frac{M_{AB} + M_{BA}}{3}$$

$$= -\frac{1.333 \times 10^5(\theta_B - \Delta) + 1.333 \times 10^5(2\theta_B - \Delta)}{3}$$

$$= -1.333 \times 10^5 \theta_B + 0.8889 \times 10^5 \Delta$$

The shear at D is identical so that

$$\sum F_x = 2(-1.333 \times 10^5 \theta_B + 0.8889 \times 10^5 \Delta) = 0$$

Moment equilibrium at joint B gives

$$\sum M_B = M_{BA} + M_{BC} = 0$$
$$2.66 \times 10^5 \theta_B - 1.333 \times 10^5 \Delta + 8 \times 10^5 \theta_B - 8 = 0$$

The two equilibrium equations solved give

$$\theta_B = 9.23 \times 10^{-6} \text{ rad}$$
$$\Delta = 1.385 \times 10^{-5} \text{ m}$$

Then substitution in the slope deflection equations gives

$$M_{AB} = 1.333 \times 10^5(9.23 \times 10^{-6}) - 1.333 \times 10^5(1.385 \times 10^{-5})$$
$$= -0.615 \text{ kN} \cdot \text{m}$$

$$M_{BA} = 2.667 \times 10^5(9.23 \times 10^{-6}) - 1.333 \times 10^5(1.385 \times 10^{-5})$$
$$= +0.615 \text{ kN} \cdot \text{m}$$

$$M_{BC} = 8 \times 10^5(9.23 \times 10^{-6}) - 8$$
$$= -0.615 \text{ kN} \cdot \text{m}$$

The remaining end moments are then

$$M_{DC} = M_{AB} = -0.615 \text{ kN} \cdot \text{m}$$
$$M_{CD} = M_{BA} = +0.615 \text{ kN} \cdot \text{m}$$
$$M_{CB} = M_{BC} = -0.615 \text{ kN} \cdot \text{m}$$

Since large reductions in effort can come from the use of symmetry and antisymmetry, it is desirable to use them whenever possible. There are cases where they can be used, but it is not evident from a glance at a sketch of the structure and its loads. These cases are more easily developed using moment distribution as a basis, so we will delay their development until Chapters 11 and 12. They are mentioned here to alert readers that there is more to follow.

10.4 MULTIPLE DEGREES OF SIDESWAY

Buildings several stories high can have several degrees of sidesway, one for each story (unless braced against it). In addition to the moment equilibrium equations taken at the joints, the slope deflection analysis requires a force equilibrium for each degree of sidesway.

EXAMPLE 10.4.1

Determine the end moments for $ABCDEF$.

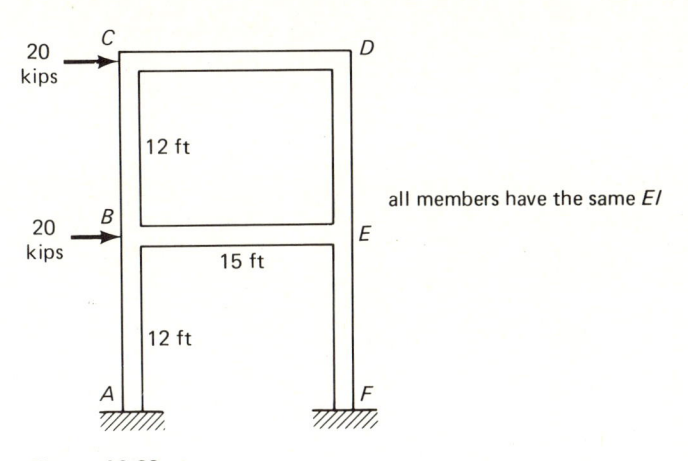

Figure 10.32

all members have the same EI

An added feature in this example is that the forces are applied at the joints. When so applied, they are not included in member fixed-end moments. They enter the solution through the force equilibrium equations. We will assume, without justification, that antisymmetry applies in this case. Justification of that assumption is left until Chapter 12.

226 STATICALLY INDETERMINATE STRUCTURES: THE SLOPE DEFLECTION METHOD

A probable elastic curve is

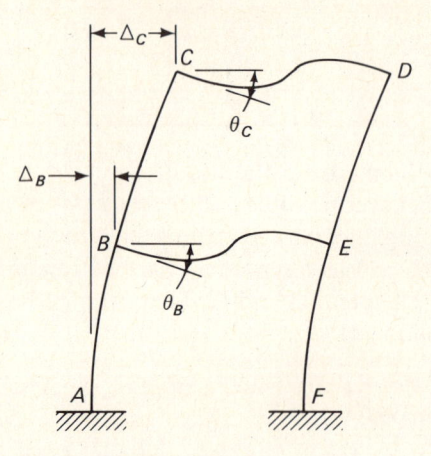

Figure 10.33

Since no displacement values are prescribed, the end moments will be independent of EI, and we can assign any value that is convenient. The value $EI = 12 \times 15 = 180$ is chosen in anticipation of lengths of 12 and 15 to appear in the slope deflection equation denominators.

$$M_{AB} = \frac{2(180)}{12}\theta_B - \frac{6(180)}{12^2}\Delta_B$$

$$= 30\theta_B - 7.5\Delta_B$$

$$M_{BA} = 60\theta_B - 7.5\Delta_B$$

$$M_{BC} = 60\theta_B + 30\theta_C - 7.5(\Delta_C - \Delta_B)$$

$$= 60\theta_B + 30\theta_C - 7.5\Delta_2$$

where $\Delta_2 = \Delta_C - \Delta_B$.

$$M_{CB} = 30\theta_B + 60\theta_C - 7.5\Delta_2$$

$$M_{CD} = \frac{4(180)}{15}\theta_C + \frac{2(180)}{15}\theta_C$$

$$= 72\theta_C$$

Antisymmetry assumes $\theta_C = \theta_D$.

$$M_{BE} = 72\theta_B$$

Moment equilibrium equations are

$$\sum M_B = M_{BA} + M_{BC} = 0$$

$$60\theta_B - 7.5\Delta_B + 60\theta_B + 30\theta_C - 7.5\Delta_2 + 72\theta_B = 0$$

or

$$192\theta_B + 30\theta_C - 7.5\Delta_B - 7.5\Delta_2 = 0$$

$$\sum M_C = M_{CB} + M_{CD} = 0$$
$$30\theta_B + 60\theta_C - 7.5\Delta_2 + 72\Delta_C = 0$$

or

$$30\theta_B + 132\theta_C - 7.5\Delta_2 = 0$$

Preparatory to writing the force equilibrium equations, we calculate the needed shear forces.

Figure 10.34

$$\sum M_C = M_{CB} + M_{BC} + 12V_{BC} = 0$$

$$V_{BC} = -\frac{M_{CB} + M_{BC}}{12}$$

$$= -\frac{30\theta_B + 60\theta_C - 7.5\Delta_2 + 60\theta_B + 30\theta_C - 7.5\Delta_2}{12}$$

$$= -7.5\theta_B - 7.5\theta_C + 1.25\Delta_2$$

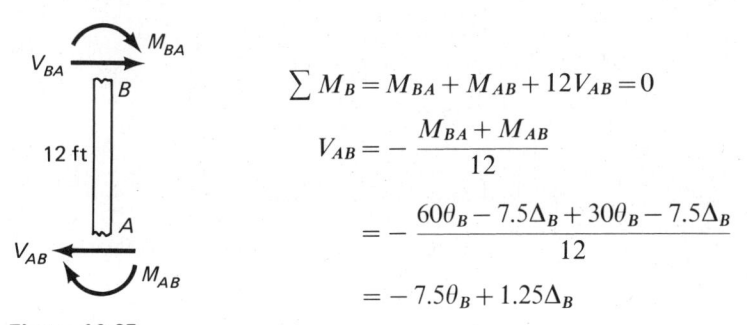

Figure 10.35

$$\sum M_B = M_{BA} + M_{AB} + 12V_{AB} = 0$$

$$V_{AB} = -\frac{M_{BA} + M_{AB}}{12}$$

$$= -\frac{60\theta_B - 7.5\Delta_B + 30\theta_B - 7.5\Delta_B}{12}$$

$$= -7.5\theta_B + 1.25\Delta_B$$

Now two force equilibrium equations are written. First cutting just above joints B and E, a free-body diagram showing only the horizontal forces is

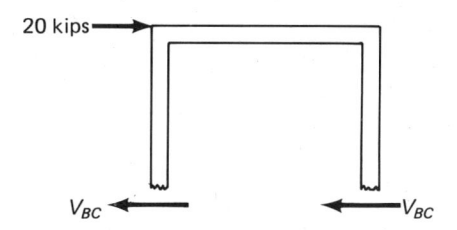

Figure 10.36

$$\sum F_x = -2V_{BC} + 20 = 0$$
$$V_{BC} = 10 \text{ kips}$$
$$-7.5\theta_B - 7.5\theta_C + 1.25\Delta_2 = 10$$

Next cutting just above A and F,

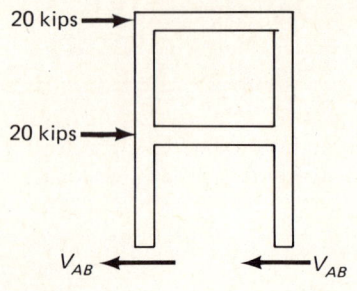

Figure 10.37

$$\pm \sum F_x = -2V_{AB} + 2(20) = 0$$
$$V_{AB} = 20 \text{ kips}$$
$$-75\theta_B + 1.25\Delta_B = 20 \text{ kips}$$

Thus the system of four equilibrium equations is

$$192\theta_B + 30\theta_C - 7.5\Delta_B - 7.5\Delta_2 = 0$$
$$30\theta_B + 132\theta_C - 7.5\Delta_2 = 0$$
$$-7.5\theta_B - 7.5\theta_C + 1.25\Delta_2 = 10$$
$$-7.5\theta_B + 1.25\Delta_B = 20$$

Solved, the values are

$$\theta_B = 1.915$$
$$\theta_C = 1.0198$$
$$\Delta_B = 27.5$$
$$\Delta_2 = 25.6$$

These are not actual displacements, since a convenient EI value was assumed. The resulting end moments are independent of EI. They are

$$M_{AB} = 30(1.915) - 7.5(27.5) = -149 \text{ kip-ft}$$
$$M_{BA} = 60(1.915) - 7.5(27.5) = -91.4 \text{ kip-ft}$$
$$M_{BC} = 60(1.915) + 30(1.0198) - 7.5(25.6) = -46.5 \text{ kip-ft}$$
$$M_{BE} = 72(1.915) = 137.9 \text{ kip-ft}$$
$$M_{CB} = 30(1.915) + 60(1.0198) - 7.5(25.6) = -73.4 \text{ kip-ft}$$
$$M_{CD} = 72(1.0198) = 73.4 \text{ kip-ft}$$

From antisymmetry, the corresponding end moments are the same, such as

$$M_{FE} = M_{AB} = -149 \text{ kip-ft}$$

The number of degrees of joint translation does not necessarily equal the number of joints that may translate. Rather, it is the number of *independent* joint translations that are possible. When a joint translates as the necessary result of another joint's translation, it is a dependent translation. Such dependent translations must be expressed as functions of the independent joint translations.

EXAMPLE 10.4.2

Determine the end moments for gabled structure $ABCDE$ loaded as shown.

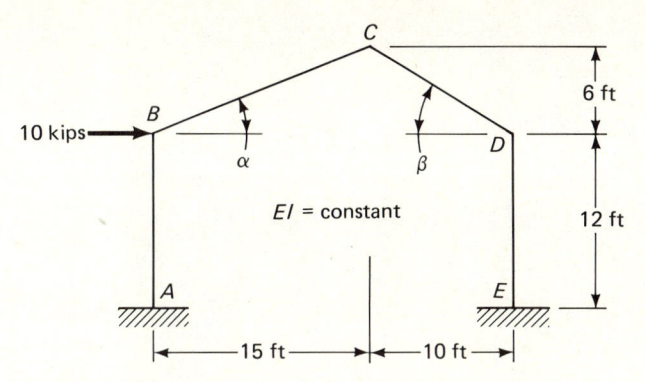

Figure 10.38

Joints B and D can be displaced horizontally independently, that is, either can move while the other does not. Thus they are independent joint translations. On the other hand, if either B or D translates, joint C must also translate. Thus the translation of joint C is dependent on that of B and D. Assume, for example, that B and D are displaced Δ_B and Δ_D to new positions B' and D'. If member lengths do not change, C moves to C'.

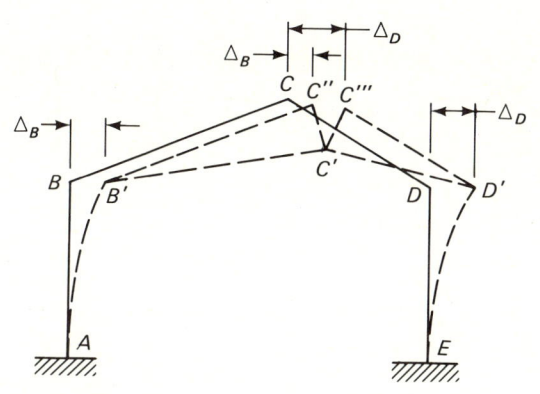

Figure 10.39

Member BC's motion can be considered in two parts: translation to $B'C'''$ and rotation about B' to $B'C'$. In the rotational part, $C''C'$ is the relative transverse end displacement of BC. Similarly, DC's motion is a translation to $D'C'''$ and a rotation about D' to $D'C'$ in which $C'''C'$ is the

230 STATICALLY INDETERMINATE STRUCTURES: THE SLOPE DEFLECTION METHOD

Steel frame of the main office building for the Georgia Railroad Bank and Trust Company. (Courtesy of Bethlehem Steel Corporation).

relative transverse end displacement of *CD*. Notice for the assumed positive Δ_B and Δ_D illustrated that $C''C'$ and $C'''C'$ correspond to clockwise and counterclockwise rotations of *BC* and *CD*, respectively. Thus they are considered positive and negative Δ's, respectively.

We work with triangle $C'C''C'''$ to establish the dependency equations,

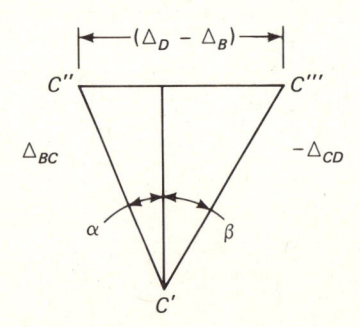

Figure 10.40

$$\Delta_{BC} \cos \alpha = (-\Delta_{CD}) \cos \beta$$
$$\Delta_{BC} \sin \alpha + (-\Delta_{CD}) \sin \beta = \Delta_D - \Delta_B$$

MULTIPLE DEGREES OF SIDESWAY

For structure $ABCDE$, $\alpha = 21.80°$ and $\beta = 30.96°$, giving

$$\Delta_{BC} = 1.077(\Delta_D - \Delta_B)$$
$$\Delta_{CD} = 1.166(\Delta_B - \Delta_D)$$

Now we are ready to write the slope deflection equations,

$$M_{AB} = \frac{2EI}{12}\theta_B - \frac{6EI}{12^2}\Delta_B$$

$$M_{BA} = \frac{4EI}{12}\theta_B - \frac{6EI}{12^2}\Delta_B$$

$$M_{BC} = \frac{4EI}{16.16}\theta_B + \frac{2EI}{16.16}\theta_C - \frac{6EI}{16.16^2}(1.077)(\Delta_D - \Delta_B)$$

$$M_{CB} = \frac{2EI}{16.16}\theta_B + \frac{4EI}{16.16}\theta_C - \frac{6EI}{16.16^2}(1.077)(\Delta_D - \Delta_B)$$

$$M_{CD} = \frac{4EI}{11.66}\theta_C + \frac{2EI}{11.66}\theta_D - \frac{6EI}{11.66^2}(1.166)(\Delta_B - \Delta_D)$$

$$M_{DC} = \frac{2EI}{11.66}\theta_C + \frac{4EI}{11.66}\theta_D - \frac{6EI}{11.66^2}(1.166)(\Delta_B - \Delta_D)$$

$$M_{DE} = \frac{4EI}{12}\theta_D - \frac{6EI}{12^2}\Delta_D$$

$$M_{ED} = \frac{2EI}{12}\theta_D - \frac{6EI}{12^2}\Delta_D$$

Moment summations about joints B, C, and D give three equations,

$$\sum M_B = M_{BA} + M_{BC}$$
$$= 0.5809EI\theta_B + 0.1238EI\theta_C - 0.01692EI\Delta_B - 0.02475EI\Delta_D = 0$$

$$\sum M_C = M_{CB} + M_{CD}$$
$$= 0.1238EI\theta_B + 0.5906EI\theta_C + 0.1715EI\theta_D - 0.0267EI\Delta_B$$
$$+ 0.0267EI\Delta_D = 0$$

$$\sum M_D = M_{DC} + M_{DE}$$
$$= 0.1715EI\theta_C + 0.6764EI\theta_D - 0.05146EI\Delta_B + 0.00979EI\Delta_D = 0.$$

and from equilibrium for the individual members AB and ED we get

$$V_{AB} = -\tfrac{1}{12}(M_{AB} + M_{BA})$$
$$V_{ED} = -\tfrac{1}{12}(M_{DE} + M_{ED})$$

Thus horizontal force equilibrium may be applied.

232 STATICALLY INDETERMINATE STRUCTURES: THE SLOPE DEFLECTION METHOD

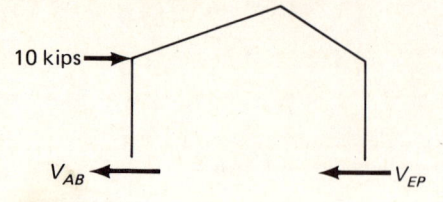

Figure 10.41

$$\sum F_X = -\tfrac{1}{12}(M_{AB} + M_{BA}) - \tfrac{1}{12}(M_{DE} + M_{ED}) + 10 = 0$$

which, upon substitution and simplification, becomes

$$0.04167 EI\theta_B + 0.04167 EI\theta_D - 0.006944 EI\Delta_B - 0.006944 EI\Delta_D = -10$$

For the fifth equilibrium equation needed to solve for the five unknowns, we use the equilibrium of member DC to get

$$V_{DC} = -\frac{1}{11.66}(M_{CD} + M_{DC})$$

and then take moments about Q for the section $ABCD$ shown.

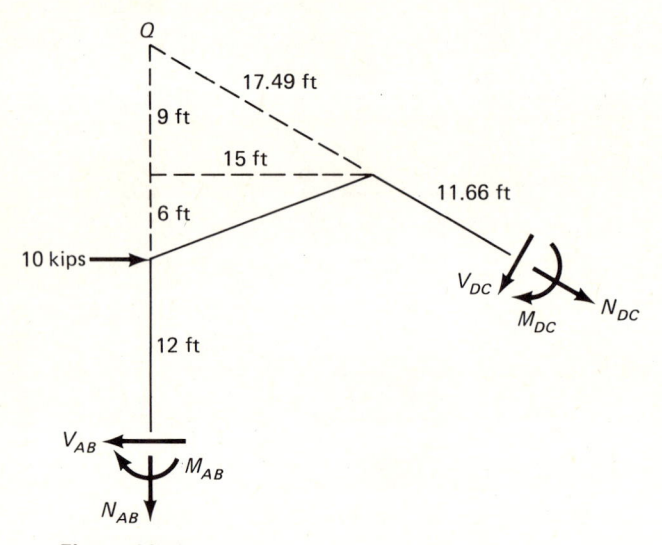

Figure 10.42

$$\sum M_Q = -15(10) + 27 V_{AB} + M_{AB} + (17.49 + 11.66) V_{DC} + M_{DC} = 0$$

Substitution and simplification gives

$$-0.9583 EI\theta_B - 1.115 EI\theta_C - 0.9434 EI\theta_D + 0.3517 EI\Delta_B$$
$$-0.2058 EI\Delta_D = 150 \text{ kip-ft}$$

Simultaneous solution of the five equations gives

$$EI\theta_B = 102.4$$
$$EI\theta_C = 35.14$$

$EI\theta_D = 111.6$

$EI\Delta_B = 1569.2$

$EI\Delta_D = 1154.9$

When these are substituted in the slope deflection equations, we get

$M_{AB} = -48.3$ kip-ft

$M_{BA} = -M_{BC} = -31.25$ kip-ft

$M_{CB} = -M_{CD} = 14.23$ kip-ft

$M_{DC} = -M_{DE} = 10.93$ kip-ft

$M_{ED} = -29.52$ kip-ft

PROBLEMS

Sequenced Problem 3, parts (m), (n), and (r).
Sequenced Problem 4, parts (m), (n), and (s).
Sequenced Problem 8, parts (a), and (b).

10.1 Use slope deflection to determine support reactions in each of the following beam structures. Draw their shear and moment diagrams.

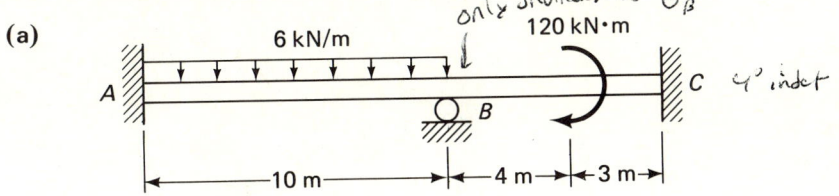

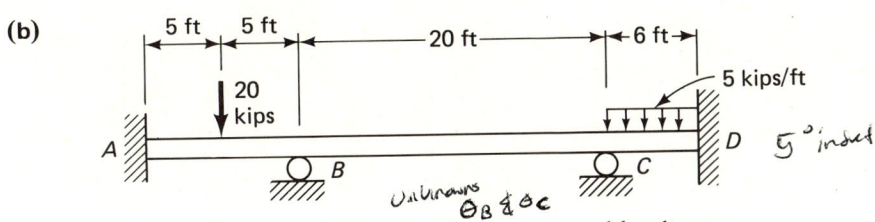

(a)

(b)

(c) Take advantage of the symmetry of structure and loading.

(c)

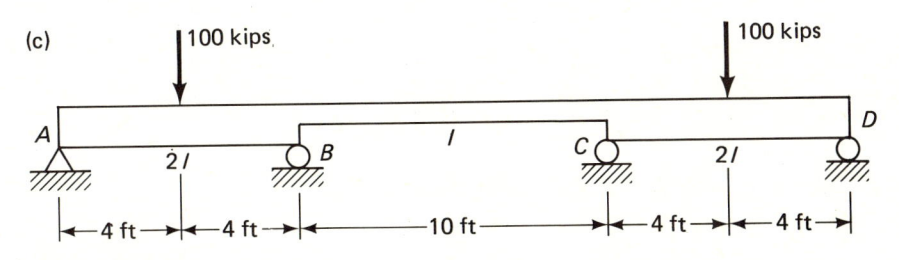

Figure P10-1 A-C

234 STATICALLY INDETERMINATE STRUCTURES: THE SLOPE DEFLECTION METHOD

(d)

40 kN/m

$EI_{AB} = 1.42 \times 10^5$ kN·m²
$EI_{BC} = 1.75 \times 10^5$ kN·m²
$EI_{CD} = 2.15 \times 10^5$ kN·m²

(e)

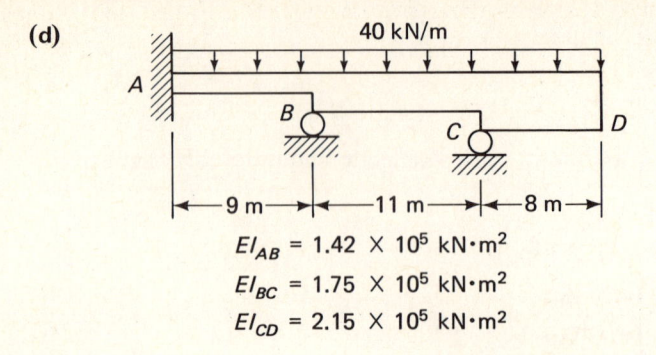

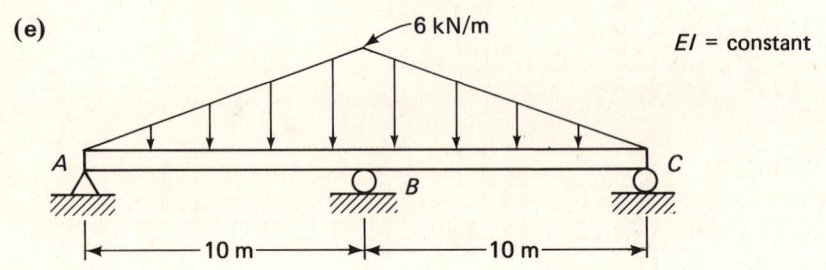

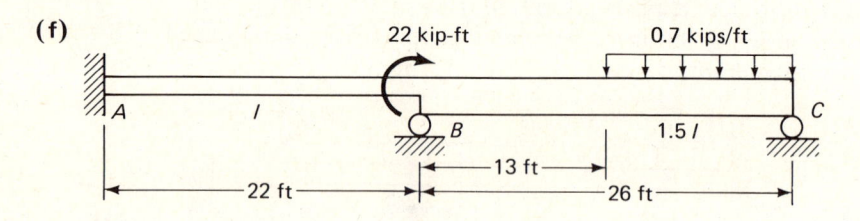

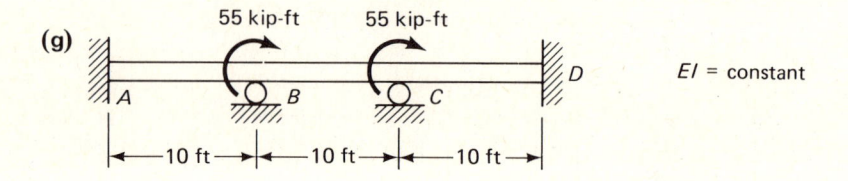

6 kN/m

EI = constant

(f)

22 kip-ft

0.7 kips/ft

(g)

55 kip-ft   55 kip-ft

EI = constant

Figure P10-1F-G

10.2

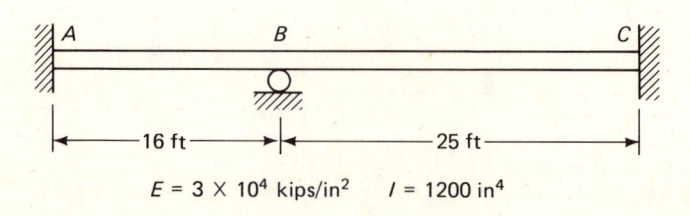

$E = 3 \times 10^4$ kips/in² $I = 1200$ in⁴

Figure P10-2

Use slope deflection to determine the end moments that are produced if support B settles $\frac{5}{8}$ in. Then calculate the support reactions and draw the shear and moment diagrams.

10.3

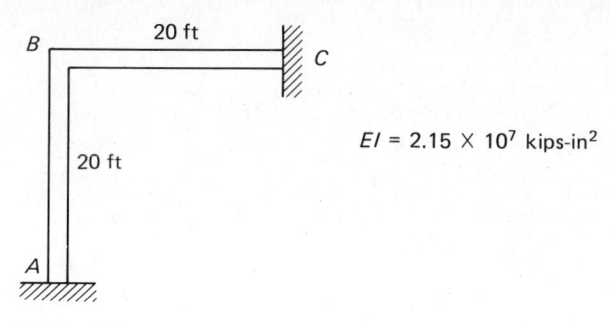

Figure P10-3

$EI = 2.15 \times 10^7$ kips-in²

Use slope deflection to determine the end moments that are produced by support A settling downward 0.25 in and rotating 1° clockwise if support C remains undisturbed. Calculate the remaining support reactions and draw the shear and moment diagrams.

10.4 Use slope deflection to determine the end moments that arise for the support settlements given. Then calculate the support reactions and draw the shear and moment diagrams.

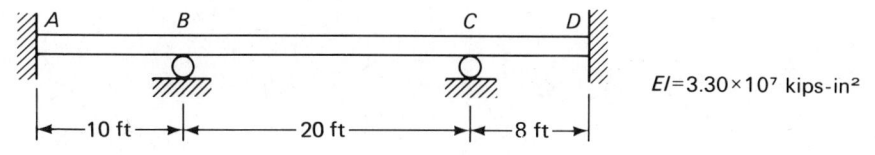

Figure P10.4

$EI = 3.30 \times 10^7$ kips-in²

(a) Support B moves downward 0.75 in, while the remaining supports remain undisturbed.
(b) Support B moves downward 0.5 in and C moves downward 0.9 in, while the others remain undisturbed.
(c) End A rotates clockwise 5°, while the other supports are undisturbed.

10.5

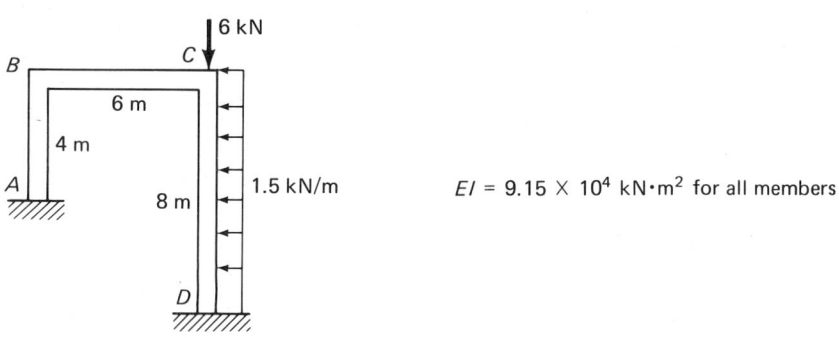

Figure P10.5

$EI = 9.15 \times 10^4$ kN·m² for all members

Use slope deflection to determine the end moments and then calculate the member end shears and the vertical support reactions, and draw the shear and moment diagrams.

10.6 Repeat Problem 10.5, but with the clamped support at *D* replaced with a pin support.

10.7

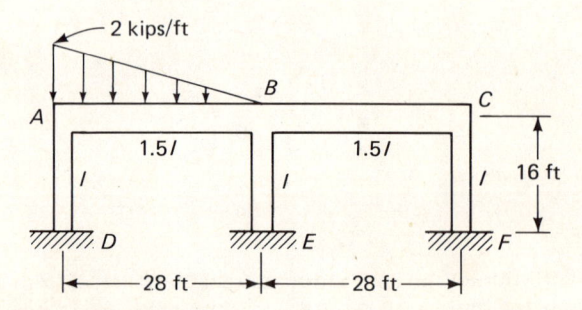

Figure P10.7

Use slope deflection to determine the end moments: then calculate the end shears and the support reactions, and draw the shear and moment diagrams.

10.8

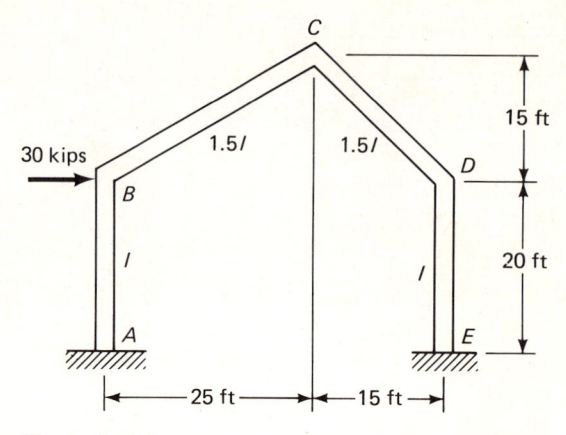

Figure P10.8

Use slope deflection to determine the end moments.

10.9

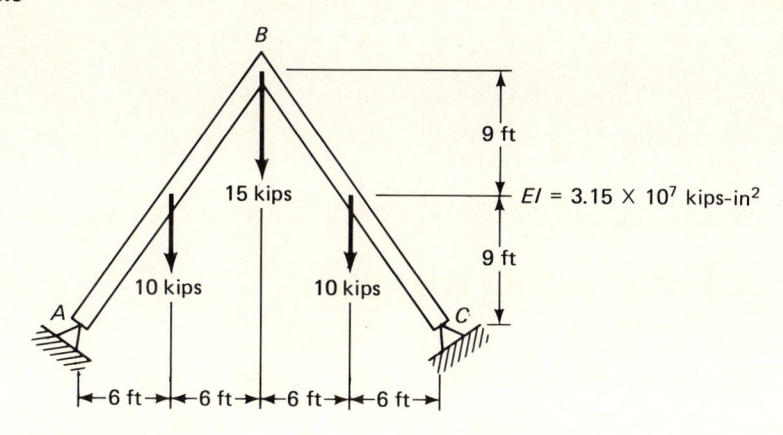

Figure P10.9

Use slope deflection to determine the end moments. Then determine the support reactions. *Hint:* Express the support reactions in components parallel and perpendicular to the members.

10.10 Repeat Problem 10.9, but with clamped supports at A and C.

10.11

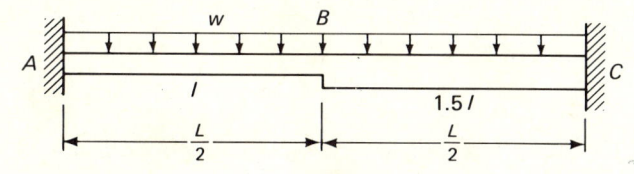

Figure P10.11

Use slope deflection to calculate the end moments. *Hint:* Treat the structure as composed of two members, AB and BC, with the joint at B subject to both rotation θ_B and translation Δ.

Moment Distribution: Structures Without Sidesway

"But what is a caucus race?" asked Alice.
"Why," said the Dodo, "the best way to explain it is to do it."
First, it marked out a race course in a sort of circle.
"The exact shape doesn't matter," it said.
Then all the party was placed along the course, here and there. There was no "one, two, three and away," but they began running when they liked and left off when they liked, so that it was not easy to know when the race was over. However, when they had been running half an hour or so the Dodo suddenly called out, "The race is over!" and they all crowded round, panting and asking, "But who has won?"
—Lewis Carroll, *Alice in Wonderland*

11.0 INTRODUCTION

Since its development and publication by Professor Hardy Cross in 1930,[1] moment distribution has been the principal tool for analyzing structures with members that are subjected primarily to bending. This may surprise you after you have first examined it, since it may seem at first to be rather disorganized. There is no exact place to begin and no clear time "when it is over," so it may seem very much like a caucus race. But order will emerge and the value of method will become evident.

If there is no sidesway (joint translation), moment distribution will give the end moments directly, avoiding the solution of simultaneous equations. If there is sidesway, moment distribution may involve the solution of simultaneous equations, but the equations will be fewer in number than would be needed for a slope deflection solution.

Moment distribution is an iterative method that is based on the slope deflection equations. The iterations are steps where, analytically, all the joints of the structure are "locked" against rotation, and then, one at a time, are "unlocked and relocked." Usually this unlocking and relocking is done several times (iterations), with each iteration bringing the deformed shape of the structure closer to that of the real structure. Done enough times, they match.

Since moment distribution is based on the slope deflection equations, it uses the slope deflection sign conventions: *clockwise* end moments, end rotations, and member rotations are *positive*.

11.1 STIFFNESS, DISTRIBUTION FACTORS, AND CARRY-OVERS

Rotational stiffness is defined as the moment that, when applied at the end of a beam, causes a unit rotation of that end, if the beam is constrained to prevent all other end displacements. For member AB, moment S_{AB} is the rotational stiffness since it causes $\theta_A = 1$, while $\theta_B = 0$ and $\Delta_{AB} = 0$.

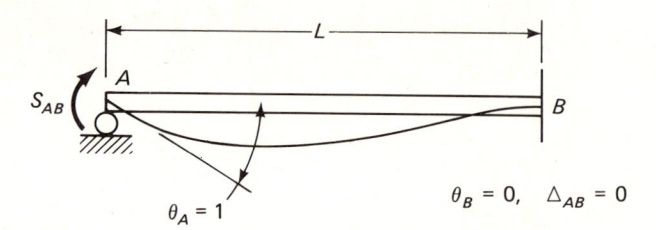

Figure 11.1

If the member is prismatic, that is, has a constant EI, the slope deflection equation (10.1) may be used to calculate the rotational stiffness,

$$M_{AB} = \frac{4EI}{L}\theta_A + \frac{2EI}{L}\theta_B - \frac{6EI}{L^2}\Delta + \text{FEM}_{AB}$$

---

[1] Cross, Hardy, "Analysis of Continuous Frames by Distributing Fixed End Moments": Proceedings, American Society of Civil Engineers, May 1930.

Hence,

$$S_{AB} = \frac{4EI}{L}(1) + 0 - 0 + 0$$

$$= \frac{4EI}{L} \tag{11.1}$$

Now consider prismatic member AB once again, with the same roller and fixed supports. If a moment M_{AB} is applied at A, the fixed support at B must apply a moment M_{BA} to prevent rotation at B. For prismatic members, the ratio of these moments is constant. Thus

$$C_{AB} = \frac{M_{BA}}{M_{AB}} \tag{11.2}$$

C_{AB} is called the carry-over factor from A to B, and M_{BA} is called the carry-over moment. Schematically,

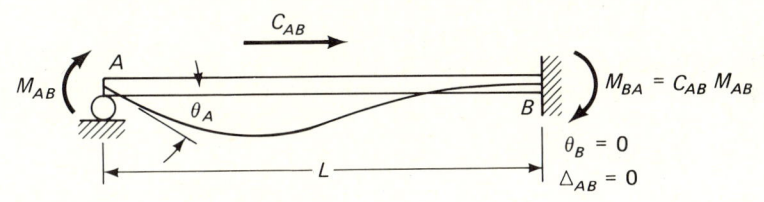

Figure 11.2

From the slope deflection equations,

$$M_{AB} = \frac{4EI}{L}\theta_A$$

$$M_{BA} = \frac{2EI}{L}\theta_A$$

showing that $C_{AB} = \frac{1}{2}$ for prismatic members.

Now consider system $ABCD$, where joint B is a rigid connection, and M is an external moment applied at joint B,

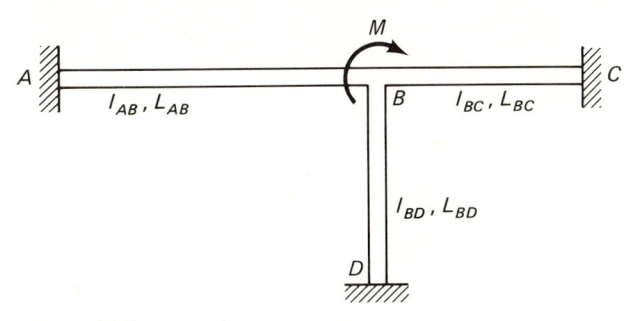

Figure 11.3

The three members are prismatic, each with different lengths and moments of inertia.

242 MOMENT DISTRIBUTION: STRUCTURES WITHOUT SIDESWAY

Responding to M, joint B and each member end at B rotates θ_B. From slope deflection,

$$M_{BA} = \frac{4EI_{AB}}{L_{AB}} \theta_B = S_{BA} \theta_B$$

$$M_{BC} = \frac{4EI_{BC}}{L_{BC}} \theta_B = S_{BC} \theta_B$$

$$M_{BC} = \frac{4EI_{BD}}{L_{BD}} \theta_B = S_{BD} \theta_B$$

Sketching the free-body diagram[2] of joint B and writing moment equilibrium, we get

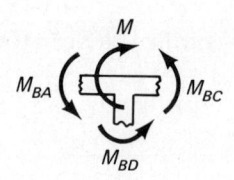

Figure 11.4

$$+\circlearrowleft \sum M_B = M - M_{BA} - M_{BC} - M_{BD} = 0$$

$$M = M_{BA} + M_{BC} + M_{BD}$$

$$= (S_{BA} + S_{BC} + S_{BD})\theta_B$$

Thus

$$\theta_B = \frac{M}{S_{BA} + S_{BC} + S_{BD}} = \frac{M}{\sum_i S_{Bi}}$$

Substituting back this value of θ_B,

$$M_{BA} = \frac{S_{BA}}{\sum_i S_{Bi}} M = D_{BA} M$$

$$M_{BC} = \frac{S_{BC}}{\sum_i S_{Bi}} M = D_{BC} M$$

$$M_{BD} = \frac{S_{BD}}{\sum_i S_{Bi}} M = D_{BD} M$$

The coefficients D_{BA}, D_{BC}, and D_{BD} are called distribution factors. They are equal to the individual member rotational stiffnesses divided by the sum of the rotational stiffnesses for all the members connected at B. Thus

$$D_{BA} = \frac{S_{BA}}{\sum_i S_{Bi}} \tag{11.3}$$

---

[2] End shear forces have been omitted on the free-body diagram. Since only moment equilibrium is to be considered, and the joint is negligible in size, these shear forces need not be considered.

and so on, where

$$\sum_i S_{Bi} = S_{BA} + S_{BC} + S_{BD}$$

that is, the sum of all the member rotational stiffnesses at B.

The common factor $4E$ cancels when calculating distribution factors for *prismatic members*, leaving a simpler form

$$D_{BA} = \frac{I_{BA}/L_{BA}}{\sum_i (I_{Bi}/L_{Bi})} \tag{11.4}$$

The ratios I/L are called *relative stiffnesses*. Their use, while convenient, is limited to cases where all the members connected at the joint are prismatic.

We illustrate a simple use of distribution factors and carry-over factors with the following example.

EXAMPLE 11.1.1

Determine all end moments for the bent ABC.

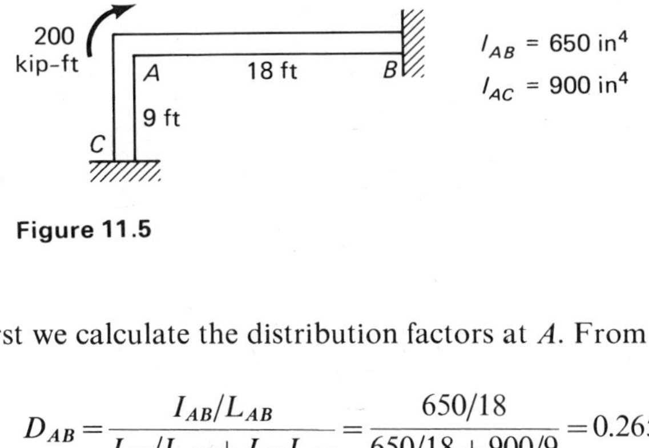

200 kip-ft, A, 18 ft, B, 9 ft, C
$I_{AB} = 650$ in^4
$I_{AC} = 900$ in^4

Figure 11.5

First we calculate the distribution factors at A. From Eq. (11.4),

$$D_{AB} = \frac{I_{AB}/L_{AB}}{I_{AB}/L_{AB} + I_{AC}L_{AC}} = \frac{650/18}{650/18 + 900/9} = 0.265$$

$$D_{AC} = \frac{I_{AC}/L_{AC}}{I_{AB}/L_{AB} + I_{AC}/L_{AC}} = \frac{900/9}{650/18 + 900/9} = 0.735$$

These distribution factors indicate how a moment applied at A is shared by AB and AC. For the 200-kip-ft moment,

$M_{AB} = 200(0.265) = 53.0$ kip-ft

$M_{AC} = 200(0.735) = 147.0$ kip-ft

Next, using the carry-over factor $\frac{1}{2}$,

$M_{BA} = 0.5(53.0) = 26.5$ kip-ft

$M_{CA} = 0.5(147.0) = 73.5$ kip-ft

11.2 FIXED END MOMENTS

Moment distribution analysis begins with every joint locked to prevent rotation. Thus at the start each member has both ends fixed, and the loads they carry produce fixed end moments. They are calculated exactly as they were for slope deflection.

11.3 MOMENT DISTRIBUTION PROCEDURE

Example 11.3.1 will illustrate the ideas underlying moment distribution: the analytical locking, unlocking, and relocking of joints. In the format of the example, the method may seem laborious. The same problem will be solved a second time in Example 11.3.2 using the usual moment distribution format. Its efficiency should become evident then, and the reasons underlying the different steps should become clear when the two examples are compared step by step.

EXAMPLE 11.3.1

Find the end moments for beam $ABCD$.

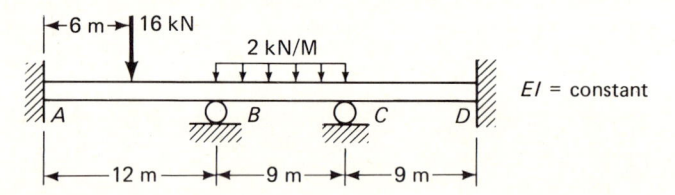

Figure 11.6

As a preliminary step we calculate the fixed end moments, applying cases 1 and 3 of Table 10.1.

$$\text{FEM}_{AB} = -\text{FEM}_{BA} = -\frac{(16 \text{ kN})(12 \text{ m})}{8} = -24 \text{ kN} \cdot \text{m}$$

$$\text{FEM}_{BC} = -\text{FEM}_{CB} = -\frac{(2 \text{ kN/m})(9^2 \text{ m}^2)}{12} = -13.5 \text{ kN} \cdot \text{m}$$

Then, using Eq. (11.4), the distribution factors are found, first at B,

$$D_{BA} = \frac{I/12}{I/12 + I/9} = 0.429$$

$$D_{BC} = \frac{I/9}{I/12 + I/9} = 0.571$$

then at C,

$$D_{CB} = \frac{I/9}{I/9 + I/9} = 0.5 = D_{CD}$$

The total of the distribution factors at each joint must always be 1, and is so here, giving a simple check against careless arithmetic.

Now we lock joints B and C against rotation, shown in the diagram by cross-hatching over the joint.

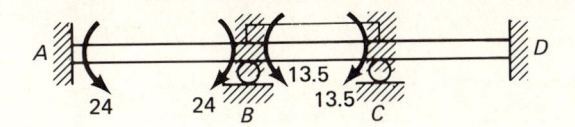

Figure 11.7

The fixed end moments are shown in their proper directions: clockwise for +, counterclockwise for −. There are net moments of +10.5 kN·m at B and +13.5 kN·m at C, provided by the artificial constraints that locked the joints. Since these constraints do not exist in the real structure, we remove them, one at a time, by adding equal, opposite moments, −10.5 kN·m at B and −13.5 kN·m at C, thereby canceling the imbalances.

At B the −10.5-kN·m is distributed −10.5(0.429) = −4.5 kN·m to BA and −10.5(0.571) = −6.0 kN·m to BC, as illustrated.

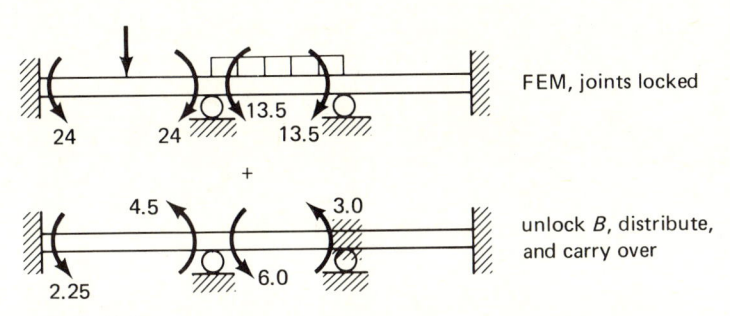

Figure 11.8

In the illustration the carry-over moments of −2.25 kN·m and −3.0 kN·m are also shown. The total of these two parts is

Figure 11.9

Joint B is balanced now, but joint C is not, now with a net +10.5 kN·m. We relock B and add −10.5 kN·m to C, distributed as −10.5(0.5) = −5.25 kN·m each to CB and CD, with half carried over. Thus

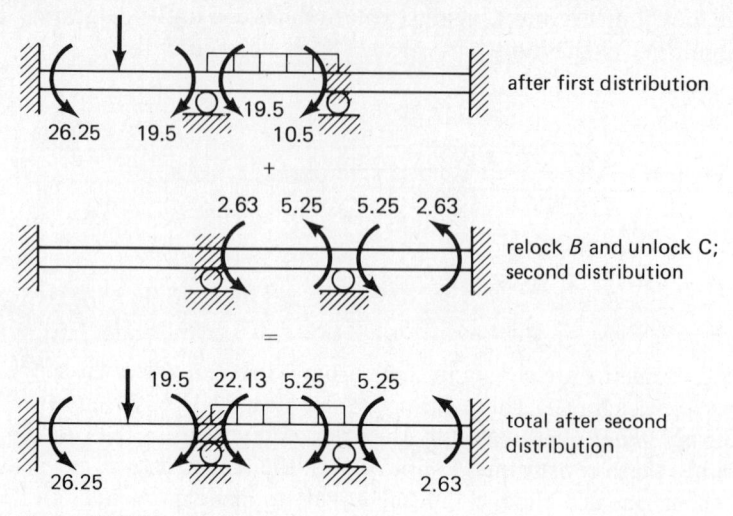

Figure 11.10

Now joint *C* is balanced, but the carry-over to *BC* has unbalanced joint *B* again, this time with a net −2.63 kN·m. So we relock *C* and unlock *B* by applying +2.63 kN·m there, distributed appropriately and carried over.

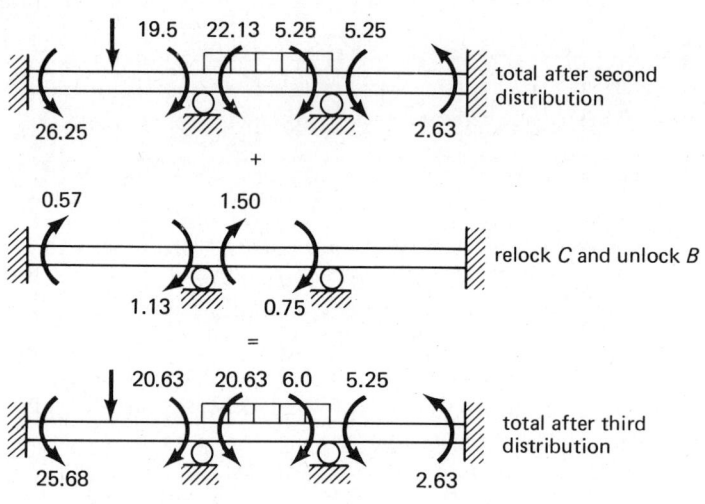

Figure 11.11

The process continues in the same way until the remaining imbalances are negligible. The process always converges, usually quite rapidly. In this example, the initial imbalances of 10.5 and 13.5 kN·m were reduced to a single 0.75 kN·m imbalance in only three distributions.

EXAMPLE 11.3.2

Repeat Example 11.3.1, using a moment distribution table.
Preliminary steps in this solution would include calculation of the

fixed end moments and distribution factors, had they not been calculated in Example 11.3.1. These were

$$\text{FEM}_{AB} = -\text{FEM}_{BA} = -24 \text{ kN} \cdot \text{m}$$
$$\text{FEM}_{BC} = -\text{FEM}_{CB} = -13.5 \text{ kN} \cdot \text{m}$$
$$D_{BA} = 0.429$$
$$D_{BC} = 0.571$$
$$D_{CB} = 0.5$$
$$D_{CD} = 0.5$$

The distribution table is arranged as illustrated.

AB		BA	BC		CB	CD		DC
	0.5 →			0.5 →			0.5 →	
0		0.429	0.571		0.5	0.5		0.0
−24		+24	−13.5		+13.5			

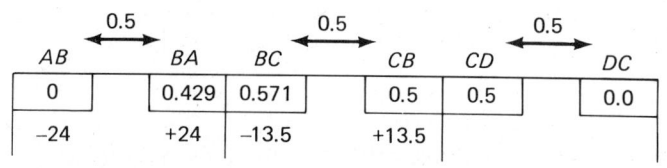

Figure 11.12

There is a column for each end moment, with the columns grouped by joints. For example, those at B, namely, BA and BC, are abutting columns in the table. A small space separates BA from its far end AB, but AB and BA are adjacent columns.

Similarly, BC and its far end CB are adjacent but separated by a space. There are boxes directly under each column label, enclosing the distribution factors. The arrows above the table show the carry-over factors from one member end to the other. Here they are all $\frac{1}{2}$, but other values will occur when modified stiffnesses (Section 11.4) are used or when there are nonprismatic members (Chapter 13).

Notice that the distribution factors are zero for AB and DC, where there are clamped ends. Moments applied at these ends would be resisted entirely by the rigid support, *none* by the more compliant member. Hence the zero distribution factor for the member end.

Directly below the distribution factor boxes are the fixed end moments. By adding these for each joint, we determine the joint imbalances. For example, at joint B, BA and BC total $+10.5$ kN · m. That is the joint B imbalance. It is balanced by distributing -10.5 kN · m there, prorated according to the distribution factors, and half is carried over to each far end.

The table at that point is

AB		BA	BC		CB	CD		DC
	0.5 →			0.5 →			0.5 →	
0		0.429	0.571		0.5	0.5		0
−24		+24	−13.5		+13.5			
−2.25 ←		−4.5	−6.0	→	−3.0			

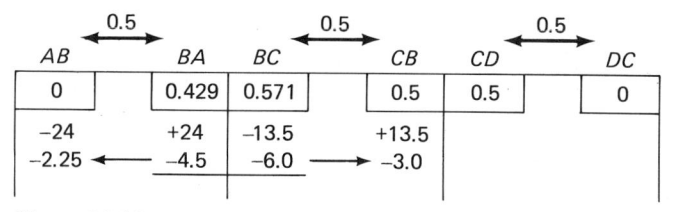

Figure 11.13

The arrows inside the table indicate a carry-over moment, and the line drawn across BA and BC indicates that joint B has been balanced for all entries above that line.

Next a moment -10.5 kN·m is distributed at C, -5.25 kN·m to each of CB and CD, and half is carried over. Then the table is

AB	BA	BC	CB	CD	DC
0	0.429	0.571	0.5	0.5	0
−24	+24	−13.5	+13.5		
−2.25 ←	−4.5	−6.0 →	−3.0		
		−2.63 ←	−5.25	−5.25 →	−2.63

(0.5 carry-over factors shown between BA↔AB, CB↔BC, DC↔CD)

Figure 11.14

At this point readers should compare the first few steps of Example 11.3.1 to the tabular solution developing here.

The distribution table, continued to its completion, is

AB	BA	BC	CB	CD	DC
0	0.429	0.571	0.5	0.5	0
−24	+24	−13.5	+13.5		
−2.25 ←	−4.5	−6.0 →	−3.0		
		−2.63 ←	−5.25	−5.25 →	−2.63
+0.56 ←	+1.13	+1.50 →	+0.75		
		−0.19 ←	−0.38	−0.37 →	−0.18
+0.04 ←	+0.08	+0.11 →	+0.05		
			−0.02	−0.03 →	−0.02
Σ −25.65	+20.71	−20.71	+5.65	−5.65	−2.83

Figure 11.15

The process was ended when all of the joints were balanced and the carry-over moments had reduced to a size considered negligible in comparison with the original fixed end moments. The final end moments were calculated by superposing the fixed end, carry-over, and distributed moments for each member end, that is, by adding each column.

A small saving in computational effort could have been made in columns AB and DC, columns that have zero for their distribution factors. Aside from their fixed end moments, these columns had only carry-over moments as entries. Rather than carry over half of each distribution in BA and CD individually, it is easier to do all the distributions first, add them, and then carry over half of their total.

MOMENT DISTRIBUTION PROCEDURE 249

Example 11.3.3 illustrates several other features of moment distribution. We apply it to a frame that has joints rigidly connecting more than two members, has a pin-ended member, and a member that, by itself, is statically determinate.

EXAMPLE 11.3.3

Calculate the end moments for frame *ABCDEF*

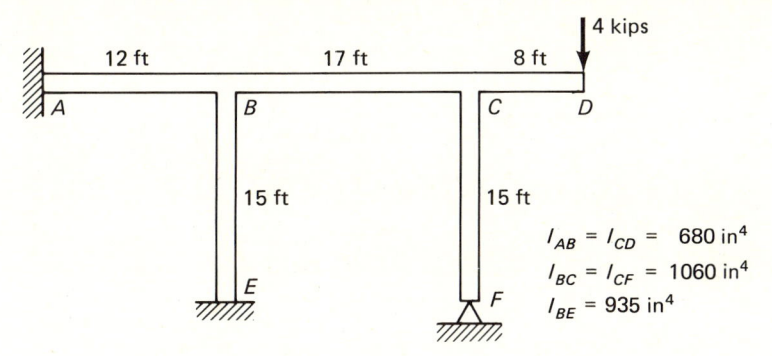

[Figure showing frame ABCDEF with horizontal member from A (fixed) through B, C, to D with 4 kips load at D. Spans: A-B = 12 ft, B-C = 17 ft, C-D = 8 ft. Vertical member B-E = 15 ft (E fixed), vertical member C-F = 15 ft (F pinned).]

$I_{AB} = I_{CD} = 680$ in^4
$I_{BC} = I_{CF} = 1060$ in^4
$I_{BE} = 935$ in^4

Figure 11.16

Since member *CD* is a statically determinate member, we can calculate its end moment *CD* by statics alone. The end moment *CD* can be considered a fixed-end moment. Thus

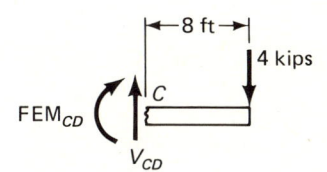

[Figure showing cantilever at C with 8 ft to 4 kip load, FEM$_{CD}$ and V$_{CD}$ reactions at C.]

Figure 11.17

$+\circlearrowleft \sum M_C = \text{FEM}_{CD} + (4 \text{ kips})(8 \text{ ft}) = 0$

$\text{FEM}_{CD} = -32$ kip-ft

Then the distribution factors are calculated using Eq. (11.4), first at *B*,

$$D_{BA} = \frac{I_{BA}/L_{BA}}{\sum (I/L)}$$

$$= \frac{680/12}{680/12 + 1060/17 + 935/15} = 0.312$$

$$D_{BC} = \frac{1060/17}{680/12 + 1060/17 + 935/15} = 0.344$$

$$D_{BE} = \frac{935/15}{680/12 + 1060/17 + 935/15} = 0.344$$

and then at C.

First, though, we observe that member CD is cantilevered from C and has no other restraints. If joint C rotates, member CD rotates with it freely, offering no resistance to rotation of C. Therefore we take its I/L to be zero.

$$D_{CD} = \frac{0}{1060/17 + 1060/15 + 0} = 0$$

$$D_{CB} = \frac{1060/17}{1060/17 + 1060/15 + 0} = 0.469$$

$$D_{CF} = \frac{1060/15}{1060/17 + 1060/15 + 0} = 0.531$$

The table has several features not found in the earlier example.

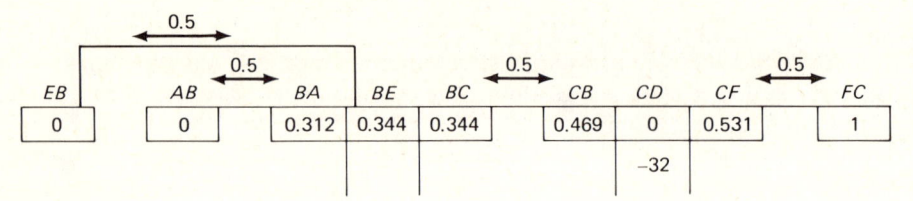

EB	AB	BA	BE	BC	CB	CD	CF	FC
0	0	0.312	0.344	0.344	0.469	0	0.531	1
						−32		

Figure 11.18

Columns for the three end moments at joint B must be grouped together. Furthermore, each should have a column for its far end located adjacent to it and spaced a small distance away. With more than two members connected at B, that is impossible. So, BE, with its far end EB having a zero distribution factor, is chosen as the center member. Its distributions will be totaled and carried over after all distributions have been made. AB and BA would have served as well.

Similarly, CD, with a zero distribution factor, was chosen as the center member at C, since no distributions will be made to it. If no zero distribution factors appear to aid in selecting interior members for groups, then a remote far end column must be established, connected by the carry-over arrow.

Finally, the support at F is a pin connection. Any external moment applied at F is absorbed totally by that member end. Hence its distribution factor is 1.

The completed table is shown.

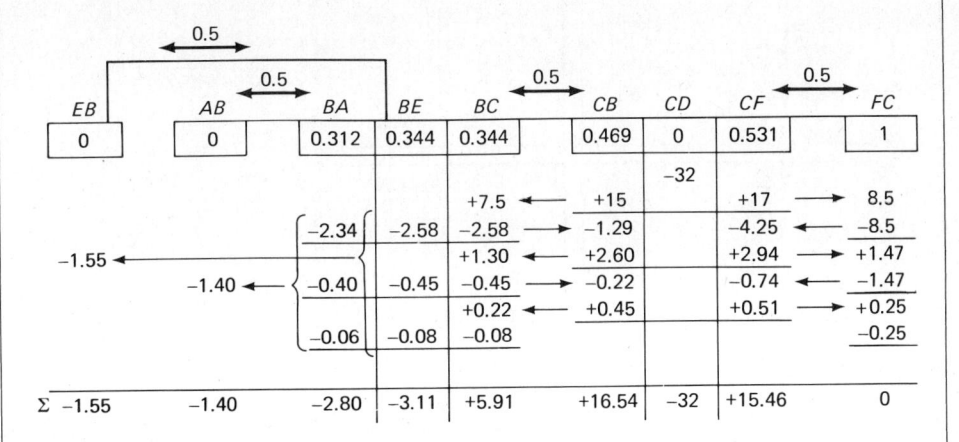

EB	AB	BA	BE	BC	CB	CD	CF	FC
0.5 →	0.5 →			0.5 →			0.5 →	
0	0	0.312	0.344	0.344	0.469	0	0.531	1
					−32			
				+7.5 ←	+15		+17 →	8.5
		−2.34	−2.58	−2.58 →	−1.29		−4.25 ←	−8.5
−1.55 ←				+1.30 ←	+2.60		+2.94 →	+1.47
	−1.40 ←	−0.40	−0.45	−0.45 →	−0.22		−0.74 ←	−1.47
				+0.22 ←	+0.45		+0.51 →	+0.25
		−0.06	−0.08	−0.08				−0.25
Σ −1.55	−1.40	−2.80	−3.11	+5.91	+16.54	−32	+15.46	0

Figure 11.19

The structure with its end moments is illustrated.

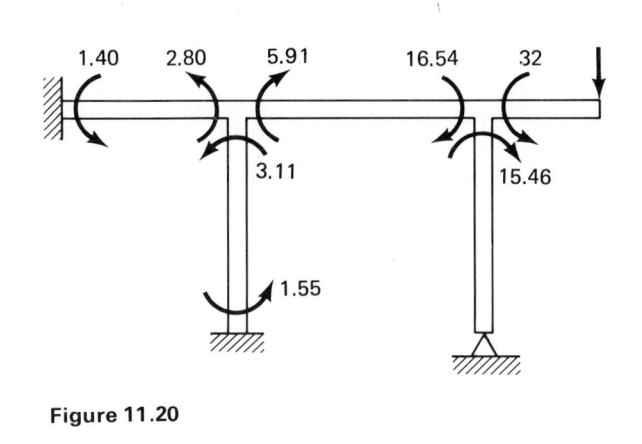

Figure 11.20

11.4 MODIFIED STIFFNESS

Three frequently encountered structural arrangements can be treated with greatly simplified calculations by using modified stiffnesses. They are for pin (or roller) ends and for symmetric and antisymmetric members.

Consider the four prismatic members shown, all rigidly connected at joint J.

The far end A of JA is clamped, so JA is similar to the members considered previously. The three other members, however, are different. The far end B of JB

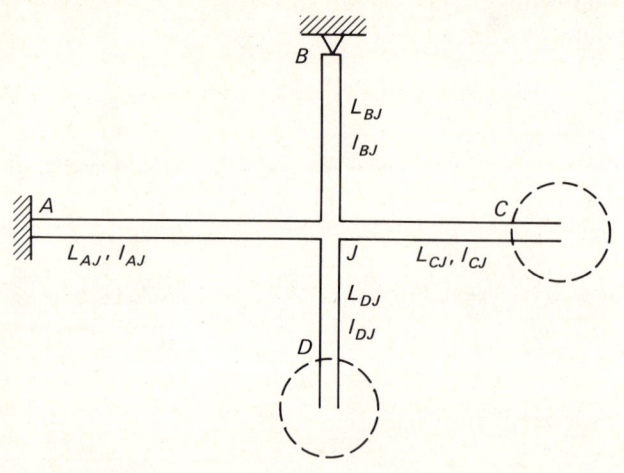

252 MOMENT DISTRIBUTION: STRUCTURES WITHOUT SIDESWAY

Figure 11.21

is pin supported so that $M_{BJ}=0$. Members JC and JD are arranged so that

$$\theta_C = -\theta_J$$
$$\theta_D = +\theta_J$$

that is, CJ and DJ are symmetric and antisymmetric, respectively. Details of the support and loading that produce symmetry and antisymmetry will be discussed later in this section.

We write and manipulate some of the slope deflection equations. First, for member JA,

$$M_{JA} = \frac{4EI}{L_{AJ}}\theta_C = S_{JA}\theta_J$$

showing that JA has the same rotational stiffness as before. Then for BJ,

$$M_{BJ} = \frac{2EI_{BJ}}{L_{BJ}}\theta_J + \frac{4EI_{BJ}}{L_{BJ}}\theta_B = 0$$

which reduces to

$$\theta_B = -\frac{\theta_J}{2}$$

so that

$$M_{JB} = \frac{4EI_{BJ}}{L_{BJ}}\theta_J + \frac{2EI_{BJ}}{L_{BJ}}\left(-\frac{\theta_J}{2}\right)$$

$$= \frac{3EI_{BJ}}{L_{BJ}}\theta_J = S^*_{JB}\theta_J$$

S^*_{JB} is the *modified stiffness for the far end pinned* and is $\tfrac{3}{4}$ of the regular stiffness.

$$S^*_{JB} = \tfrac{3}{4}S_{JB} \quad \text{far end pinned} \tag{11.5}$$

Use of this modified stiffness presumes that when any moment is applied at JB, the far end moment M_{BJ} remains zero.

Next consider member CJ,

$$M_{JC} = \frac{4EI_{JC}}{L_{JC}}\theta_J + \frac{2EI_{JC}}{L_{JC}}(-\theta_J)$$

$$= \frac{2EI_{JC}}{L_{JC}}\theta_J = S^*_{JC}\theta_J$$

S^*_{JC} is the *modified stiffness for a symmetric member* and is one half the regular stiffness.

$$S^*_{JC} = \tfrac{1}{2} S_{JC} \qquad \text{symmetric member} \tag{11.6}$$

The slope deflection equation for the far end moment M_{CJ} is

$$M_{CJ} = \frac{4EI_{JC}}{L_{JC}}(-\theta_J) + \frac{2EI_{JC}}{L_{JC}}\theta_J$$

$$= -\frac{2EI_{JC}}{L_{JC}}\theta_J$$

showing that when the symmetric member modified stiffness is used, the far end moment is automatically equal and opposite to the near end moment applied.

Now for member JD,

$$M_{JD} = \frac{4EI}{L_{JD}}\theta_J + \frac{2EI}{L_{JD}}\theta_J$$

$$= \frac{6EI}{L_{JD}}\theta_J = S^*_{JD}\theta_J$$

S^*_{JD} is the *modified stiffness for an antisymmetric member* and is $\tfrac{3}{2}$ the regular stiffness.

$$S^*_{JD} = \tfrac{3}{2} S_{JD} \qquad \text{antisymmetric member} \tag{11.7}$$

In this case the far end moment is equal to the near end moment and has the same sign.

Whenever symmetry or antisymmetry is used, it will be for the center member of a structure. The structure's end moments at image locations will have the same values and the same or opposite signs as appropriate. Thus we need work with only half the structure. Further, when unlocking a joint, members whose distribution factors are based on modified stiffnesses will have their far end moments accounted for automatically. Hence there is no carry-over to their far ends.

EXAMPLE 11.4.1

Find the end moments for *ABCDEF*. Assume all members have the same *I*.

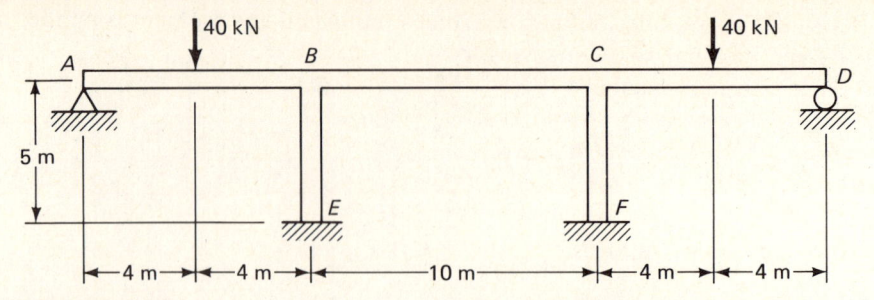

254   MOMENT DISTRIBUTION: STRUCTURES WITHOUT SIDESWAY

Figure 11.22

The structure is symmetric with respect to the midpoint of BC and is loaded symmetrically. Thus BC will deform symmetrically, permitting use of the modified stiffness for symmetric members for BC. The far end of BA is pin supported, so the far end pinned modified stiffness may be used for BA.

The distribution factors are found using Eq. (11.4), but with the modifying coefficients multiplying the relative stiffnesses where applicable.

$$D_{BA} = \frac{\frac{3}{4}(I/8)}{\frac{3}{4}(I/8) + \frac{1}{2}(I/10) + I/5} = 0.273$$

$$D_{BC} = \frac{\frac{1}{2}(I/10)}{\frac{3}{4}(I/8) + \frac{1}{2}(I/10) + I/5} = 0.145$$

$$D_{BE} = \frac{I/5}{\frac{3}{4}(I/8) + \frac{1}{2}(I/10) + I/5} = 0.582$$

A check shows that their summation is 1.

Fixed end moments are found for AB and BA, but are not needed for CD and DC, since we are working with only the left half of the structure.

$$\text{FEM}_{AB} = -\text{FEM}_{BA} = -\frac{40(8)}{8} = -40 \text{ kN} \cdot \text{m}$$

Our table is for only half of the structure.

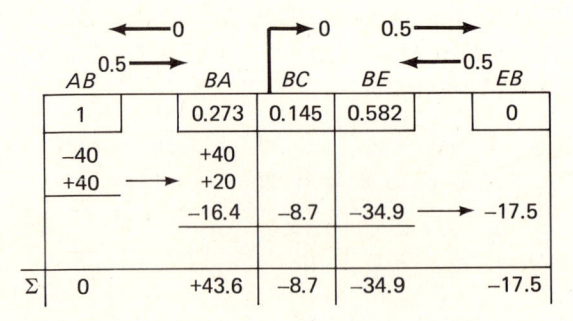

Figure 11.23

Notice that in member AB no carry-over was used from BA to AB, nor was there a carry-over from BC. In these cases a modified stiffness was used to calculate the near end distribution factors. In the remaining cases distribution factors were based on regular stiffnesses requiring that regular carry-overs remain in use. Note the difference in the carry-over factor from AB to BA as compared to that from BA to AB.

The complete set of end moments is taken from the table with equal, opposite end moments at image locations shown. The end moments are in kN·m.

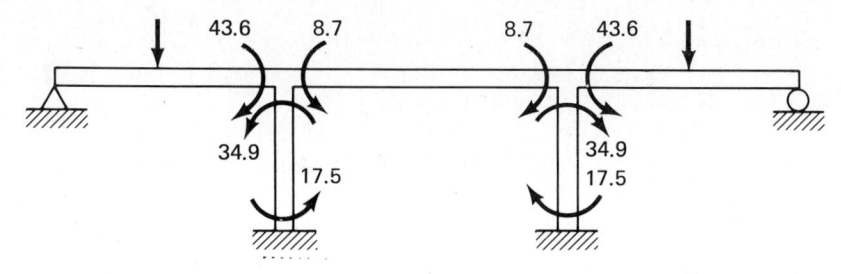

Figure 11.24

EXAMPLE 11.4.2

The structure of Example 11.4.1 is subjected to the wind load shown. Calculate the end moments.

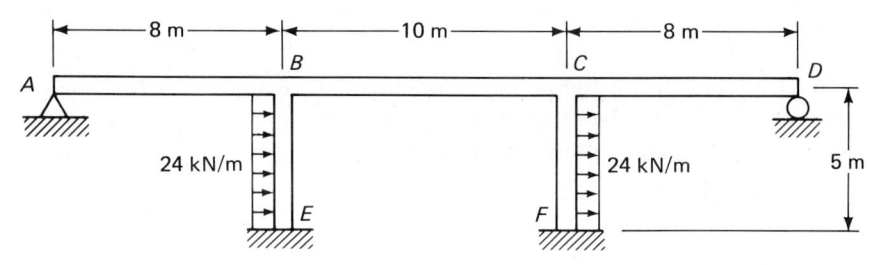

Figure 11.25

Member BC is antisymmetric for this loading, a fact that may not be obvious. Let us proceed as if the antisymmetry had not been noticed, treating BC as a regular member. Then the only modified stiffness to be used will be for the far end pinned for BA.

The distribution factors at B are

$$D_{BA} = \frac{\frac{3}{4}(I/8)}{\frac{3}{4}(I/8) + I/5 + I/10} = 0.238$$

$$D_{BC} = \frac{I/10}{\frac{3}{4}(I/8) + I/5 + I/10} = 0.254$$

$$D_{BE} = \frac{I/5}{\frac{3}{4}(I/8) + I/5 + I/10} = 0.508$$

All three differ from those in Example 11.4.1.

The distribution factors for joint C are identical to those at B.

$D_{CD} = D_{BA} = 0.238$
$D_{CF} = D_{BE} = 0.508$
$D_{CB} = D_{BC} = 0.254$

The fixed end moments are

$$FEM_{EB} = FEM_{FC} = -FEM_{BE} = -FEM_{CF}$$
$$= -\frac{24(5)^2}{12} = -50 \text{ kN} \cdot \text{m}$$

We set up the distribution table.

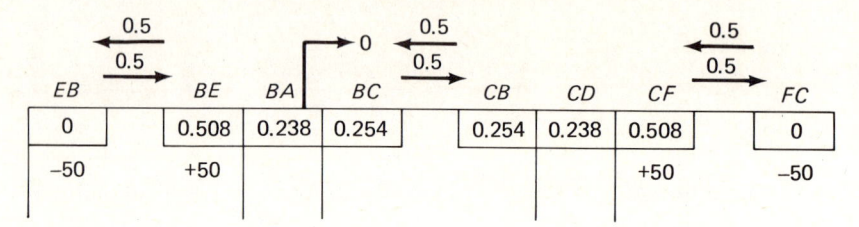

EB	BE	BA	BC	CB	CD	CF	FC
0	0.508	0.238	0.254	0.254	0.238	0.508	0
−50	+50					+50	−50

Carry-over factors above the table: 0.5 (EB←BE), 0.5 (BE→EB), 0 (BA/BC), 0.5 (CB→BC), 0.5 (BC←CB), 0.5 (CF→FC), 0.5 (FC←CF).

Figure 11.26

At this point the antisymmetry of BC becomes obvious: the distribution factors and carry-over factors are, item by item, identical at image locations. Furthermore, the fixed end moments are equal in magnitude and sign, item by item, at image locations. These two conditions are necessary and sufficient conditions for use of antisymmetry.[3]

Having discovered the antisymmetry, it is economical in effort to recalculate the distribution factors and set up a new table, rather than continue the one we have just begun. Thus

$$D_{BA} = \frac{\frac{3}{4}(I/8)}{\frac{3}{4}(I/8) + \frac{3}{2}(I/10) + I/5} = 0.211$$

$$D_{BC} = \frac{\frac{3}{2}(I/10)}{\frac{3}{4}(I/8) + \frac{3}{2}(I/10) + I/5} = 0.338$$

$$D_{BE} = \frac{I/5}{\frac{3}{4}(I/8) + \frac{3}{2}(I/10) + I/5} = 0.451$$

All differ from the ones found before, and from those of Example 11.4.1.
The table is simple and quick to converge.

---

[3] If the fixed end moments had been equal, but opposite in sign at the image locations, we would have concluded that BC is symmetric.

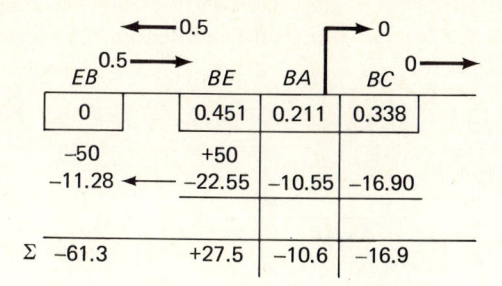

	←0.5			0 →
0.5→				0→
EB	BE	BA	BC	
0	0.451	0.211	0.338	
−50	+50			
−11.28 ←	−22.55	−10.55	−16.90	
Σ −61.3	+27.5	−10.6	−16.9	

Figure 11.27

Readers are urged to complete the table that was discarded earlier to verify that the end moments found are the same, and to observe the extent of the extra effort required in not using modified stiffness.

The end moments for the entire structure are

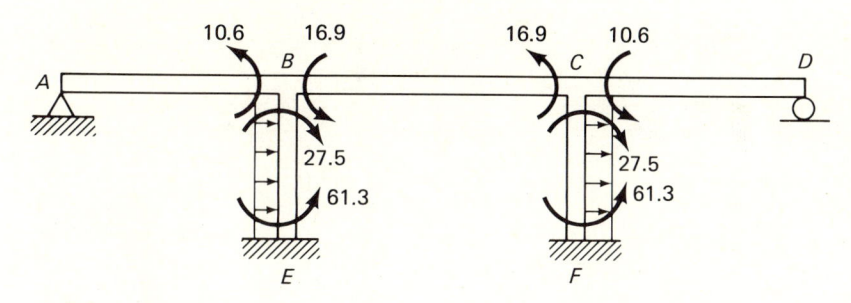

Figure 11.28

where the end moments are in kN · m.

PROBLEMS

Sequenced Problem 3, parts (o) and (p).
Sequenced Problem 4, parts (o) and (p).
Sequenced Problem 6, parts (a) and (b).
Sequenced Problem 8, parts (a) and (b).
11.1 Solve Problems 10.1, parts (a)–(g), using moment distribution.
11.2

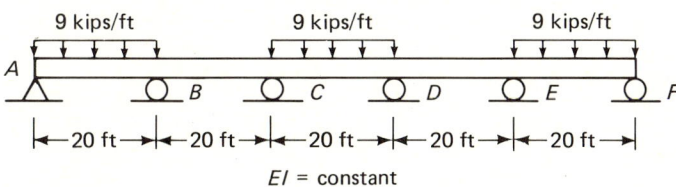

Figure P11.2

(a) Use moment distribution to solve for all of the end moments. Do *not* use any modified stiffnesses.

(b) Use moment distribution to solve for all of the end moments. Take advantage of all applicable modified stiffnesses.

11.3

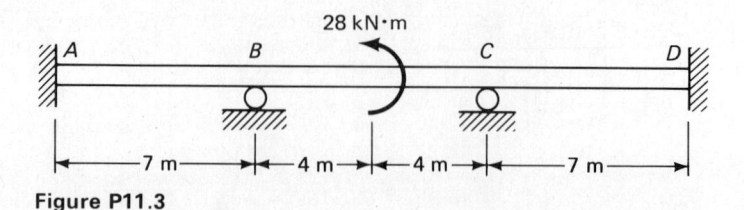

Figure P11.3

Use moment distribution to determine the end moments, then calculate the vertical support reactions and draw the shear and moment diagrams.

11.4

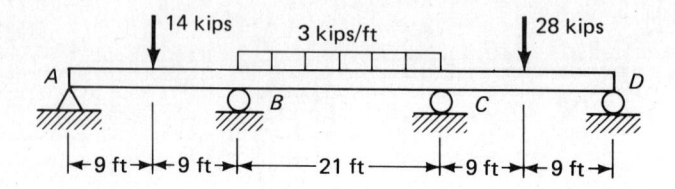

The structure can be considered the superposition of the two following cases:

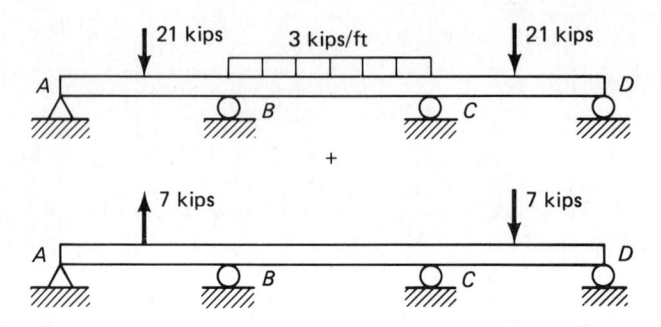

Figure P11.4

One is symmetric, the other is antisymmetric. Take advantage of this to solve the parts, using moment distribution and appropriate modified stiffnesses, and then superpose the results.

11.5

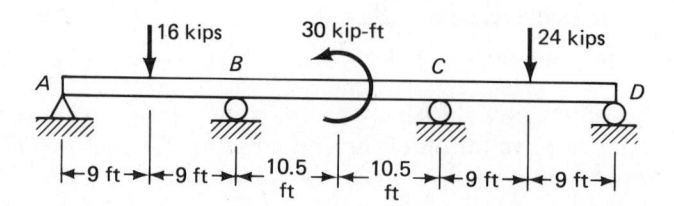

The structure can be considered the superposition of the two following cases:

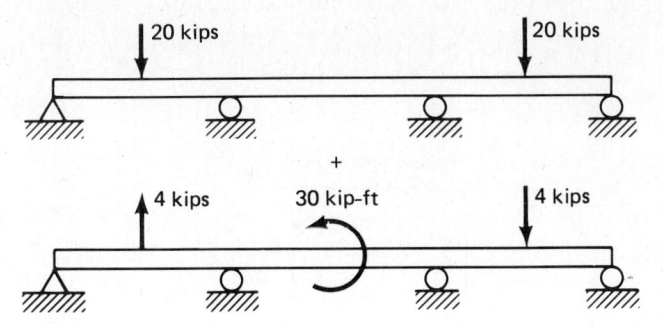

Figure P11.5

One is symmetric, the other is antisymmetric. Take advantage of this to solve the parts using moment distribution and appropriate modified stiffnesses. Then superpose the results.

11.6 Solve Problem 10.4c, by moment distribution.

11.7

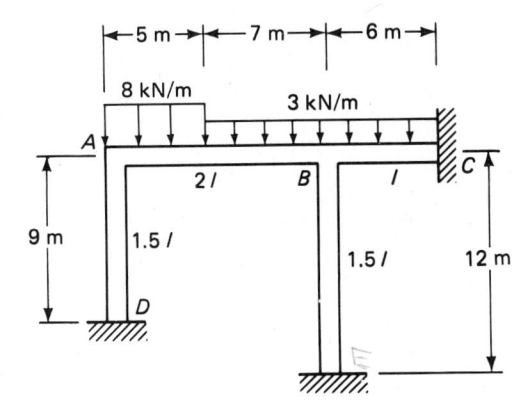

Figure P11.7

Use moment distribution to determine the end moments.

11.8

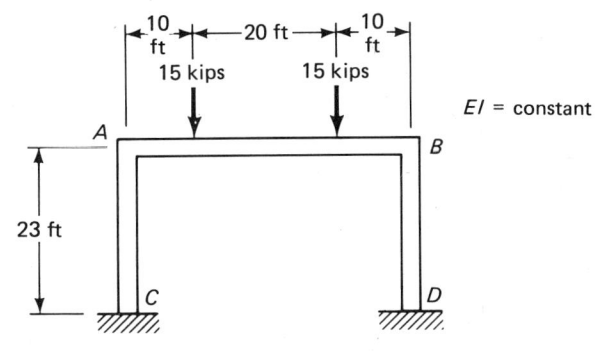

EI = constant

Figure P11.8

Use moment distribution to determine the end moments. Then evaluate the support reactions and draw the shear and moment diagrams.

11.9

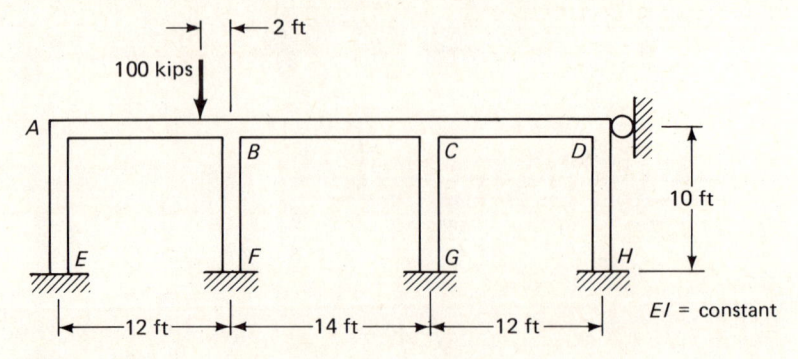

Figure P11.9

EI = constant

Use moment distribution to determine the end moments.

11.10

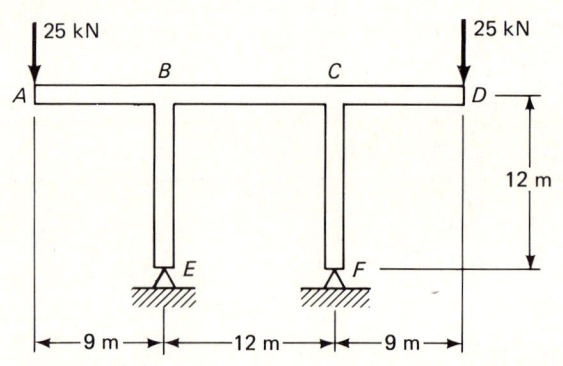

Figure P11.10

Use moment distribution to determine the end moments.

Moment Distribution: Structures With Joint Translation

The second grade was holding a spelling bee. After a number of rounds only Betsy and Albert remained. When Betsy missed the word "farm" Albert became excited. He had heard "farm" spelled many times— he was going to win. Stepping forward proudly, he said "Farm: E-I-E-I-O."

12.0 INTRODUCTION

By now you probably feel comfortable with moment distribution; perhaps you do it almost automatically. But all the structures that we have considered so far have been braced against sidesway (joint translation). In reality many structures are not so braced. It would be easy to overlook the sidesway and analyze them with an automatic "E-I-E-I-O," but that would be a dangerous mistake. Sidesway may have a sizeable effect on the end moments. Therefore, it must be included in the analysis. This is done in two parts—one related to the loaded structure but with added supports preventing sidesway, and the other where loads are applied at the added support points that cause the unbraced structure to undergo sidesway. The first part, where sidesway is prevented, is treated using the techniques of Chapter 11. Thus we proceed directly to the second part, sidesway induced.

12.1 SIDESWAY INDUCED FIXED END MOMENTS AND THEIR DISTRIBUTION

Consider prismatic member AB with transverse relative end displacement Δ, supported so that end rotations are prevented.

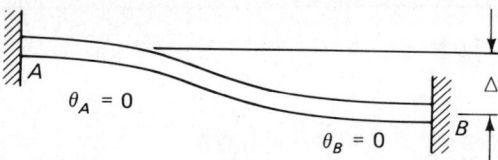

Figure 12.1

Since rotations are prevented, we can consider the end moments to be fixed end moments and evaluate them using the slope deflection equation (10.1). Therefore

$$FEM_{AB} = FEM_{BA} = -\frac{6EI}{L^2}\Delta \tag{12.1}$$

We illustrate their use with the following example.

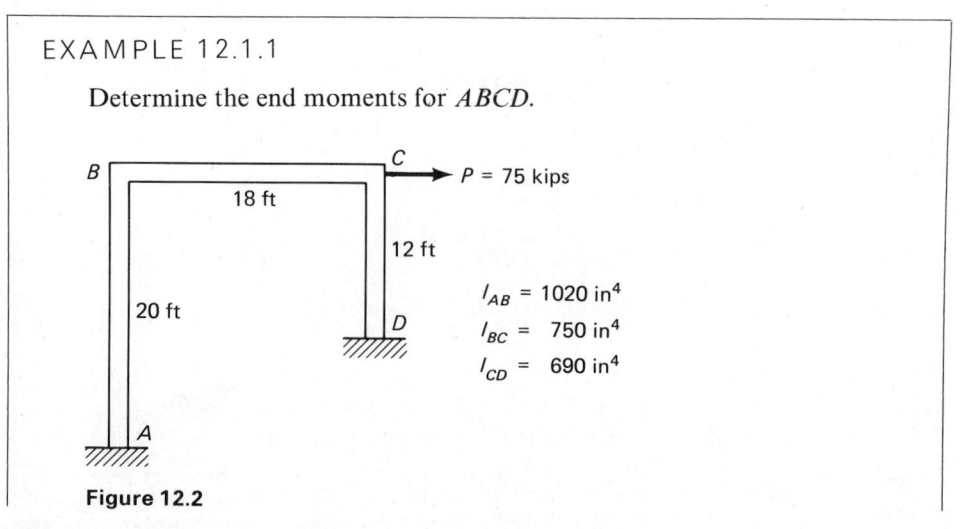

EXAMPLE 12.1.1

Determine the end moments for $ABCD$.

$P = 75$ kips

18 ft
12 ft
20 ft

$I_{AB} = 1020$ in^4
$I_{BC} = 750$ in^4
$I_{CD} = 690$ in^4

Figure 12.2

SIDESWAY INDUCED FIXED END MOMENTS AND THEIR DISTRIBUTION

First we calculate the distribution factors. There being no applicable modified stiffnesses, Eq. (11.4) applies directly,

$$D_{BA} = \frac{1020/20}{1020/20 + 750/18} = 0.550$$

$$D_{BC} = \frac{750/18}{1020/20 + 750/18} = 0.450$$

$$D_{CB} = \frac{750/18}{750/18 + 690/12} = 0.420$$

$$D_{CD} = \frac{690/12}{750/18 + 690/12} = 0.580$$

We assume that members AB and CD do not change their lengths and conclude that the only possible motions of B and C are horizontal, when members AB and CD bend. Further, we assume that BC does not change its length and conclude that the horizontal motions of B and C are identical. Initially we lock joints B and C to prevent rotation, giving the following initial deformed shape.

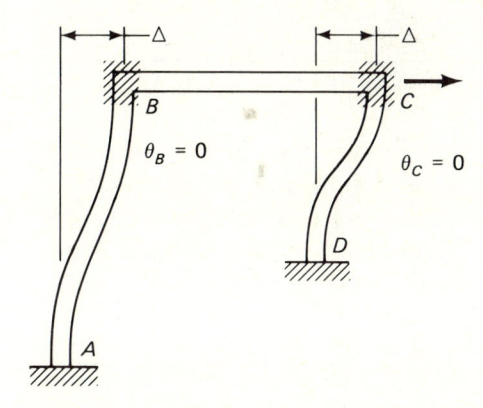

Figure 12.3

The cross-hatching at B and C, as in Chapter 11, implies the presence of fictitious constraints that prevent rotation. They will be removed by the successive unlocking and relocking of these joints as before.

From Eq. (12.1) the fixed end moments for the deformed structure are

$$\text{FEM}_{AB} = \text{FEM}_{BA} = -\frac{6E(1020)}{20^2}\Delta = -15.30E\Delta$$

$$\text{FEM}_{CD} = \text{FEM}_{DC} = -\frac{6E(690)}{12^2}\Delta = -28.75E\Delta$$

Since member BC translates axially without deforming, its fixed end moments are zero.

At this point there is no known value of Δ, and our procedure requires us to *assume* a convenient value. If E is the same for all members, we can assume a convenient value for the product $E\Delta$, say $E\Delta = 7$. That was an

264 MOMENT DISTRIBUTION: STRUCTURES WITH JOINT TRANSLATION

arbitrary choice made to set values of the fixed end moments around 100 or 200, numbers conveniently handled in a distribution table. It is important that whatever $E\Delta$ value is used, it must be used for all of the fixed end moments since their relative values and signs are critical.

$$FEM_{AB} = FEM_{BA} = -15.3(7) = -107.1 \text{ kip-ft}$$
$$FEM_{CD} = FEM_{DC} = -28.75(7) = -201.2 \text{ kip-ft}$$

The distribution table is arranged and completed as before, but starts with the fixed end moments for the assumed sidesway.

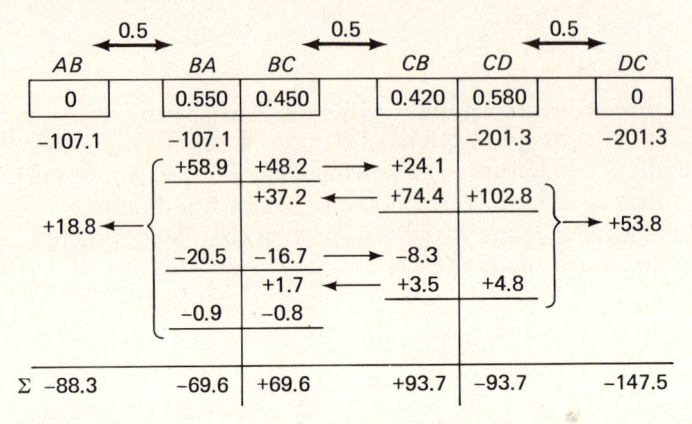

AB	BA	BC	CB	CD	DC
0	0.550	0.450	0.420	0.580	0
−107.1	−107.1			−201.3	−201.3
	+58.9	+48.2 →	+24.1		
		+37.2 ←	+74.4	+102.8	
+18.8 ←					→ +53.8
	−20.5	−16.7 →	−8.3		
		+1.7 ←	+3.5	+4.8	
	−0.9	−0.8			
Σ −88.3	−69.6	+69.6	+93.7	−93.7	−147.5

Carry-over factors 0.5 between AB–BA, BC–CB, CD–DC.

Figure 12.4

In the distribution, joint B was unlocked first by distributing $+107.1$ as $+58.9$ and $+48.2$ to BA and BC, and carrying over $48.2/2 = 24.1$ to CB. The carry-over to AB was delayed to the end. Joint B was relocked and C unlocked by distributing $+177.2$ (from -201.3 and $+24.3$) to CB and CD. The distribution continued until the carry-overs were negligible.

It may be tempting to assume that the end moments found are those from $P = 75$ kips at C. They are not, for they came from the *arbitrary* choice of $E\Delta = 7$. We can determine what the value of P must be to produce the end moments found. We do so by summing the horizontal forces acting on the structure, but first we need the end shears at AB and DC.

For member AB,

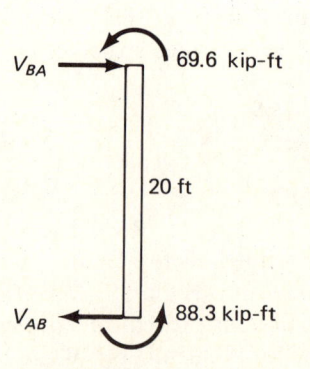

V_{BA} ← 69.6 kip-ft (top)

20 ft

V_{AB} ← 88.3 kip-ft (bottom)

Figure 12.5

SIDESWAY INDUCED FIXED END MOMENTS AND THEIR DISTRIBUTION 265

$$+\circlearrowright \sum M_B = 20V_{AB} - 69.6 - 88.3 = 0$$
$$V_{AB} = 7.90 \text{ kips}$$

For member DC,

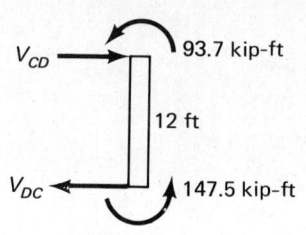

Figure 12.6

$$+\circlearrowright \sum M_C = 12V_{DC} - 93.7 - 147.5 = 0$$
$$V_{DC} = 20.1 \text{ kips}$$

A partial free-body diagram of $ABCD$, showing only horizontal forces, is sketched and used to write $\sum F_x = 0$.

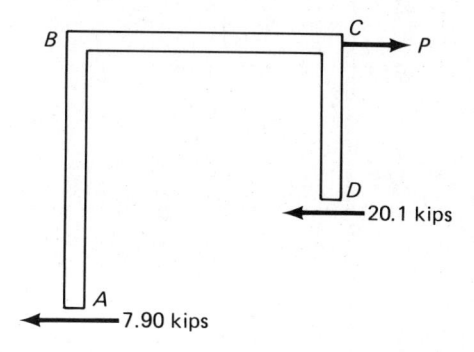

Figure 12.7

$$\pm \sum F_x = P - 7.90 - 20.1 = 0$$
$$P = 28.0 \text{ kips}$$

Thus the end moments found are for $P = 28$ kips. The end moments for $P = 75$ kips would be $75/28$ times as big. Thus

$$M_{AB} = -88.3 \left(\frac{75}{28}\right) = -236.5 \text{ kip-ft}$$

$$M_{BA} = -M_{BC} = -69.6 \left(\frac{75}{28}\right) = -186.4 \text{ kip-ft}$$

$$M_{CB} = -M_{DC} = +93.7 \left(\frac{75}{28}\right) = +251 \text{ kip-ft}$$

$$M_{DC} = -147.5 \left(\frac{75}{28}\right) = -395.1 \text{ kip-ft}$$

Thus

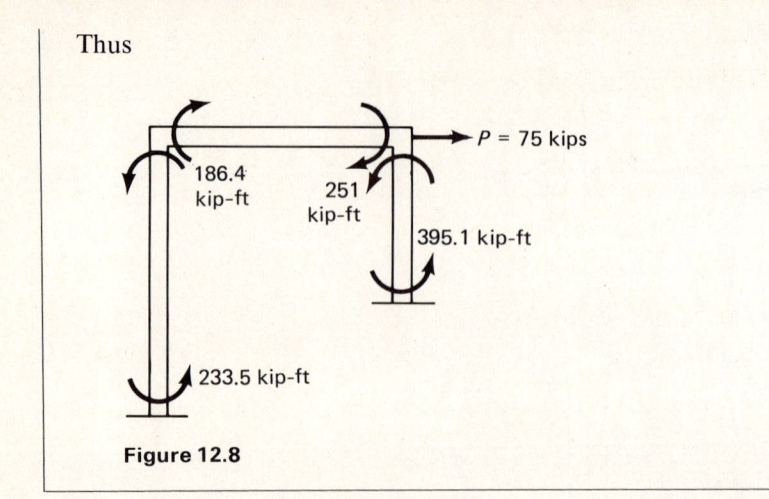

Figure 12.8

12.2 SUPERPOSITION OF SIDESWAY PREVENTED AND SIDESWAY INDUCED

We start with Example 12.2.1, using the structure of Example 12.1.1, but subjected to different loading.

EXAMPLE 12.2.1

Determine the end moments.

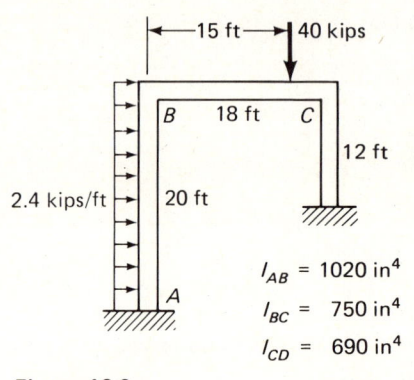

$I_{AB} = 1020 \text{ in}^4$
$I_{BC} = 750 \text{ in}^4$
$I_{CD} = 690 \text{ in}^4$

Figure 12.9

We will consider the structure and its loading in two parts as illustrated. Figure 12.10(a) is the structure with its original loading but constrained by force P to *prevent* sideway. Figure 12.10(b) has sideway induced by force P, equal and opposite to the force P of Figure 12.10(a).

We consider Figure 12.10(a) first, starting with the fixed end moments for its member loading, using formulas from Table 10.1.

$$\text{FEM}_{AB} = -\text{FEM}_{BA} = -\frac{2.4(20)^2}{12} = -80 \text{ kip-ft}$$

SUPERPOSITION OF SIDESWAY PREVENTED AND SIDESWAY INDUCED

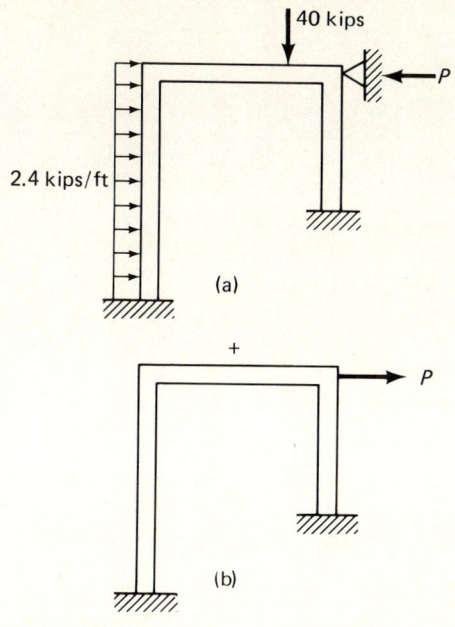

Figure 12.10

$$\text{FEM}_{BC} = -\frac{40(15)(3)^2}{18^2} = -16.67 \text{ kip-ft}$$

$$\text{FEM}_{CB} = +\frac{40(3)(15)^2}{18^2} = +83.3 \text{ kip-ft}$$

The distribution factors for this structure were found in Example 12.1.1, so we are ready to establish and use a distribution table for Figure 12.10(a).

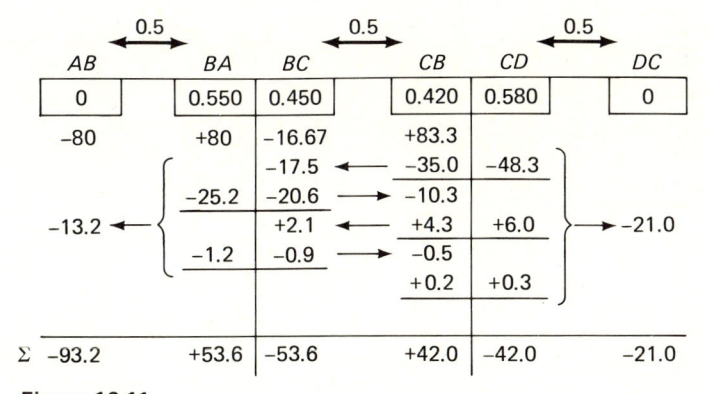

	AB	BA	BC	CB	CD	DC
	0	0.550	0.450	0.420	0.580	0
	−80	+80	−16.67	+83.3		
			−17.5	−35.0	−48.3	
		−25.2	−20.6	−10.3		
	−13.2		+2.1	+4.3	+6.0	−21.0
		−1.2	−0.9	−0.5		
				+0.2	+0.3	
Σ	−93.2	+53.6	−53.6	+42.0	−42.0	−21.0

Figure 12.11

With the end moments of Figure 12.10(a) known we can evaluate the force P that prevents the sidesway. This process requires a horizontal force summation and begins with the calculation of the end shears at A and D.

For AB,

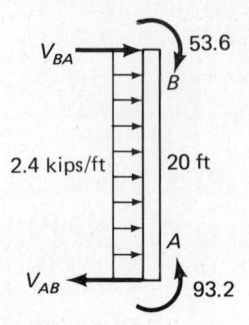

Figure 12.12

$$+\circlearrowleft \sum M_B = 20V_{AB} - 93.2 + 53.6 - 2.4(20)\left(\frac{20}{2}\right) = 0$$

$$V_{AB} = 26.0 \text{ kips}$$

For CD,

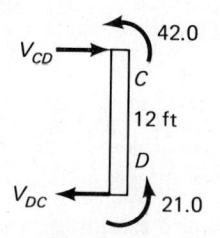

Figure 12.13

$$+\circlearrowleft \sum M_C = 12V_{DC} - 21 - 42 = 0$$

$$V_{DC} = 5.25 \text{ kips}$$

A free-body diagram of the structure showing only the horizontal forces is drawn, and a force summation is taken.

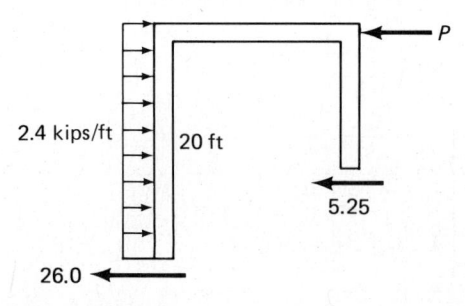

Figure 12.14

$$\pm \sum F_x = 2.4(20) - 26 - 5.25 - P = 0$$

$$P = 16.75 \text{ kips}$$

SUPERPOSITION OF SIDESWAY PREVENTED AND SIDESWAY INDUCED 269

Figure 12.10(b), sidesway induced, was solved in Example 12.1.1, but with one exception: the end moments found were for $P = 28$ kips and were scaled for $P = 75$ kips. Here we need $P = 16.75$ kips and scale the end moments accordingly. Thus

$$M_{AB} = -88.3 \left(\frac{16.75}{28}\right) = -52.8 \text{ kip-ft}$$

$$M_{BA} = -M_{BC} = -69.6 \left(\frac{16.75}{28}\right) = -41.6 \text{ kip-ft}$$

$$M_{CB} = -M_{CD} = +93.7 \left(\frac{16.75}{28}\right) = +56.1 \text{ kip-ft}$$

$$M_{DC} = -147.5 \left(\frac{16.75}{28}\right) = -88.2 \text{ kip-ft}$$

Now we can superpose Figure 12.15(a) and (b).

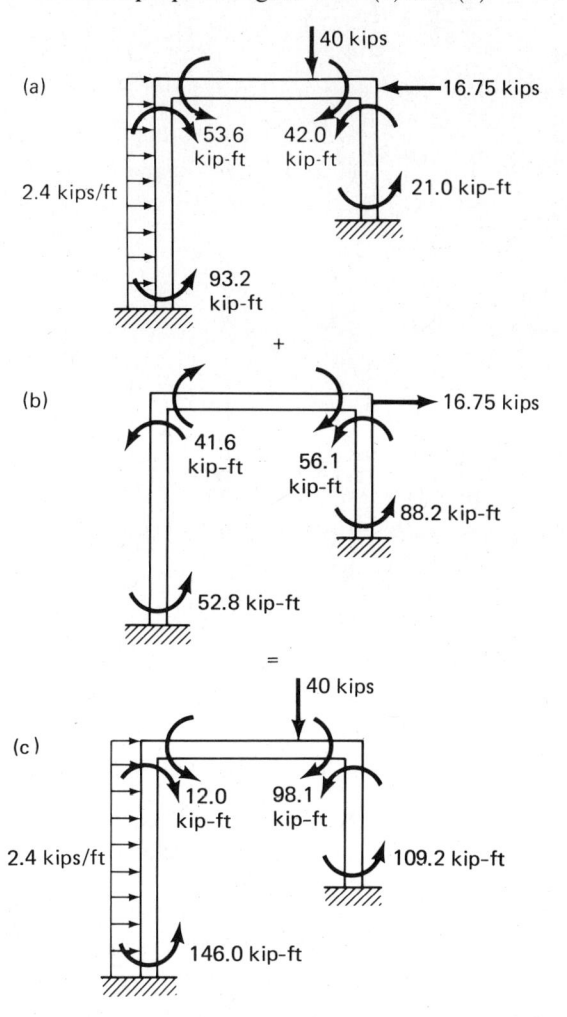

Figure 12.15

Notice that in the superposition the two forces at C cancel, leaving no net force at C. Since there is no provision for a force at C, this is a necessary result.

12.3 SIDESWAY INDUCED FIXED END MOMENTS; MEMBERS WITH FAR END PINNED

If a member subjected to sidesway has a pin support at one end, we can shorten our work by using a modifier. It will set $M=0$ at the pin and change the fixed end moment to a new value.

Consider prismatic member AB with transverse relative end displacement Δ, supported so there is no rotation of end B and $M=0$ at end A.

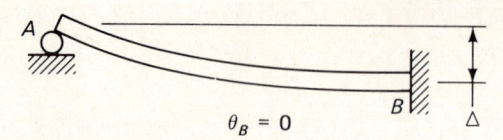

Figure 12.16

Based on the slope deflection equations (10.1),

$$M_{AB} = 0 = \frac{4EI}{L}\theta_A - \frac{6EI}{L^2}\Delta$$

showing that

$$\theta_A = \frac{3}{2}\left(\frac{\Delta}{L}\right)$$

and that

$$M_{BA} = \frac{2EI}{L}\theta_A - \frac{6EI}{L^2}\Delta = \frac{2EI}{L}\left(\frac{3}{2}\right)\left(\frac{\Delta}{L}\right) - \frac{6EI}{L^2}\Delta = -\frac{3EI}{L^2}\Delta$$

Thus for a member with one end pinned, the sidesway induced fixed end moment is

$$\text{FEM}_{BA} = -\frac{3EI}{L^2}\Delta \tag{12.2}$$

at the end away from the pin, while at the pin it is zero.

12.4 SIDESWAY INVOLVING SYMMETRY OR ANTISYMMETRY

Symmetry and antisymmetry are found quite frequently in structures with sidesway. Use of their modified stiffness produces the same savings in effort as in structures without sidesway. Care must be exercised in their application, for it is not uncommon for symmetry or antisymmetry to apply in one part while neither applies in the other part. Clearly, different sets of distribution factors would be used for the two parts. Example 12.4.1 illustrates such a case and also illustrates the use of the sidesway induced fixed end moments for far end pinned.

SIDESWAY INDUCED FIXED END MOMENTS; MEMBERS WITH FAR END PINNED   271

EXAMPLE 12.4.1

Find the end moments for the loaded structure, assuming that all members have the same EI.

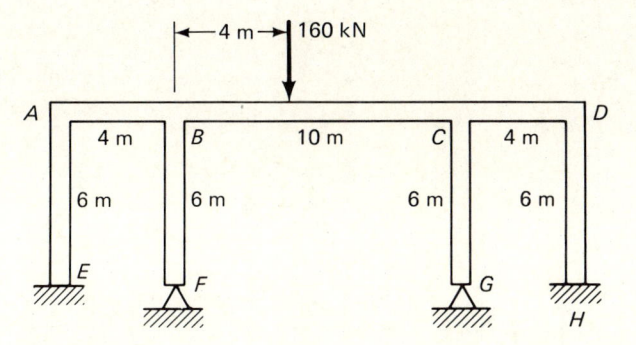

Figure 12.17

First we prevent sidesway by adding a horizontal restraint at D. Then we calculate the load induced fixed end moments, using formulas from Table 10.1.

$$FEM_{BC} = -\frac{160(4)(6^2)}{10^2} = -230.4 \text{ kN} \cdot \text{m}$$

$$FEM_{CB} = +\frac{160(6)(4^2)}{10^2} = +153.6 \text{ kN} \cdot \text{m}$$

The distribution factors are

$$D_{AE} = D_{DH} = \frac{I/6}{I/6 + I/4} = 0.400$$

$$D_{AB} = D_{DC} = \frac{I/4}{I/6 + I/4} = 0.600$$

$$D_{BA} = D_{CD} = \frac{I/4}{I/4 + I/10 + \frac{3}{4}(I/6)} = 0.526$$

$$D_{BF} = D_{CG} = \frac{\frac{3}{4}(I/6)}{I/4 + I/10 + \frac{3}{4}(I/6)} = 0.263$$

$$D_{BC} = D_{CB} = \frac{I/10}{I/4 + I/10 + \frac{3}{4}(I/6)} = 0.211$$

Only one modified stiffness was used, far end pinned for D_{BF} and D_{CG}. The distribution table is

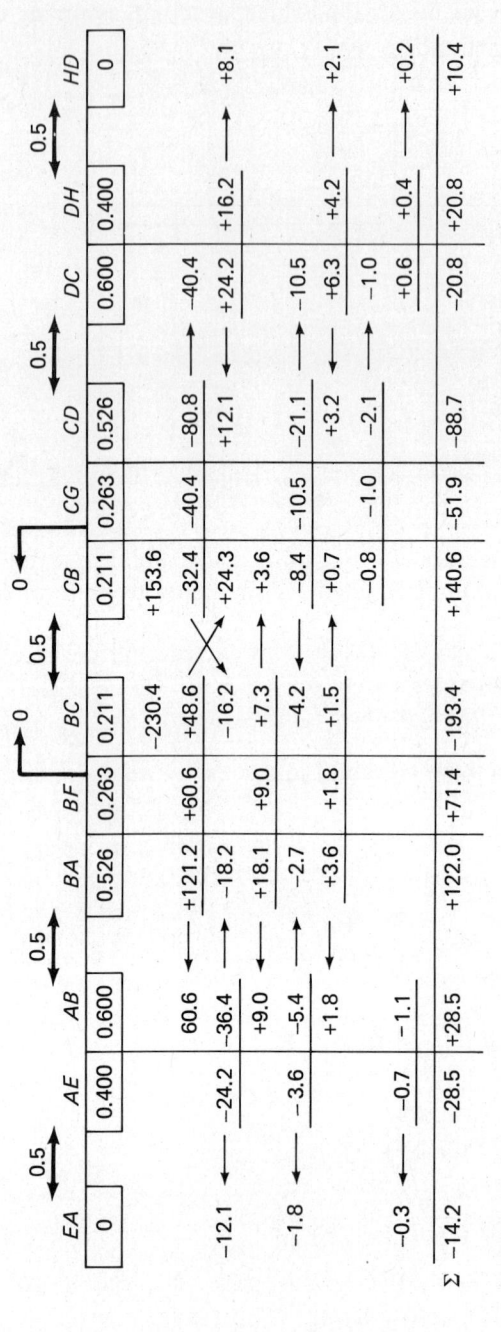

272 MOMENT DISTRIBUTION: STRUCTURES WITH JOINT TRANSLATION

	EA	AE	AB	BA	BF	BC	CB	CG	CD	DC	DH	HD
	0	0.400	0.600	0.526	0.263	0.211	0.211	0.263	0.526	0.600	0.400	0
		−24.2	60.6	+121.2	+60.6	−230.4 +48.6	+153.6 −32.4	−40.4	−80.8 +12.1	−40.4 +24.2	+16.2	+8.1
	−12.1		−36.4	−18.2		−16.2	+24.3					
		−3.6	+9.0	+18.1	+9.0	+7.3	+3.6	−10.5	−21.1 +3.2	−10.5 +6.3	+4.2	+2.1
	−1.8		−5.4	−2.7		−4.2	−8.4					
		−0.7	+1.8	+3.6	+1.8	+1.5	+0.7	−1.0	−2.1	−1.0 +0.6	+0.4	+0.2
	−0.3		−1.1			−0.8						
Σ	−14.2	−28.5	+28.5	+122.0	+71.4	−193.4	+140.6	−51.9	−88.7	−20.8	+20.8	+10.4

Figure 12.18

SIDESWAY INDUCED FIXED END MOMENTS; MEMBERS WITH FAR END PINNED 273

The resulting end moments for sidesway prevented are displayed.

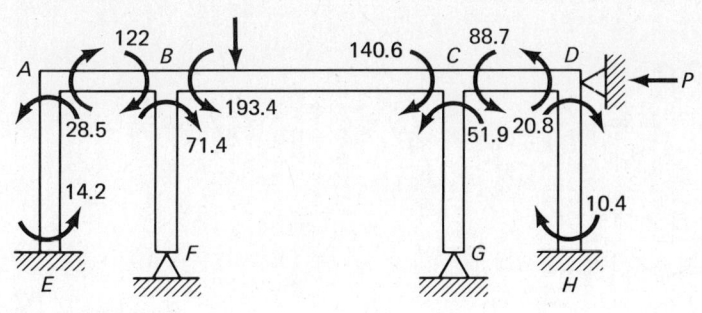

Figure 12.19

The end shears at the supports must be calculated in order to sum forces horizontally and to determine the force P that prevents sidesway.

For AE,

$$+\circlearrowleft \Sigma M_A = 6V_{EA} - 14.2 - 28.5 = 0$$
$$V_{EA} = 7.1 \text{ kN}$$

Figure 12.20

For BF,

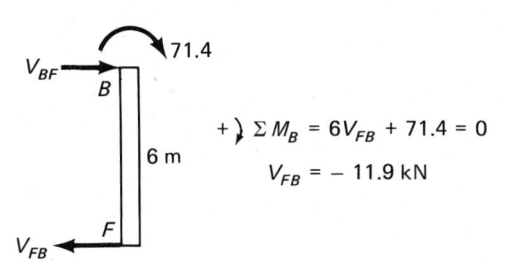

$$+\circlearrowleft \Sigma M_B = 6V_{FB} + 71.4 = 0$$
$$V_{FB} = -11.9 \text{ kN}$$

Figure 12.21

For CG,

$$+\circlearrowleft \Sigma M_C = 6V_{GC} - 51.9 = 0$$
$$V_{GC} = 8.6 \text{ kN}$$

Figure 12.22

For *DH*,

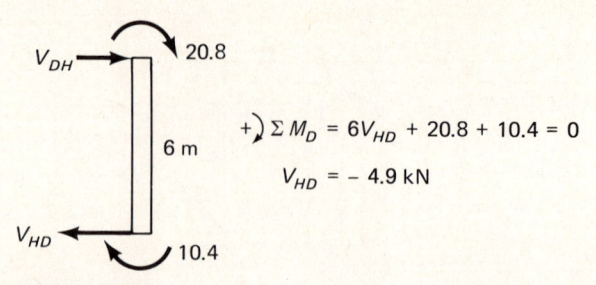

$$+\circlearrowright \Sigma M_D = 6V_{HD} + 20.8 + 10.4 = 0$$
$$V_{HD} = -4.9 \text{ kN}$$

Figure 12.23

From a free-body diagram showing only the horizontal forces, we apply $\Sigma F_x = 0$.

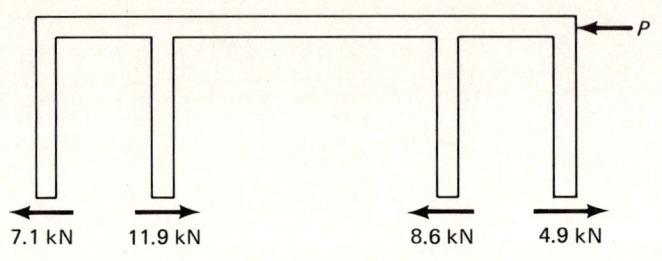

Figure 12.24

$$\pm \Sigma F_x = -7.1 + 11.9 - 8.6 + 4.9 - P = 0$$
$$P = 1.1 \text{ kN}$$

Now we induce sidesway.

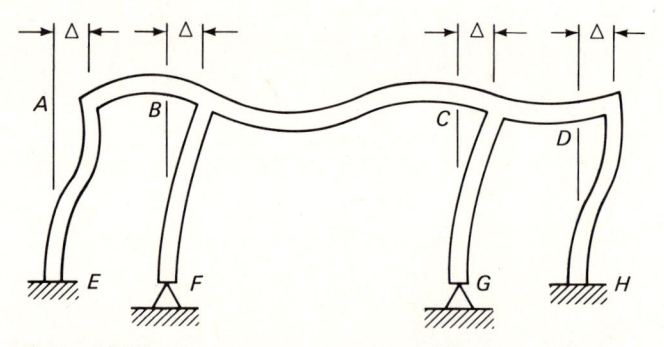

Figure 12.25

Assuming no changes in member lengths, we conclude that each joint may move only horizontally, each the distance Δ.
The sidesway induced fixed end moments are

$$\text{FEM}_{AE} = \text{FEM}_{EA} = \text{FEM}_{DH} = \text{FEM}_{HD} = -\frac{6EI}{6^2}\Delta = -\frac{EI\Delta}{6}$$

SIDESWAY INDUCED FIXED END MOMENTS; MEMBERS WITH FAR END PINNED

$$\text{FEM}_{BF} = \text{FEM}_{CG} = -\frac{3EI}{6^2}\Delta = -\frac{EI\Delta}{12}$$

$$\text{FEM}_{FB} = \text{FEM}_{GC} = 0$$

Those for *BF* and *CG* were based on the far end pinned formulation, Eq. (12.2), the others on Eq. (12.1). We assign a convenient value of $EI\Delta$, say 600, to give manageable values.

$$\text{FEM}_{AE} = \text{FEM}_{EA} = \text{FEM}_{DH} = \text{FEM}_{HD} = -100$$

$$\text{FEM}_{BF} = \text{FEM}_{CG} = -50$$

$$\text{FEM}_{FB} = \text{FEM}_{GC} = 0$$

The unloaded structure is symmetric, a fact confirmed by the distribution factors. The sidesway fixed end moments are antisymmetric, since they have equal values and the same signs at image points. Thus member *BC* is antisymmetric. Recognizing that there will be a sizable savings in effort, we recalculate the joint *B* distribution factors, intending to work with just one half of the structure.

$$D_{BA} = \frac{I/4}{I/4 + \frac{3}{2}(I/10) + \frac{3}{4}(I/6)} = 0.476$$

$$D_{BC} = \frac{\frac{3}{2}(I/10)}{I/4 + \frac{3}{2}(I/10) + \frac{3}{4}(I/6)} = 0.286$$

$$D_{BF} = \frac{\frac{3}{4}(I/6)}{I/4 + \frac{3}{2}(I/10) + \frac{3}{4}(I/6)} = 0.238$$

The distribution table is

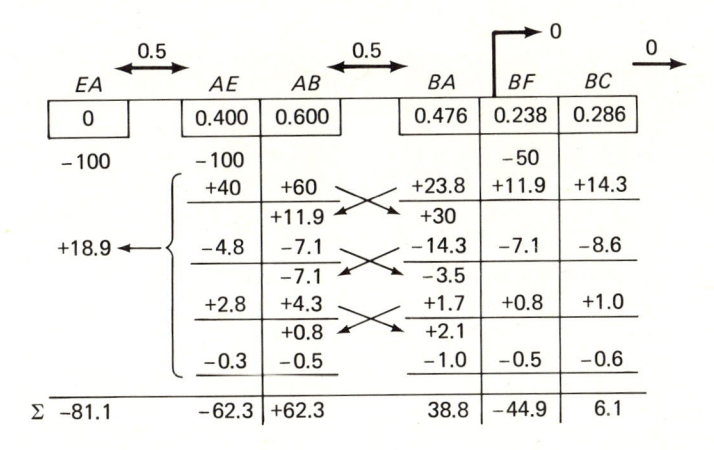

	EA	AE	AB	BA	BF	BC
	0	0.400	0.600	0.476	0.238	0.286
	−100	−100		−50		
		+40	+60	+23.8	+11.9	+14.3
			+11.9	+30		
+18.9		−4.8	−7.1	−14.3	−7.1	−8.6
			−7.1	−3.5		
		+2.8	+4.3	+1.7	+0.8	+1.0
			+0.8	+2.1		
		−0.3	−0.5	−1.0	−0.5	−0.6
Σ −81.1		−62.3	+62.3	38.8	−44.9	6.1

(Arrows: 0.5 between EA and AE; 0.5 between AB and BA; 0 at BF; 0 at BC)

Figure 12.26

The resulting end moments for sidesway (not yet scaled) are displayed, where equal moments with the same signs are taken at image locations.

276 MOMENT DISTRIBUTION: STRUCTURES WITH JOINT TRANSLATION

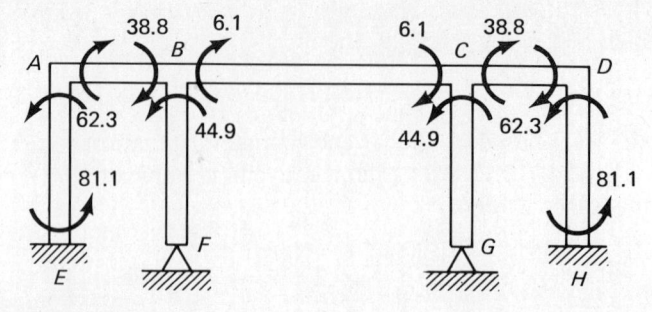

Figure 12.27

Two end shears are calculated.
For AE,

Figure 12.28

$$+\circlearrowleft \sum M_A = 6V_{EA} - 81.1 - 62.3 = 0$$
$$V_{EA} = 23.9 \text{ kN}$$

For BF,

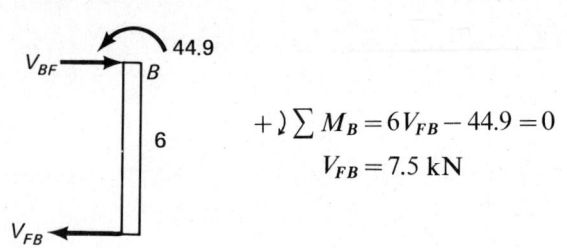

$$+\circlearrowleft \sum M_B = 6V_{FB} - 44.9 = 0$$
$$V_{FB} = 7.5 \text{ kN}$$

Figure 12.29

so that for horizontal forces,

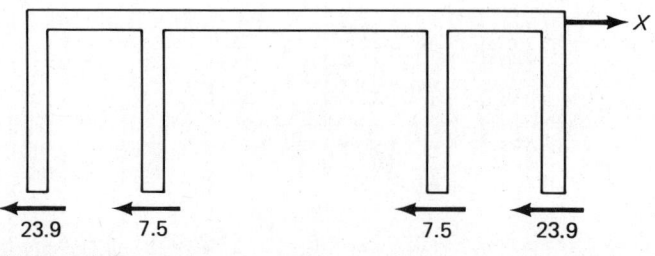

Figure 12.30

$$\pm \sum F_x = X - 23.9 - 7.5 - 7.5 - 23.9 = 0$$
$$X = 62.8 \text{ kN}$$

SIDESWAY INDUCED FIXED END MOMENTS; MEMBERS WITH FAR END PINNED 277

Since the sideway prevented part required a horizontal force $P=1.1$ kN to prevent sidesway, we scale the sidesway end moments to a value $X = 1.1$. Thus

$$M_{EA} = M_{HD} = -81.1\left(\frac{1.1}{62.8}\right) = -1.4 \text{ kN} \cdot \text{m}$$

$$M_{AE} = M_{DH} = -M_{AB} = -M_{DC} = -62.3\left(\frac{1.1}{62.8}\right) = -1.1 \text{ kN} \cdot \text{m}$$

$$M_{BA} = M_{CD} = +38.8\left(\frac{1.1}{62.8}\right) = +0.7$$

$$M_{BF} = M_{CG} = -44.9\left(\frac{1.1}{62.8}\right) = -0.8 \text{ kN} \cdot \text{m}$$

$$M_{BC} = M_{CB} = 6.1\left(\frac{1.1}{62.8}\right) = 0.1 \text{ kN} \cdot \text{m}$$

The sidesway prevented and induced parts are superposed.

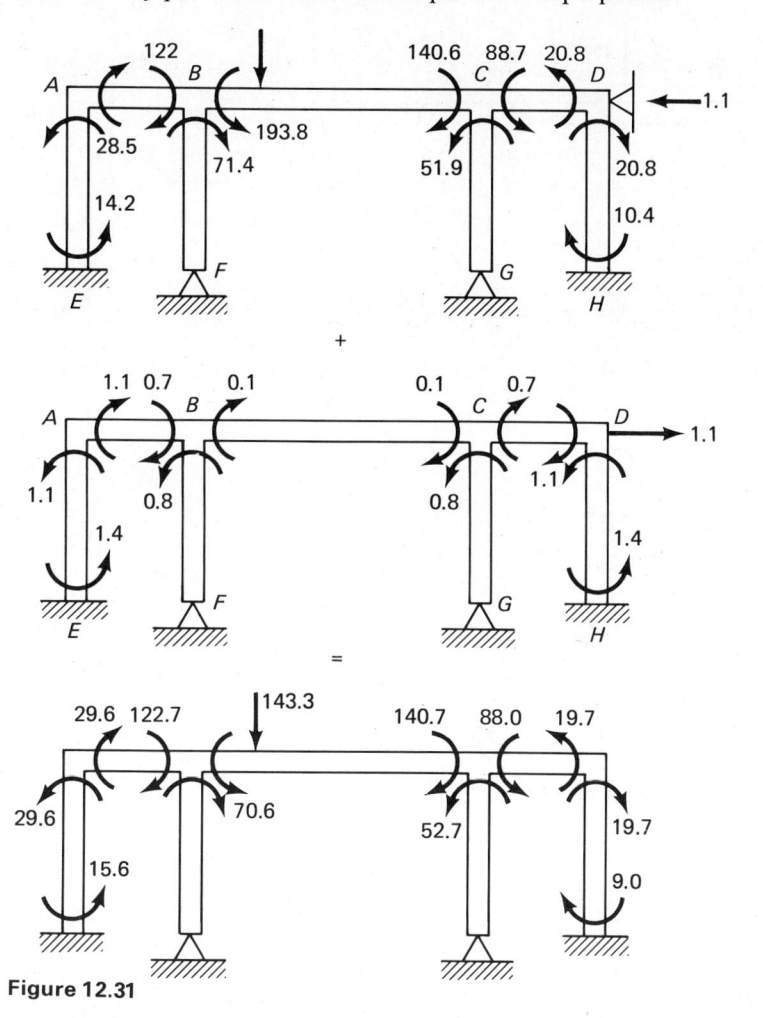

Figure 12.31

278 MOMENT DISTRIBUTION: STRUCTURES WITH JOINT TRANSLATION

A geometrically symmetric structure that is loaded symmetrically will not require a force to prevent sidesway. Hence there would be no need for the sidesway induced part in the analysis. But do not be misled. If the structure has both symmetric loading and symmetric distribution factors, there still may be sidesway if the structure is not geometrically symmetric. A sidesway analysis is then necessary. Consider the following example.

EXAMPLE 12.4.2

Determine the end moments of $ABCD$.

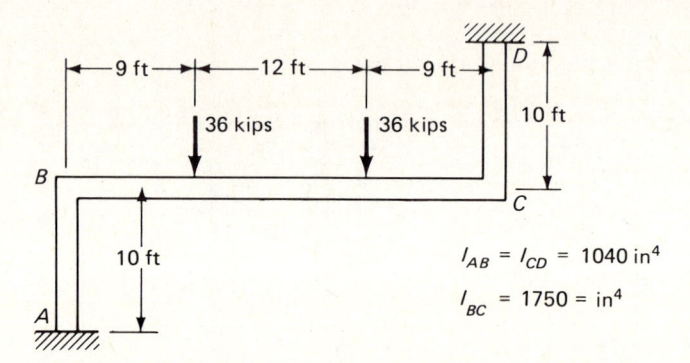

Figure 12.32

$I_{AB} = I_{CD} = 1040\ in^4$

$I_{BC} = 1750 = in^4$

We consider first the end moments for sidesway prevented, presuming a lateral brace at C. The load related fixed end moments, from Table 10.1, are

$$\text{FEM}_{BC} = -\text{FEM}_{CB} = -\frac{36(9)(21)}{30}$$

$$= -226.8\ \text{kip-ft}$$

They are symmetric. It is easily seen that the distribution factors at B will match those at C, verifying that the modified stiffness for symmetry may be used for member BC. We calculate the distribution factors at B using the symmetric modified stiffness.

$$D_{BC} = \frac{\frac{1}{2}(1750/30)}{\frac{1}{2}(1750/30) + 1040/10} = 0.219$$

$$D_{BA} = \frac{1040/10}{\frac{1}{2}(1750/30) + 1040/10} = 0.781$$

Now the distribution table for the left half of the structure is established and completed.

SIDESWAY INDUCED FIXED END MOMENTS; MEMBERS WITH FAR END PINNED 279

```
           0.5                0
   AB   ←─────    BA     BC   ────→
    0             0.781  0.219
                        -226.8
  +88.5  ←──── +177.1   +49.7
  ─────────────────────────────
Σ +88.5        +177.1  -177.1
```

Figure 12.33

The loaded structure, with sideswayprevented, is sketched with its end moments, having used symmetry to establish those for the right half.

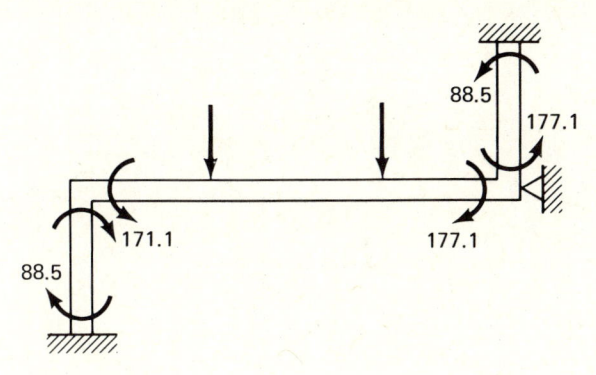

Figure 12.34

The end shears at A and D are calculated.
For AB,

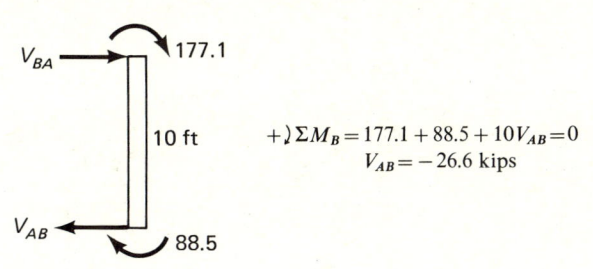

$$+\circlearrowleft \Sigma M_B = 177.1 + 88.5 + 10 V_{AB} = 0$$
$$V_{AB} = -26.6 \text{ kips}$$

Figure 12.35

For CD,

$$+\circlearrowright \Sigma M_C = 88.5 + 177.1 - 10 V_{DC} = 0$$
$$V_{DC} = -26.6 \text{ kips}$$

Figure 12.36

280 MOMENT DISTRIBUTION: STRUCTURES WITH JOINT TRANSLATION

Now the free-body diagram is drawn, showing only the horizontal forces.

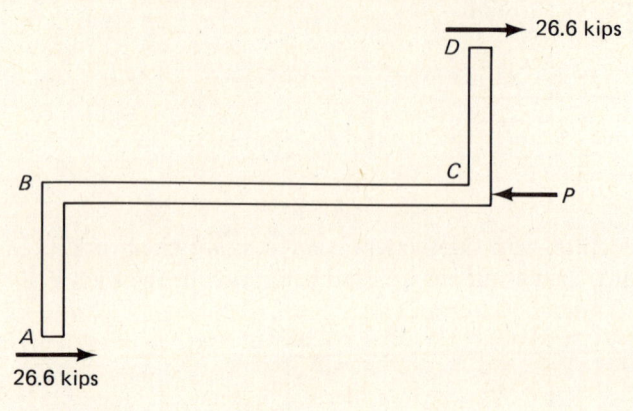

Figure 12.37

From which

$$\pm\sum F_x = 26.6 + 26.6 - P = 0$$
$$P = 53.2 \text{ kips}$$

Thus 53.2 kips directed to the left is required to prevent sidesway.
Now we induce sidesway, as suggested by the broken line.

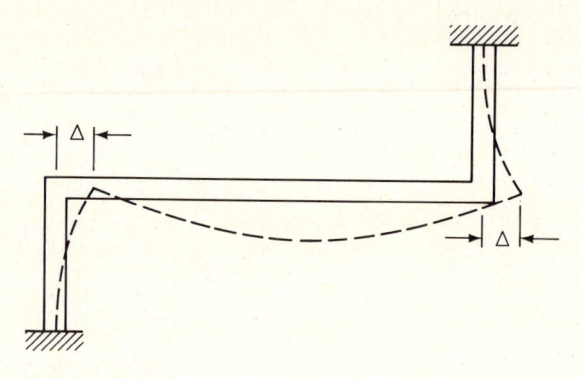

Figure 12.38

Only horizontal displacements of B and C are possible, each assumed to be Δ. The sidesway induced fixed end moments are

$$\text{FEM}_{AB} = \text{FEM}_{BA} = -\frac{6E(1040)}{10^2}\Delta$$

$$\text{FEM}_{CD} = \text{FEM}_{DC} = -\frac{6E(1040)}{10^2}(-\Delta)$$

They will all have the same value, and may be assumed to be any convenient value. But they will have opposite signs between those for AB and those for *CD*. We assume then that

$$FEM_{AB} = FEM_{BA} = -100 \text{ kip-ft}$$
$$FEM_{CD} = FEM_{DC} = +100 \text{ kip-ft}$$

Thus these fixed end moments are symmetric, too, permitting use of the same distribution factors as before. The distribution table is established and completed.

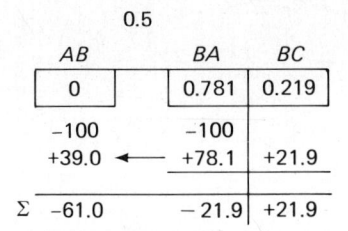

	0.5	
AB	BA	BC
0	0.781	0.219
−100	−100	
+39.0 ←	+78.1	+21.9
Σ −61.0	−21.9	+21.9

Figure 12.39

The sidesway induced structure with end moments is

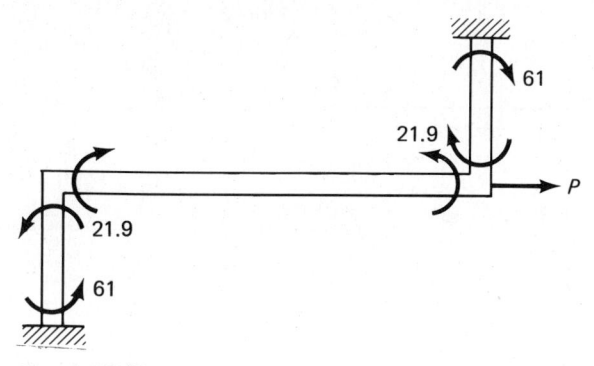

Figure 12.40

The end shears at *A* and *D* are calculated as before, both with values of 8.29 kips, and both to the left, from which the force inducing sidesway is 16.6 kips to the left. We scale to match the force found that prevents sidesway, 53.2 kips, giving

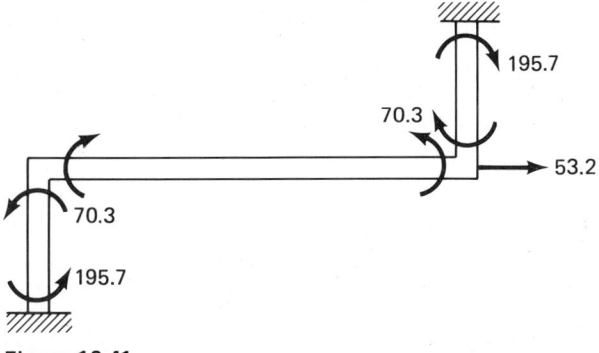

Figure 12.41

Superposing this with the sidesway prevented part,

Figure 12.42

gives

Figure 12.43

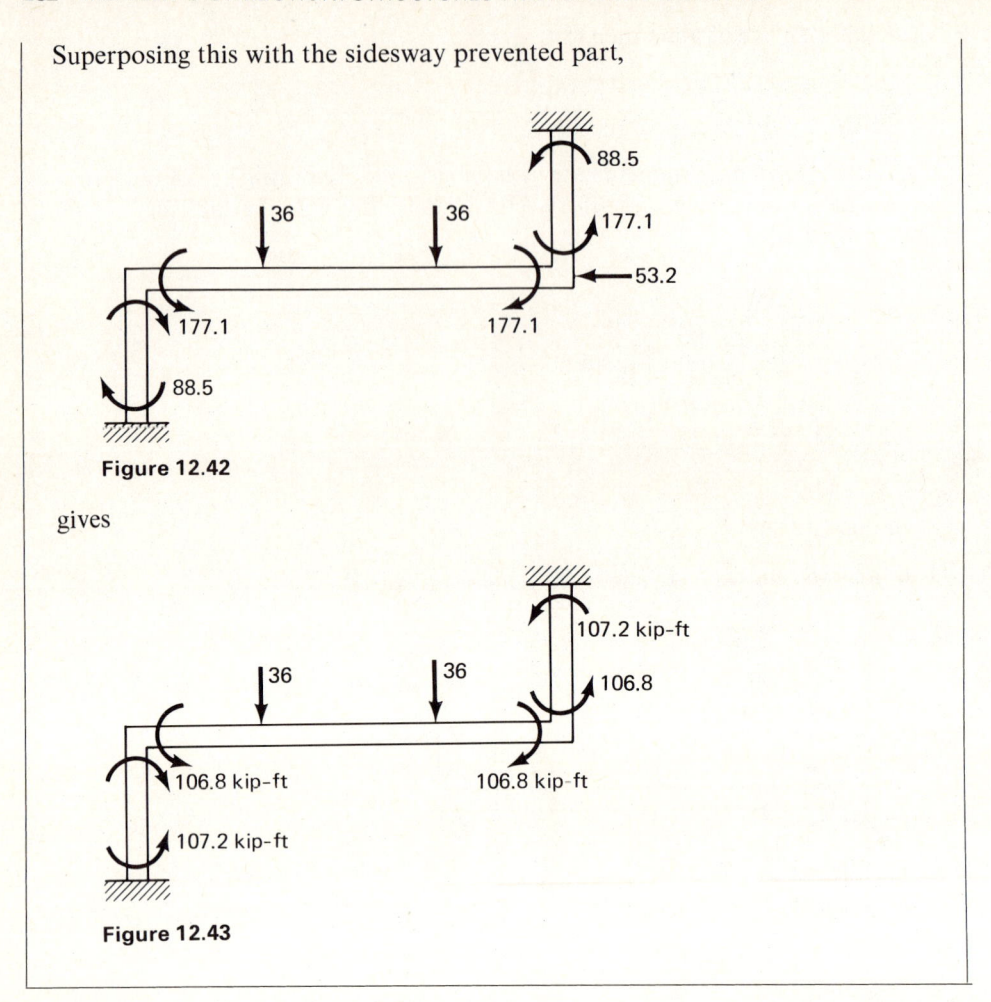

12.5 MULTIPLE DEGREES OF SIDESWAY

The procedures used for a single degree of sidesway are readily extended to multiple degrees of sidesway. The loaded structure, artificially braced against sidesway, is analyzed, and the forces that prevent each sidesway are calculated. Then each sidesway is induced separately. In each, end moments and end shear forces are found that are used to calculate the force causing the induced sidesway as well as those forces that prevent other sidesways. As a calculational expedient, each of the forces causing a sidesway is scaled to an algebraic value $X_i (i = 1, 2, \ldots, n)$, and the end moments and the forces that prevent the other sidesways are scaled to match.

We impose the restriction that all the forces acting at one joint, either inducing or preventing sidesway, must have a total of zero. That serves as the basis for writing a set of simultaneous equations solved for the unknown sidesway forces X_i. Then the end moments for each sidesway case are rescaled to these values, and all are superposed along with those for sidesway prevented.

The following example illustrates the procedure for a two-degree sideway structure.

EXAMPLE 12.5.1

Determine the end moments for the structure shown.

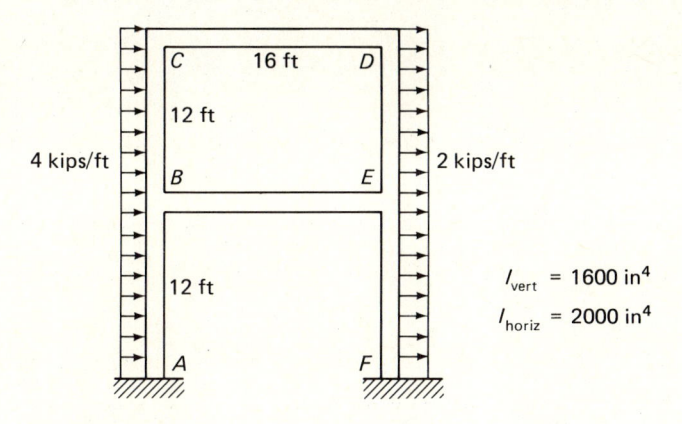

$I_{vert} = 1600\ in^4$
$I_{horiz} = 2000\ in^4$

Figure 12.44

First, for sideway prevented, we find the distribution factors by using regular stiffnesses since there are no applicable modified stiffnesses.

$$D_{ED} = D_{EF} = D_{BC} = D_{BA} = \frac{1600/12}{(1600/12)(2) + 2000/16} = 0.340$$

$$D_{EB} = D_{BE} = \frac{2000/16}{(1600/12)(2) + 2000/16}$$

$$D_{CD} = D_{DC} = \frac{2000/16}{2000/16 + 1600/12} = 0.484$$

$$D_{CB} = D_{DE} = \frac{1600/12}{2000/16 + 1600/12} = 0.516$$

The fixed end moments are

$$\text{FEM}_{AB} = \text{FEM}_{BC} = -\text{FEM}_{BA} = -\text{FEM}_{CB} = -\frac{4(12^2)}{12} = -48\ \text{kip-ft}$$

$$\text{FEM}_{FE} = \text{FEM}_{ED} = -\text{FEM}_{EF} = -\text{FEM}_{DE} = -\frac{2(12^2)}{12} = -24\ \text{kip-ft}$$

The distribution table follows.

284 MOMENT DISTRIBUTION: STRUCTURES WITH JOINT TRANSLATION

Figure 12.45

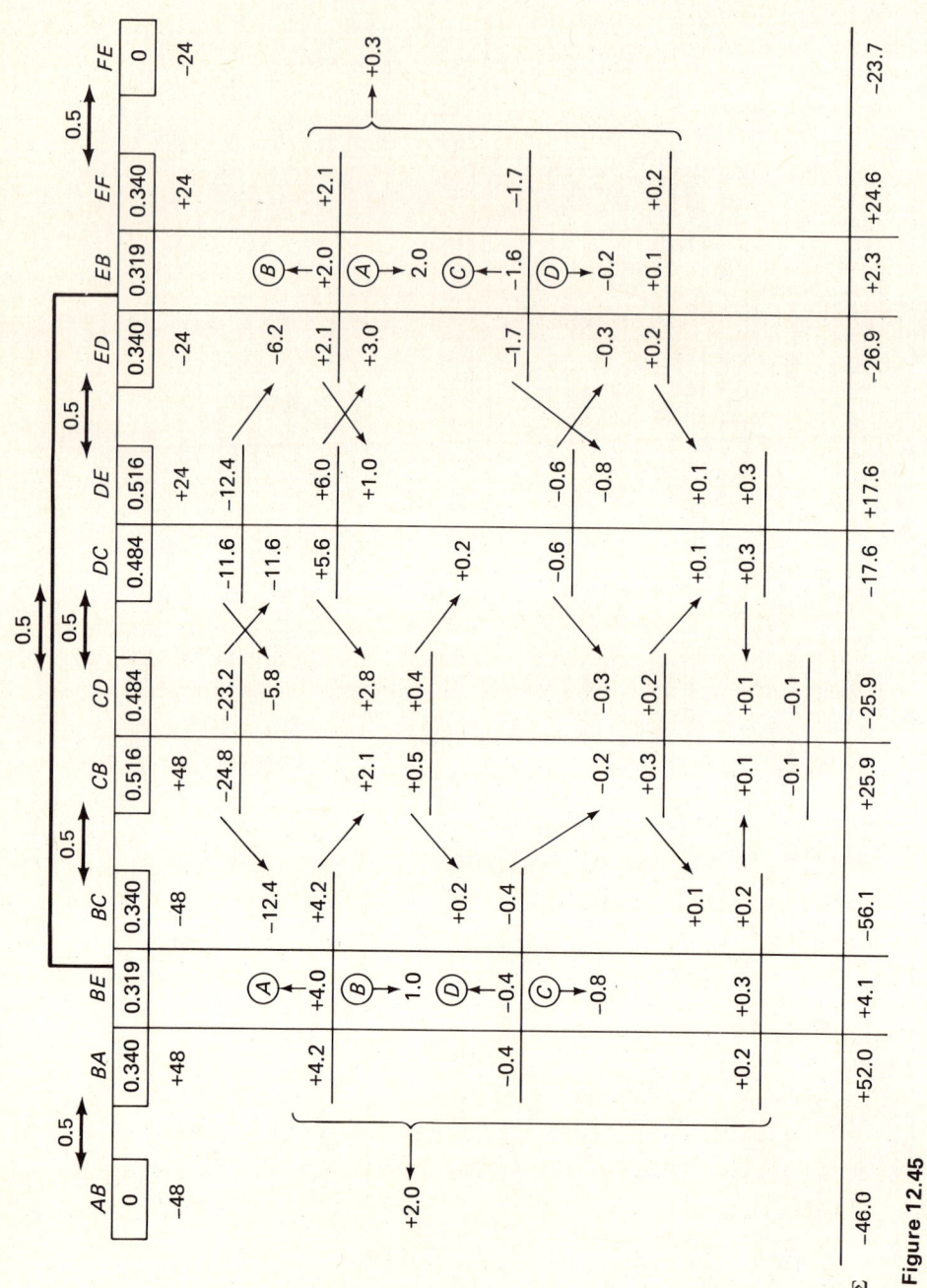

The arrangement of the table, particularly with regard to member EB, is worthy of close examination. End moment columns BE and EB are not adjacent to each other in the table since it is not possible to arrange adjacent far end columns for more than two members connected at a joint. With three members connected at B and E, one member at each joint must have its far end located remotely on the table. The carry-overs are indicated by the line across the top of the table joining columns BE and EB. It shows the carry-over factors with the arrows next to the line. The analyst keeps track of what carry-overs have been completed by using small arrows within the table. These arrows point away from a distribution and toward a carry-over moment, with matching circled letters to show corresponding parts.

With the end moments known we can calculate certain end shears, anticipating the summations of horizontal forces. For the shear at the bottom of BC we draw the free-body diagram of BC.

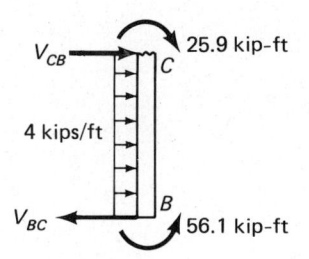

Figure 12.46

$$+\circlearrowleft \sum M_C = 12 V_{BC} + 25.9 - 56.1 - 4(12)(6) = 0$$
$$V_{BC} = 26.5 \text{ kips}$$

A similar process gives the shear at the bottom of DE as

$$V_{ED} = +12.8 \text{ kips}$$

so that we can now sketch a free-body diagram of the top half of the structure and sum forces horizontally.

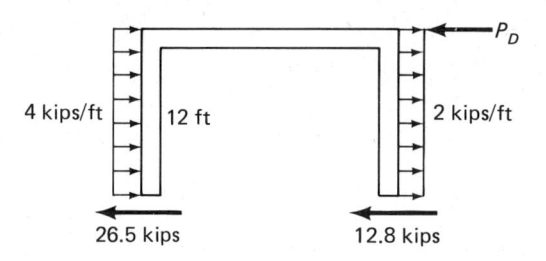

Figure 12.47

$$\pm \sum F_x = 4(12) - 26.5 + 2(12) - 12.8 - P_D = 0$$
$$P_D = 32.7 \text{ kips}$$

where force $P_D = 32.7$ kips is the force preventing sidesway at D.

The same process is used with member AB to get the shear at its bottom. Thus

$V_{AB} = 23.5$ kips

Again, for member EF we get

$V_{EF} = 11.9$ kips

so we can now draw the free-body diagram of the whole structure (showing only the horizontal forces) and solve for P_E.

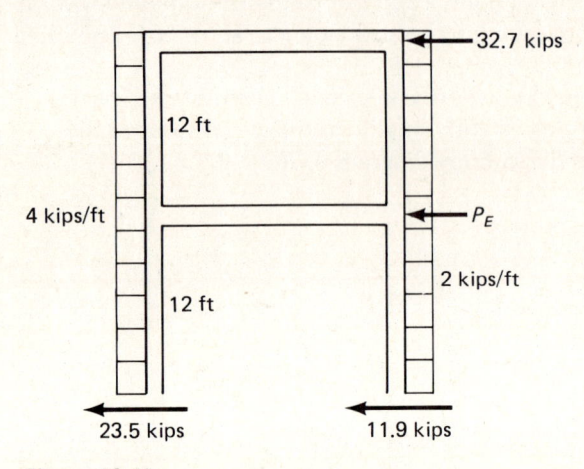

Figure 12.48

$$+\sum F_x = 4(24) + 2(24) - 23.5 - 11.9 - 32.7 - P_E = 0$$
$$P_E = 75.9 \text{ kips}$$

where P_E is the force preventing sidesway at E.

Now we are ready to consider sidesway. This will be done in two parts: the horizontal motion of CD while BE is not displaced, and the horizontal motion of BE while CD is not displaced.

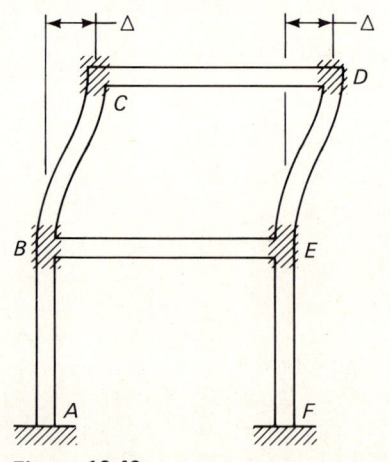

Figure 12.49

MULTIPLE DEGREES OF SIDESWAY

The diagram shows CD displaced to the right a distance Δ. Joints B, C, D, and E are locked against rotation. The sidesway induced fixed end moments for BC, CB, DE, and ED will all be the same and will all have the same sign. Thus all image locations will have equal fixed end moments with the same sign. Since the distribution factors found previously were symmetric, we conclude that antisymmetry exists for members CD and BE. We recalculate the distribution factors to account for the antisymmetry, effectively cutting in half the distribution table and the effort required to complete it.

$$D_{BC} = D_{BA} = \frac{1600/12}{1600/12 + 1600/12 + \frac{3}{2}(2000)/16}$$

$$= 0.294$$

$$D_{BE} = \frac{\frac{3}{2}(2000)/16}{1600/12 + 1600/12 + \frac{3}{2}(2000)/16}$$

$$= 0.413$$

$$D_{CD} = \frac{\frac{3}{2}(2000)/16}{\frac{3}{2}(2000)/16 + 1600/12}$$

$$= 0.584$$

$$D_{CB} = \frac{1600/12}{\frac{3}{2}(2000)/16 + 1600/12}$$

Having recognized that all the sidesway induced fixed end moments are equal and have the same sign, we are free to assume any value for them. For convenience we let

$$\text{FEM}_{CB} = \text{FEM}_{BC} = -100 \text{ kip-ft}$$

establish the distribution table and complete it.

AB		BA	BE	BC		CB	CD	
	0.5 →			← 0	0.5 →		0 →	
0		0.294	0.413	0.294		0.584	0.416	
				−100		−100		
		+29.4	+41.2	+29.4	↘	+58.4	+41.6	
				+29.2	↗	+14.7		
+10.9 ←		−8.6	−12.0	−8.6	↘	−8.6	−6.1	
				−4.3	↗	−4.3		
		+1.3	+1.7	+1.3	↘	+2.5	+1.8	
				+1.2	↗	+0.6		
		−0.4	−0.4	−0.4		−0.4	−0.2	
Σ +10.9		+21.7	+30.5	−52.2		−37.1	+37.1	

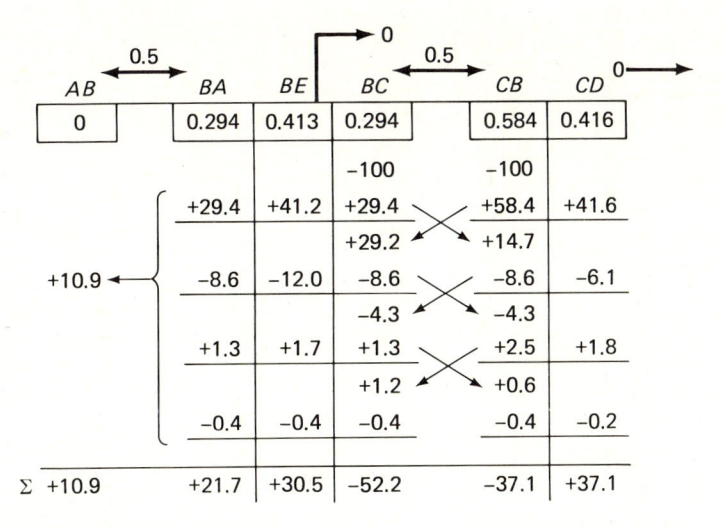

Figure 12.50

Following the procedures used before, that is, isolating individual vertical members and calculating their end shears, we get

$$V_{BC} = V_{ED} = 7.44 \text{ kips}$$

so that for the upper story

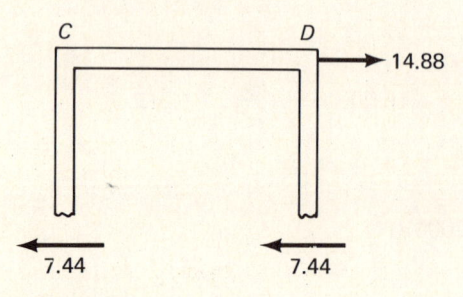

Figure 12.51

We continue to evaluate end shears

$$V_{AB} = V_{FE} = -2.72 \text{ kips}$$

Thus the complete structure has the horizontal forces shown.

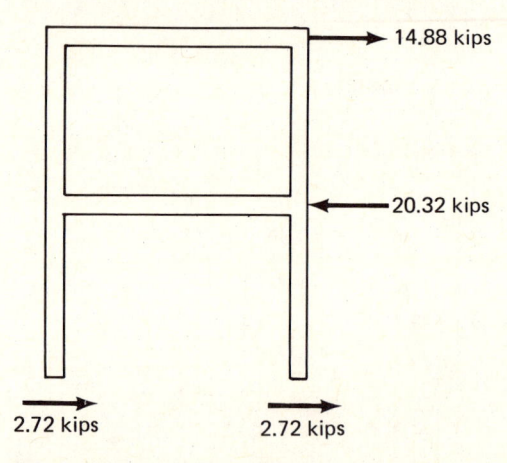

Figure 12.52

Now we scale the horizontal force at D to a value X_1 by multiplying it by $X_1/14.88$, and scale all the other forces and end moments by the same ratio. Thus for sidesway of CD we get

MULTIPLE DEGREES OF SIDESWAY

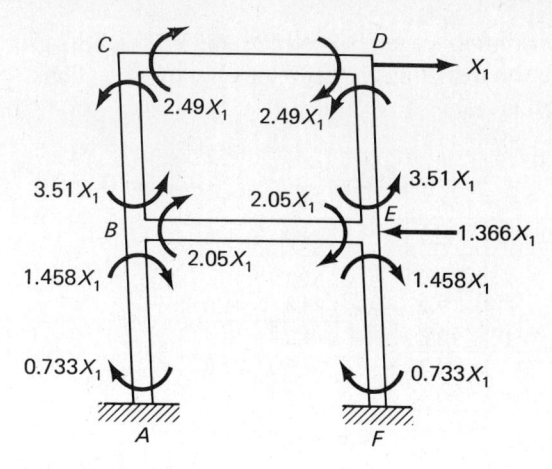

Figure 12.53

Later we will evaluate X_1 and use that value to calculate the actual end moments resulting from the sidesway of CD.

Next we introduce the second sidesway, that of BE. In this, further sidesway of CD will be prevented. Thus the assumed mode of displacement is

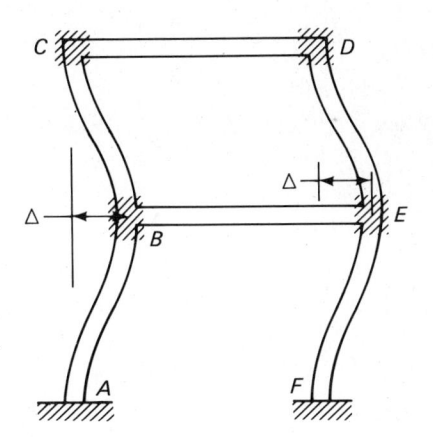

Figure 12.54

The fixed end moments will all have the same value

$$-\frac{6E(1600)}{12^2}\Delta$$

except that $+\Delta$ will apply to members AB and EF and $-\Delta$ will apply to members BC and DE, since for the latter two Δ relates to a counter-clockwise rotation of the members. We assume the convenient value of 100 kip-ft so that the FEM for AB, BA, EF, and FE are $+100$ and those for CB, BC, DE, and ED are -100 kip-ft.

Evidently we have antisymmetry again since the same value and sign occur at image points while the distribution factors are symmetric. Thus we use the same distribution factors.

	0.5				0.5		0
AB		BA	BE	BC		CB	CD
0		0.294	0.413	0.294		0.584	0.416
−100		−100		+100		+100	
				−29.2	←	−58.4	−41.6
		+8.6	+12.0	+8.6	→	+4.3	
				−1.2	←	−2.5	−1.8
+4.5		+0.4	+0.4	+0.4			
Σ −95.5		−91.0	+12.4	+78.6		+43.4	−43.4

Figure 12.55

The evaluation of end shears and the calculation of horizontal forces proceeds exactly as it did for the sidesway of CD. Thus for the top half

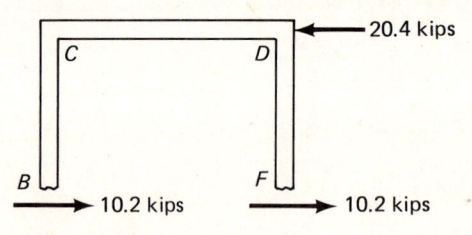

[Figure 12.56: Frame showing C, D at top with 20.4 kips arrow at D; B, F at bottom with 10.2 kips arrows at B and F]

Figure 12.56

and for the whole structure

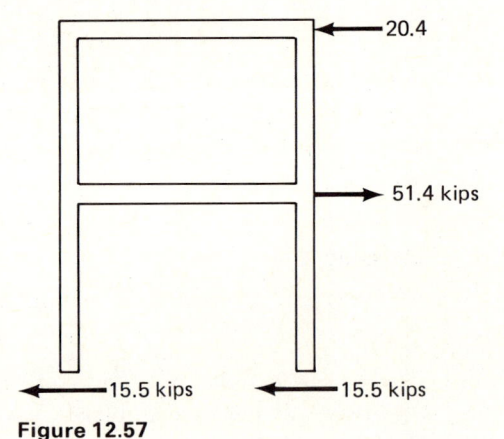

[Figure 12.57: Two-story frame with 20.4 arrow at top right, 51.4 kips arrow at middle right, and 15.5 kips arrows at bottom left corners]

Figure 12.57

As done for the sidesway of CE, the force causing the sidesway is scaled to a variable, in this case X_2. All forces and end moments are scaled to match by the factor $X_2/51.4$, giving

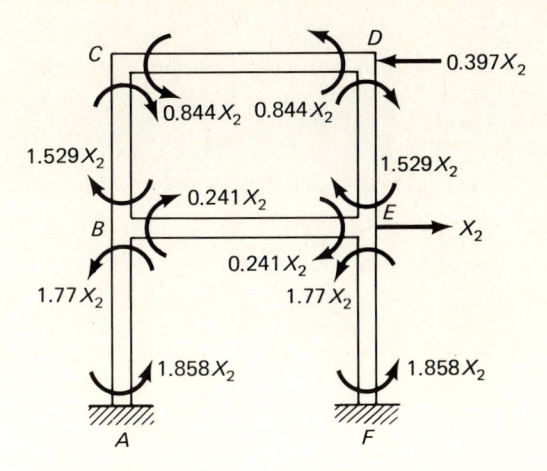

Figure 12.58

To prepare for superposing the three cases we note that at joint D we have the forces $\leftarrow 32.7$ kips, $\rightarrow X_1$ and $\leftarrow 0.397\, X_2$ from the original loading but sidesway prevented, sidesway of CD, and sidesway of BE, respectively. Because in actuality there can be no net force there, we let their sum equal zero. Thus

$$X_1 - 0.397\, X_2 - 32.7 = 0$$

Similarly, at joint E the three external forces must cancel. Thus

$$-1.366\, X_1 + X_2 - 75.9 = 0$$

Solving these two equations, we get

$X_1 = 137.2$ kips

$X_2 = 263.3$ kips

It is easiest to combine the three parts in a table, using a column for the end moments in each part (Table 12.1).

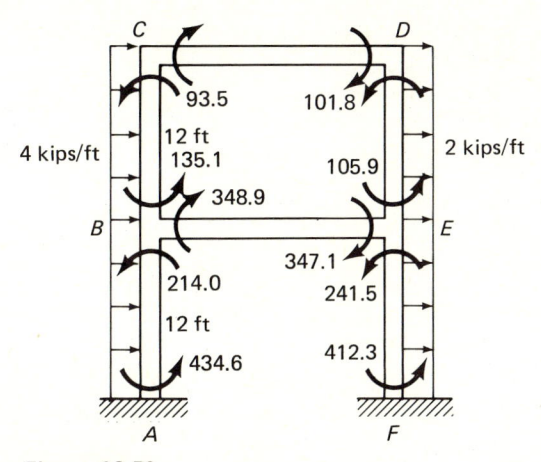

Figure 12.59

Table 12.1

Member	M Sidesway Prevented	M Sidesway of CD, $X_1 = 137.2$	M Sidesway of BE, $X_2 = 263.3$	M Total
AB	−46.0	+100.6	−489.2	−434.6
BA	+52.0	+200.0	−466.0	−214.0
BE	+4.1	+281.3	+63.5	+348.9
BC	−56.1	−481.6	+402.6	−135.1
CB	+25.9	−341.6	+222.2	−93.5
CD	−25.9	+341.6	−222.2	+93.5
DC	−17.6	+341.6	−222.2	+101.8
DE	+17.6	−341.6	+222.2	−101.8
ED	−26.9	−481.6	+402.6	−105.9
EB	+2.3	+281.3	+63.5	+347.1
EF	+24.6	+200.0	−466.0	−241.5
FE	−23.7	+100.6	−489.2	−412.3

The combined end moments from the three parts are shown in Figure 12.59.

PROBLEMS

Sequenced Problem 3, part (q).
Sequenced Problem 4, part (q).
Sequenced Problem 6, parts (c), (d), and (e).
Sequenced Problem 8, parts (c), (d), and (e).

12.1

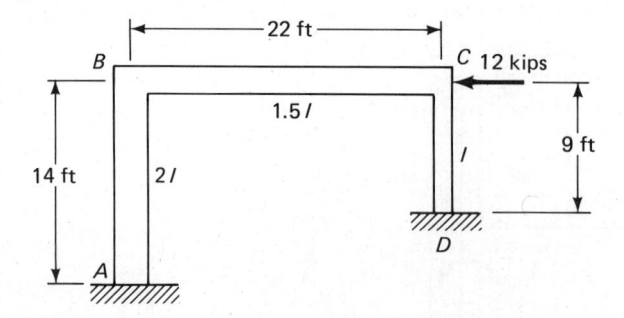

Figure P12.1

Use moment distribution to determine the end moments caused by the 12-kip force at C.

12.2 Solve Problem 10.2 by moment distribution.
12.3 Solve Problem 10.3 by moment distribution.

12.4

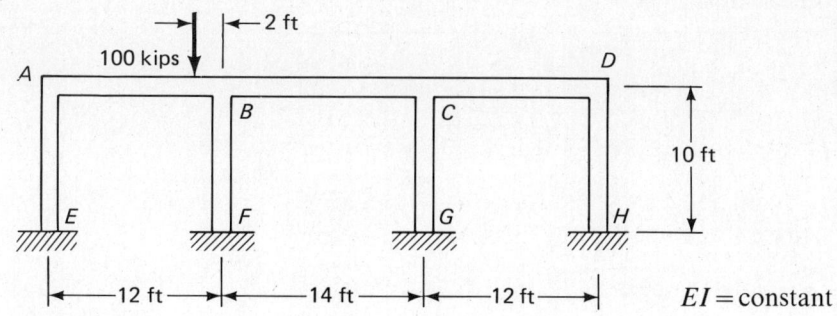

Figure P12.4

Determine the end moments using moment distribution. The sidesway prevented part for this is equivalent to Problem 11.9.

12.5 Use moment distribution to determine the end moments.

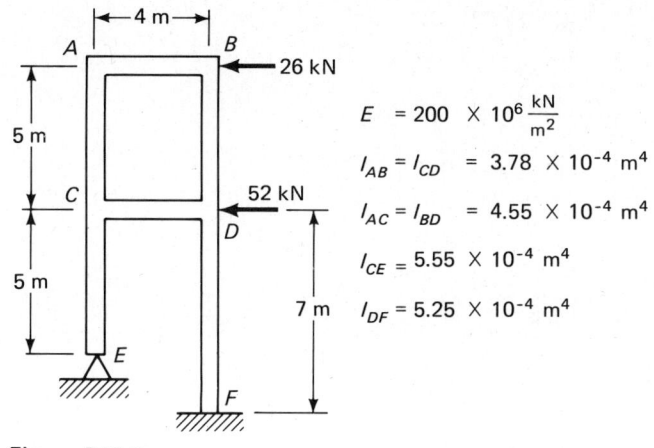

$E = 200 \times 10^6 \ \frac{kN}{m^2}$

$I_{AB} = I_{CD} = 3.78 \times 10^{-4} \ m^4$

$I_{AC} = I_{BD} = 4.55 \times 10^{-4} \ m^4$

$I_{CE} = 5.55 \times 10^{-4} \ m^4$

$I_{DF} = 5.25 \times 10^{-4} \ m^4$

Figure P12.5

12.6

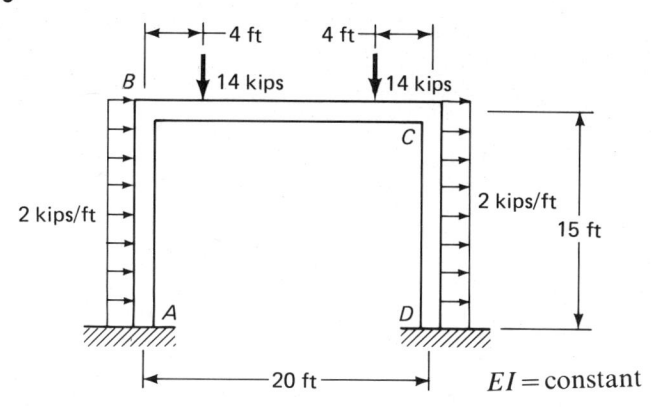

Figure P12.6

Use moment distribution to determine the end moments. *Suggestion:* For sidesway prevented, consider the external loading in two parts, one symmetric, the other antisymmetric.

Beams and Rigid Frames With Nonprismatic Members

The irascible professor had developed an ulcer that progressively worsened until he had been hospitalized for surgery. Several days into his recuperation he received a very cheery get well card from one of his classes. He enjoyed the printed sentiment and then read the written note: "By a vote of 19 to 16, your MWF 9:00 class wishes you well and expresses hope for your speedy return."[1]

---

[1] Having started each of the preceding chapters with an anecdote or cartoon, I'm afraid that it would break the pattern if I didn't start with one for this chapter, too. This story has nothing to do with nonprismatic members, but it does maintain my approach and you may even enjoy its message. If you can think of a better story, one relating to nonprismatic members, please send it to me. —*Author*

13.0 INTRODUCTION

Structures in the earlier chapters were composed almost exclusively of prismatic members, that is, members with the same cross section from end to end. But there are many structures that have one or more nonprismatic members, such as those illustrated below.

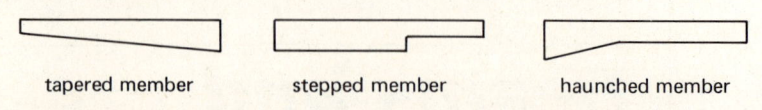

tapered member   stepped member   haunched member

Figure 13.1

This chapter develops methods for calculating the rotational stiffnesses, carry-over factors, and fixed end moments for nonprismatic members, and shows how the slope deflection and moment distribution methods are applied in structures that have nonprismatic members.

13.1 PUBLISHED PROPERTIES FOR NONPRISMATIC MEMBERS

Rotational stiffnesses, carry-over factors, and fixed end moments for some nonprismatic members can be found in engineering literature. The *Handbook of Frame Constants*[2] is found in many civil engineering design offices. It lists a variety of properties for 1320 different members, including the fixed end moments for uniformly distributed loads and for an arbitrarily placed concentrated force. Because it was developed for use in the design of reinforced concrete beams, the *Handbook of Frame Constants* concentrates on beams with rectangular cross sections and offers little useful information for the typical I-shaped steel member cross section.

The *Structural Engineering Handbook*[3] gives a few curves for rotational stiffness coefficients and, with some interpretation, carry-over factors for tapered beams.

The most abundant source of information is *Die Cross-Methode*.[4] This book is a valuable reference even though it is in German. Much of its information may be extracted by non-German readers through its wealth of diagrams.

13.2 INTEGRAL FORMULAS FOR CARRY-OVER FACTORS, ROTATIONAL STIFFNESSES, AND FIXED END MOMENTS

Consider member AB with varying EI, supported by a clamp at B and roller at A, and loaded with clockwise couple S_{AB} at A which causes a unit rotation at A.

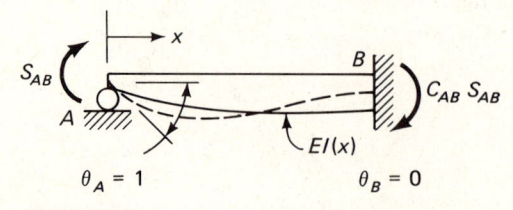

Figure 13.2

---

[2] *Handbook of Frame Constants*, Portland Cement Association, Old Orchard Road, Skokee, IL 60076, 1958.
[3] Gaylord, E. H. Jr. and Gaylord, C. N., *Structural Engineering Handbook*, 2nd ed., McGraw-Hill, New York, 1979, pp. 1–48.
[4] Guldan, R., *Die Cross-Methode*, Wien-Springer-Verlag, Vienna, Austria, 1955.

INTEGRAL FORMULAS FOR CARRY-OVER FACTORS

By definition S_{AB} is the rotational stiffness for end A. The clamped support at B exerts moment $C_{AB}S_{AB}$ in preventing rotation at B. By definition C_{AB} is the carry-over factor from A to B.

We will use conjugate beam analysis to evaluate S_{AB} and C_{AB}. We choose a simply supported beam for our primary structure and use superposition so that S_{AB} and $C_{AB}S_{AB}$ may be considered separately.

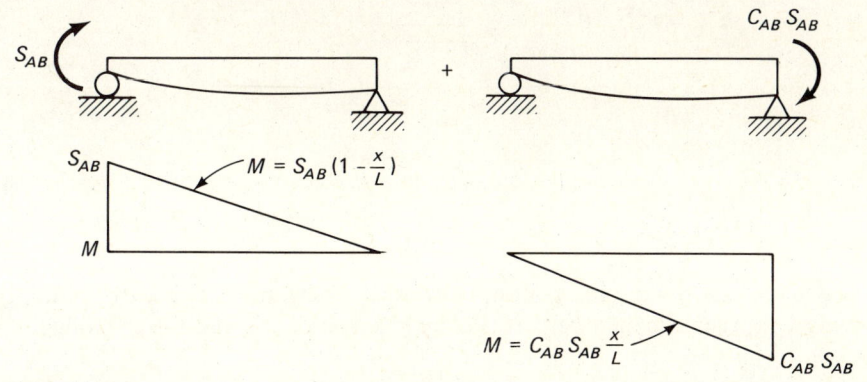

Figure 13.3

The two moment diagrams must be converted to M/EI diagrams for use as conjugate loads. The conversion is more complicated than for prismatic members since the conversion must account for the variation in EI.

Conceptually the two M/EI diagrams are shown.

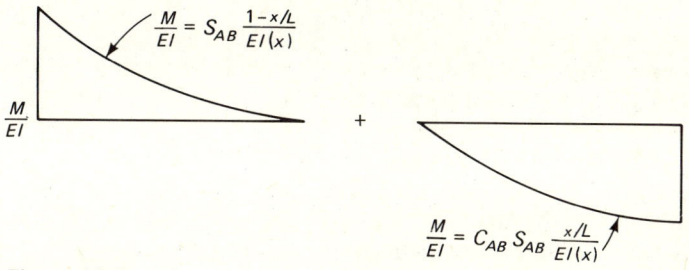

Figure 13.4

The shape of these curves will depend, of course, on how EI varies as a function of x for a particular beam.

The conjugate beam, with its loading, is drawn.

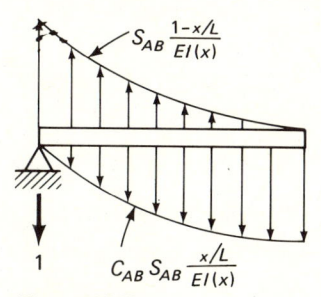

Figure 13.5

A unit downward conjugate reaction is assumed at A. It gives the conjugate beam the necessary unit negative conjugate shear at A that corresponds with the unit downward slope of the real beam there.

Equilibrium of the conjugate beam is necessary. Thus

$$+\circlearrowright\sum M_A = S_{AB}\int_0^L \frac{x(1-x/L)}{EI(x)}dx - C_{AB}S_{AB}\int_0^L \frac{x(x/L)}{EI(x)}dx = 0$$

Therefore

$$C_{AB} = \frac{\int_0^L \frac{x - x^2/L}{EI(x)}dx}{\int_0^L \frac{x^2/L}{EI(x)}dx} \qquad (13.1)$$

The two integrals in Eq. (13.1), along with one other, arise repeatedly in analyses involving nonprismatic members. It is convenient to define and use the following,

$$J_0 = \int_0^L \frac{1}{EI(x)}dx \qquad (13.2a)$$

$$J_1 = \int_0^L \frac{x}{EI(x)}dx \qquad (13.2b)$$

$$J_2 = \int_0^L \frac{x^2}{EI(x)}dx \qquad (13.2c)$$

Eq. (13.1) then may be written

$$C_{AB} = \frac{LJ_1}{J_2} - 1 \qquad (13.3)$$

Readers are encouraged to verify that Eq. (13.3) gives $C_{AB} = 0.5$ for $EI = $ constant.

We continue our use of equilibrium for the conjugate beam.

$$\sum F_y = S_{AB}\int_0^L \frac{1-x/L}{EI(x)}dx - S_{AB}C_{AB}\int_0^L \frac{x/L}{EI(x)}dx - 1 = 0$$

We substitute the expressions of Eqs. (13.2) and (13.3) to simplify this equilibrium equation. It becomes

$$S_{AB}J_0 - S_{AB}\frac{J_1}{L} - S_{AB}\left(\frac{LJ_1}{J_2} - 1\right)\frac{J_1}{L} = 1$$

which gives

$$S_{AB} = \frac{J_2}{J_0 J_2 - J_1^2} \qquad (13.4)$$

It is easily verified that Eq. (13.4) gives $S_{AB} = 4EI/L$ for a prismatic member.

A similar development gives S_{BA} and C_{BA}.

INTEGRAL FORMULAS FOR CARRY-OVER FACTORS

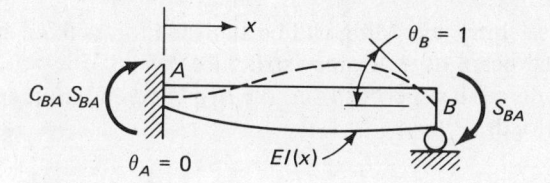

Figure 13.6

The conjugate beam with its loads is

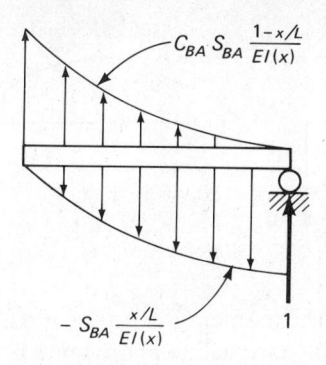

Figure 13.7

$$+\circlearrowleft \sum M_B = C_{BA} S_{BA} \int_0^L \frac{(L-x)(1-x/L)}{EI(x)} dx - S_{BA} \int \frac{(x/L)(L-x)}{EI(x)} dx = 0$$

which reduces to

$$C_{BA} = \frac{J_1 - J_2/L}{LJ_0 - 2J_1 + J_2/L} \tag{13.5}$$

$$+\uparrow \sum F_y = C_{BA} S_{BA} \int_0^L \frac{1-x/L}{EI(x)} dx - S_{BA} \int_0^L \frac{x/L}{EI(x)} dx + 1 = 0$$

which becomes

$$S_{BA} = \frac{J_2 - 2J_1 L + L^2 J_0}{J_0 J_2 - J_1^2} \tag{13.6}$$

Notice that evaluation of three integrals, those for J_0, J_1, and J_2, followed by their substitution in simple formulas is all that is necessary to determine both rotational stiffnesses and both carry-over factors.

Let us explore how we evaluate fixed end moments.

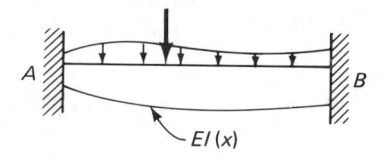

Figure 13.8

Nonprismatic member AB with its arbitrary loading will be analyzed for its fixed end moments using a simply supported beam as a primary structure.

Three separate parts are considered in superposition: the two fixed end moments and a part related to the external loads.

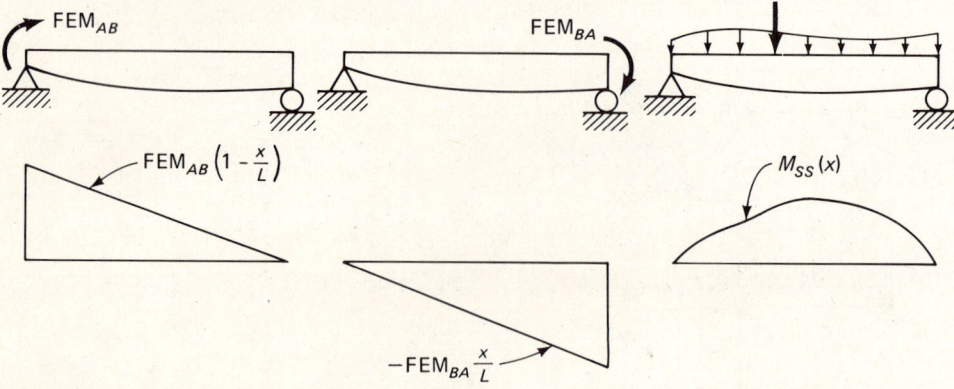

Figure 13.9

Their moment diagrams are shown, where $M_{SS}(x)$ is the moment due to the member loading, as evaluated for a *simply supported* beam. Since we will use conjugate beam analysis, the moment diagrams are converted to M/EI diagrams.

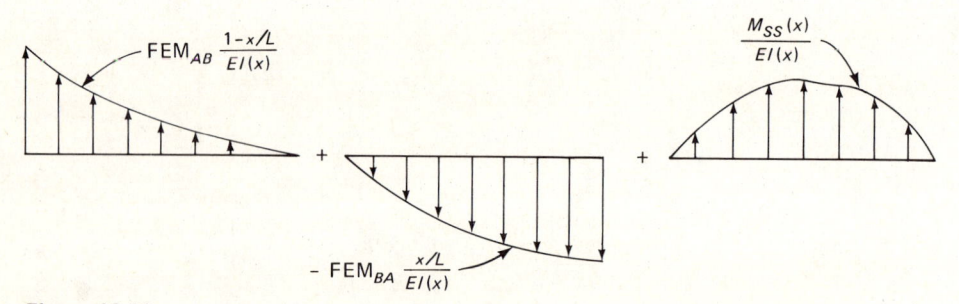

Figure 13.10

As before, their shapes depend on $EI(x)$ for the beam. The conjugate beam itself is *completely unsupported* so that its equilibrium must come from $\sum F_y = 0$ and $\sum M = 0$ for the conjugate loading *alone*. First summing forces, we get

$$\sum F_y = \text{FEM}_{AB} \int_0^L \frac{1-x/L}{EI(x)} dx - \text{FEM}_{BA} \int_0^L \frac{x/L}{EI(x)} dx + \int_0^L \frac{M_{SS}(x)}{EI(x)} dx = 0$$

Next, summing moments about A, we get

$$\sum M_A = \text{FEM}_{AB} \int_0^L \frac{x - x^2/L}{EI(x)} dx - \text{FEM}_{BA} \int_0^L \frac{x^2/L}{EI(x)} dx + \int_0^L \frac{xM_{SS}(x)}{EI(x)} dx = 0$$

Substitution of Eqs. (13.2) and solving simultaneously gives

$$\text{FEM}_{AB} = \frac{J_1 \int_0^L \frac{xM_{SS}(x)}{EI(x)} dx - J_2 \int_0^L \frac{M_{SS}(x)}{EI(x)} dx}{J_0 J_2 - J_1^2} \tag{13.7a}$$

$$\text{FEM}_{BA} = \frac{(J_0 L - J_1) \int_0^L \frac{x M_{SS}(x)}{EI(x)} dx + (J_2 - J_1 L) \int_0^L \frac{M_{SS}(x)}{EI(x)} dx}{J_0 J_2 - J_1^2} \quad (13.7b)$$

13.3 NUMERICAL PROCEDURES

With the exception of some trivial cases, the integrals of Eqs. (13.2) and (13.7) are extremely difficult to evaluate. Numerical integration using Simpson's one-third rule is an effective alternative. If we wish to evaluate

$$\int_A^B f(x) \, dx$$

we choose N, an even number of intervals of x between A and B, evaluate $f(x)$ at each, and substitute them in the formula:

$$\int_A^B f(x) \, dx = \frac{h}{3}(f_0 + 4f_1 + 2f_2 + 4f_3 + \cdots + 2f_{N-2} + 4f_{N-1} + f_N)$$

where $h = (B - A)/N$ and $f_i = f(0 + ih)$.

It is presumed that $f(x)$ and its first derivative are continuous from A to B. If there is a discontinuity in that region, say at point C, the integral must be evaluated in two parts, as follows:

$$\int_A^B f(x) \, dx = \int_A^C f(x) \, dx + \int_C^B f(x) \, dx$$

EXAMPLE 13.3.1

Evaluate the rotational stiffnesses and carry-over factors for a 30-ft-long tapered beam consisting of a 1.5-in-thick linearly tapered web, 24 in deep at A and 36 in deep at B, with top and bottom flanges each 28 in wide and 2.5 in thick continuously welded to the web.

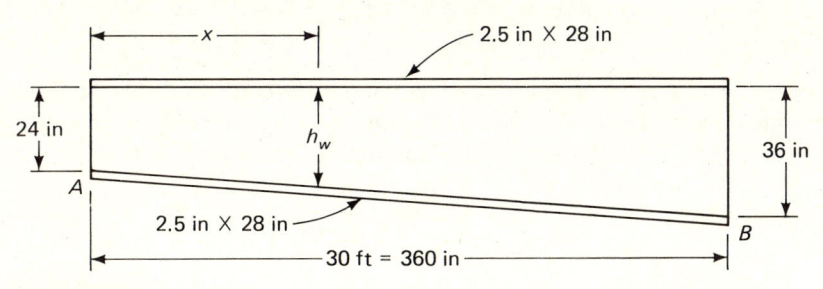

Figure 13.11

$$h_w = 24 + \frac{12}{360}x = 24 + 0.03333x \text{ in}$$

where x is measured in inches.

$$I = \tfrac{1}{12}(28)(h_w + 5)^3 - \tfrac{1}{12}(26.5)h_w^3$$

$$= 0.125 h_w^3 + 35 h_w^2 + 175 h_w + 291.66$$

Substituting $h_w = 24 + 0.03333x$, we get

$$I(x) = 26{,}380 + 69x + 0.0489x^2 + 4.63 \times 10^{-6}x^3 \text{ in}^4$$

where x is in inches.

If $E = 3 \times 10^4$ kips/in^2, the expression for $EI(x)$ becomes

$$EI(x) = 0.1389x^3 + 1467x^2 + 2.07 \times 10^6 x + 7.91 \times 10^8 \text{ kip-in}^2$$

where x is inches from end A.

To evaluate J_0, J_1, and J_2 by integration when $EI(x)$ is so complicated would challenge the most persevering of analysts, so we opt for numerical integration using Simpson's one-third rule.

For this illustration we will use $N = 6$, which will give rather crude evaluations. A larger value, say 20, for a hand calculation or 1000 for a computer calculation, would be preferred.

The interval of x is $h = (360 - 0)/6 = 60$ in. We let

$$j_0(x) = \frac{1}{EI(x)}$$

$$j_1(x) = \frac{x}{EI(x)}$$

$$j_2(x) = \frac{x^2}{EI(x)}$$

and establish Table 13.1.

Table 13.1

x	k	$kj_0(x)$	$kj_1(x)$	$kj_2(x)$
0	1	1.264×10^{-9}	0	0
60	4	4.345×10^{-9}	2.607×10^{-7}	1.564×10^{-5}
120	2	1.885×10^{-9}	2.263×10^{-7}	2.715×10^{-5}
180	4	3.30×10^{-9}	5.941×10^{-7}	1.069×10^{-4}
240	2	1.455×10^{-9}	3.493×10^{-7}	8.38×10^{-5}
300	4	2.584×10^{-9}	7.75×10^{-7}	2.326×10^{-4}
360	1	5.77×10^{-10}	2.078×10^{-7}	7.47×10^{-5}
		1.541×10^{-8}	2.413×10^{-6}	5.403×10^{-4}

Applying Simpson's one-third rule gives

$$J_0 = \tfrac{60}{3}(1.541 \times 10^{-8}) = 3.08 \times 10^{-7} \text{ kip}^{-1} \text{ in}^{-1}$$

$$J_1 = \tfrac{60}{3}(2.413 \times 10^{-6}) = 4.83 \times 10^{-5} \text{ 1/kip}$$

$$J_2 = \tfrac{60}{3}(5.403 \times 10^{-4}) = 1.081 \times 10^{-2} \text{ in/kip}$$

Now we are ready to evaluate the rotational stiffnesses and carry-over factors by substituting J_0, J_1, and J_2 in Eqs. (13.3)–(13.6).

From Eq. (13.3),

$$C_{AB} = \frac{LJ_1}{J_2} - 1$$

$$= \frac{(360 \text{ in})(4.83 \times 10^{-5} \text{ kip}^{-1})}{1.081 \times 10^{-4} \text{ in/kip}} - 1 = 0.607$$

From Eq. (13.4)

$$S_{AB} = \frac{J_2}{J_0 J_2 - J_1^2}$$

$$= \frac{1.081 \times 10^{-2} \text{ in/kip}}{(3.08 \times 10^{-7} \text{ kip}^{-1} \text{ in}^{-1})(1.081 \times 10^{-2} \text{ in/kip}) - (4.83 \times 10^{-5})^2 \text{ kip}^{-2}}$$

$$= 1.085 \times 10^7 \text{ kip-in/rad}$$

$$= 9.04 \times 10^5 \text{ kip-ft/rad}$$

From Eq. (13.5),

$$C_{BA} = \frac{J_1 - J_2/L}{LJ_0 - 2J_1 + J_2/L}$$

$$= \frac{(4.83 \times 10^{-5} \text{ kip}^{-1}) - (1.081 \times 10^{-2} \text{ in/kip})/360 \text{ in}}{360 \text{ in}(3.08 \times 10^{-7} \text{ kip}^{-1} \text{ in}^{-1}) - 2(4.83 \times 10^{-5}) \text{ kip}^{-1} + (1.081 \times 10^{-2} \text{ in/kip})/360 \text{ in}}$$

$$= 0.406$$

From Eq. (13.6),

$$S_{BA} = \frac{J_2 - 2J_1 L + L^2 J_0}{J_0 J_2 - J_1^2}$$

$$= \frac{(1.08 \times 10^{-2} \text{ in/kip}) - 2(4.83 \times 10^{-5} \text{ kip}^{-1})(360 \text{ in}) + (360^2 \text{ in}^2)(3.08 \times 10^{-7} \text{ kip}^{-1} \text{ in}^{-1})}{(3.08 \times 10^{-7} \text{ kip}^{-1} \text{ in}^{-1})(1.081 \times 10^{-2} \text{ in/kip}) - (4.83 \times 10^{-5})^2 \text{ kip}^{-2}}$$

$$= 1.595 \times 10^7 \text{ kip-in/rad}$$

$$= 1.329 \times 10^6 \text{ kip-ft/rad}$$

When making calculations like these it is good practice to compare the results found with known values that should be close. Certainly S_{AB} should be slightly larger than the S_{AB} for a prismatic member of the same length as the tapered member, and with an EI throughout equal to that of the smaller end of the tapered beam. Thus

$$S_{AB \text{ comparison}} = \frac{4(3 \times 10^4 \text{ kip/in}^2)(26,380 \text{ in}^4)}{360 \text{ in}}$$

$$= 8.79 \times 10^6 \text{ kip-in/rad}$$

The value of $S_{AB} = 1.085 \times 10^7$ kip-in/rad of the tapered member is slightly larger than that of the prismatic, as expected. Its S_{AB}, therefore, seems to be a reasonable value.

A similar comparison is made of S_{BA} for a tapered beam to that for a prismatic member with an I equal to that of the larger end. It shows that

304 BEAMS AND RIGID FRAMES WITH NONPRISMATIC MEMBERS

$$S_{BA\text{comparison}} = \frac{4(3 \times 10^4)(5.78 \times 10^4)}{360}$$

$$= 1.926 \times 10^7 \text{ kip-in/rad}$$

is slightly larger for the prismatic beam than for the tapered beam, $S_{BA} = 1.57 \times 10^7$ kip-in/rad. Again this is reasonable.

EXAMPLE 13.3.2

Determine the fixed end moments for the tapered beam of Example 13.3.1 loaded by a uniformly distributed downward force w.

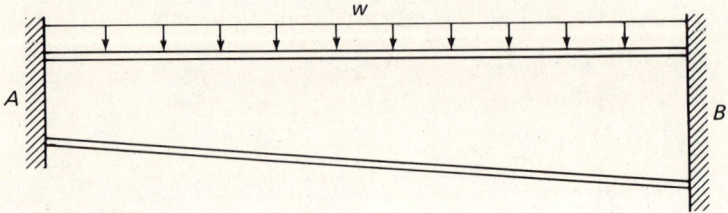

Figure 13.12

We apply Eq. (13.7),

$$\text{FEM}_{AB} = \frac{J_1 \int_0^L \frac{xM_{SS}(x)}{EI(x)}dx - J_2 \int_0^L \frac{M_{SS}(x)}{EI(x)}dx}{J_0 J_2 - J_1^2} \qquad (13.7a)$$

$$\text{FEM}_{BA} = \frac{(J_0 L - J_1)\int_0^L \frac{xM_{SS}(x)}{EI(x)}dx + (J_2 - J_1 L)\int_0^L \frac{M_{SS}(x)}{EI(x)}dx}{J_0 J_2 - J_1^2} \qquad (13.7b)$$

Close examination of these equations shows that there are two new integrals. Their integrands are $M_{SS}(x)$ multiplied by $1/EI(x)$ for one and by $x/EI(x)$ for the other, where $M_{SS}(x)$ is the load related bending moment for a simply supported beam. We evaluated $1/EI(x)$ and $x/EI(x)$ at 60-in intervals along the beam when we calculated J_0, J_1, and J_2 in Example 13.3.1. Thus we need only to multiply these known values by $M_{SS}(x)$ as found at the same locations.

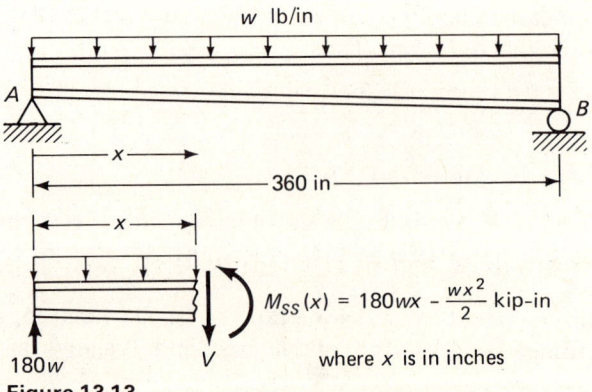

$$M_{SS}(x) = 180wx - \frac{wx^2}{2} \text{ kip-in}$$

where x is in inches

Figure 13.13

We set up Table 13.2 for Simpson's rule, repeating several columns from Table 13.1.

Table 13.2

x	k	kj_0	kj_1	$M_{SS}(x)$	$\dfrac{M_{SS}(x)kj_0}{=\dfrac{kM_{SS}(x)}{EI(x)}}$	$\dfrac{M_{SS}(x)kj_1}{=\dfrac{kxM_{SS}(x)}{EI(x)}}$
0	1	1.269×10^{-9}	0	0	0	0
60	4	4.345×10^{-9}	2.607×10^{-7}	$9 \times 10^3 w$	$3.91 \times 10^{-5}w$	$2.346 \times 10^{-3}w$
120	2	1.885×10^{-9}	2.263×10^{-7}	$1.44 \times 10^4 w$	$2.71 \times 10^{-5}w$	$3.259 \times 10^{-3}w$
180	4	3.30×10^{-9}	5.941×10^{-7}	$1.62 \times 10^4 w$	$5.346 \times 10^{-5}w$	$9.623 \times 10^{-3}w$
240	2	1.455×10^{-9}	3.493×10^{-7}	$1.44 \times 10^4 w$	$2.095 \times 10^{-5}w$	$5.0299 \times 10^{-3}w$
300	4	2.584×10^{-9}	7.75×10^{-7}	$9 \times 10^3 w$	$2.326 \times 10^{-5}w$	$6.977 \times 10^{-3}w$
360	1	5.77×10^{-10}	2.078×10^{-7}	0	0	0
					$1.639 \times 10^{-4}w$	$2.724 \times 10^{-2}w$

Thus from Simpson's rule,

$$\int_0^L \frac{xM_{SS}(x)}{EI(x)}\,dx = \tfrac{60}{3}(2.724 \times 10^{-2}w) = 5.45 \times 10^{-1}w \text{ in}$$

and

$$\int_0^L \frac{M_{SS}(x)}{EI(x)}\,dx = \tfrac{60}{3}(1.639 \times 10^{-4}w) = 3.278 \times 10^{-3}w \qquad \text{(dimensionless)}$$

We substitute these values, along with the previously calculated J_0, J_1, and J_2, into Eqs. (13.7),

$$\text{FEM}_{AB} = \frac{(4.83 \times 10^{-5} \text{ kip}^{-1})(5.45 \times 10^{-1}w \text{ in}) - (1.081 \times 10^{-2} \text{ in/kip})(3.278 \times 10^{-3}w)}{(3.08 \times 10^{-7} \text{ kip}^{-1} \text{ in}^{-1})(1.081 \times 10^{-2} \text{ in/kip}) - (4.83 \times 10^{-5})^2 \text{ kip}^{-2}}$$

$$= -9130w \text{ kip-in} \qquad (w \text{ kip/in})$$

$$\text{FEM}_{BA} = \frac{\{[(3.08 \times 10^{-7} \text{ kip}^{-1} \text{ in}^{-1})(360 \text{ in}) - (4.83 \times 10^{-5} \text{ kip}^{-1})](5.45^{-1}w \text{ kip/in}) + [(1.081 \times 10^{-2} \text{ in/kip}) - (4.83 \times 10^{-5} \text{ kip}^{-1})(360 \text{ in})](3.278 \times 10^{-3}w)\}}{(3.08 \times 10^{-7} \text{ kip}^{-1} \text{ in}^{-1})(1.081 \times 10^{-2} \text{ kip}^{-1}) - (4.83 \times 10^{-5} \text{ kip}^{-2})}$$

$$= 12{,}580w \text{ kip-in} \qquad (w \text{ kip/in})$$

For comparison, a prismatic member has $\pm wL^2/12$ for this loading. For a 360-in beam, these values are $\pm 10{,}800w$. Thus the values found seem reasonable and have the correct signs.

13.4 MOMENT DISTRIBUTION FOR STRUCTURES WITH NONPRISMATIC MEMBERS

The derivation of the expressions for distribution factors of Chapter 11 was not restricted to prismatic members. We found in Eq. (11.3) that

$$D_{BA} = \frac{S_{BA}}{\sum_i S_{Bi}} \tag{11.3}$$

where S_{BA} and S_{Bi} are rotational stiffnesses. For a prismatic member

$$S_{BA} = \frac{4EI}{L}$$

while for nonprismatic members S_{BA} is found using any of the procedures of the preceding sections of this chapter.

No changes are made in the use of carry-over factors, although the values differ from those for prismatic members. It is also important to note that C_{AB} and C_{BA} are never equal unless the member is symmetric.

We will illustrate moment distribution for a structure with nonprismatic members with the following example.

EXAMPLE 13.4.1

Determine the end moments for $ABCD$.

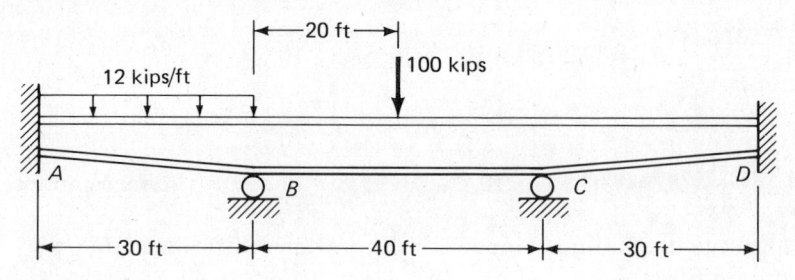

Figure 13.14

Members AB and CD are both tapered members identical to the member of Example 13.3.1, while BC is a prismatic member with $EI = 1.732 \times 10^9$ kip-in^2.

$$S_{BA} = S_{CD} = 1.595 \times 10^7 \text{ kip-in/rad} \quad \text{from Example 13.3.1}$$

$$S_{BC} = S_{CB} = \frac{4(1.732 \times 10^9 \text{ kip-in}^2)}{40 \times 12 \text{ in}} \quad \text{from Eq. (11.1)}$$

$$= 1.443 \times 10^7 \text{ kip-in/rad}$$

Applying Eq. (11.3),

$$D_{BA} = D_{CD} = \frac{1.595 \times 10^7}{1.595 \times 10^7 + 1.443 \times 10^7}$$

$$= 0.525$$

$$D_{BC} = D_{CB} = \frac{1.443 \times 10^7}{1.595 \times 10^7 + 1.443 \times 10^7}$$

$$= 0.475$$

The fixed end moments are

$$\text{FEB}_{AB} = -9130(1) = -9130 \text{ kip-in} \quad \text{from Example 13.3.1}$$
$$= -760.8 \text{ kip-ft}$$
$$\text{FEM}_{BA} = +12{,}580(1) = +12{,}580 \text{ kip-in} \quad \text{from Example 13.3.1}$$
$$= +1048.3 \text{ kip-ft}$$
$$\text{FEM}_{BC} = \text{FEM}_{CB} = -\frac{100(40)}{8} = -500 \text{ kip-ft} \quad \text{from Table 10.1}$$

Carry-over factors are

$$C_{AB} = C_{DC} = 0.607$$
$$C_{BA} = C_{CD} = 0.406$$
$$C_{BC} = C_{CB} = 0.5$$

So we have everything necessary to set up and complete the distribution table.

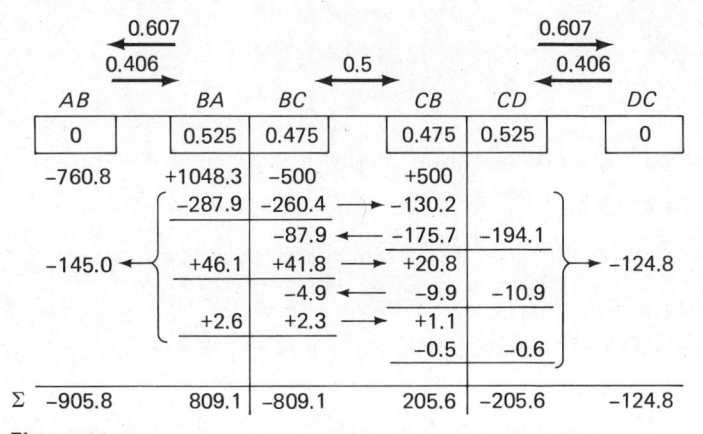

Figure 13.15

The carry-overs from BA to BC and from CD to DC were delayed until the distribution was completed. Both used the 0.607 carry-over factor rather than the 0.5 for a prismatic member.

13.5 RELATIONSHIP BETWEEN ROTATIONAL STIFFNESSES AND CARRY-OVER FACTORS

There is a relationship between the rotational stiffnesses and carry-over factors that serves as a valuable check of our calculations. It also will be used in some of the derivations for modified stiffnesses.

Consider the simply supported beam AB, shown as a single line for simplicity, but specified to be nonprismatic.

Figure 13.16

If we apply S_{AB} and $C_{AB}S_{AB}$ at A and B, respectively, it will deform as follows:

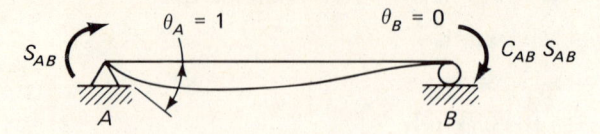

Figure 13.17

The work done externally and stored internally as strain energy is

$$W_1 = \tfrac{1}{2}S_{AB}(1)$$

Next, without removing S_{AB} and $C_{AB}S_{AB}$, we add S_{BA} and $C_{BA}S_{BA}$, deforming the beam as shown.

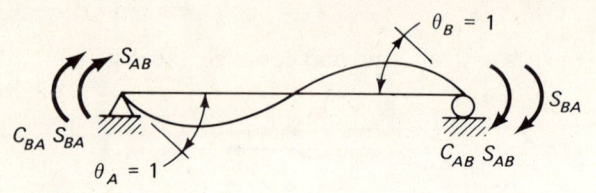

Figure 13.18

The work added and stored as strain energy in this step is

$$W_2 = \tfrac{1}{2}S_{BA}(1) + C_{AB}S_{AB}(1)$$

Thus the total work done and stored as strain energy in the two steps is

$$W = W_1 + W_2 = \tfrac{1}{2}S_{AB}(1) + \tfrac{1}{2}S_{BA}(1) + C_{AB}S_{AB}(1)$$

Now let us repeat the loading, but in reverse order, that is, we apply S_{BA} and $C_{BA}S_{BA}$ first.

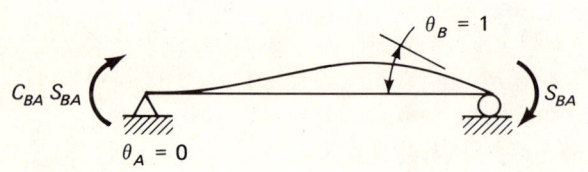

Figure 13.19

The work done is

$$W_3 = \tfrac{1}{2}S_{BA}(1)$$

Now we apply S_{AB} and $C_{AB}S_{AB}$ without removing S_{BA} and $C_{BA}S_{BA}$,

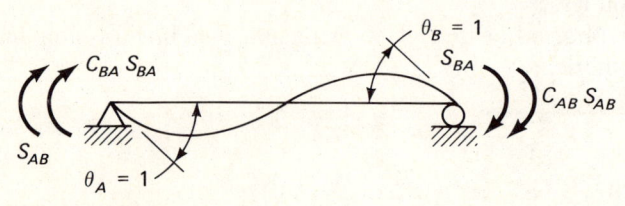

Figure 13.20

during which the work done is

$$W_4 = \tfrac{1}{2}S_{AB}(1) + C_{BA}S_{BA}(1)$$

The total work done in the two steps is

$$W = W_3 + W_4$$
$$= \tfrac{1}{2}S_{AB}(1) + \tfrac{1}{2}S_{BA}(1) + C_{BA}S_{BA}(1)$$

The total load on the beam is the same for both sequences. The deformed shape must be the same after both sequences and the total energy stored is also the same. Therefore

$$\tfrac{1}{2}S_{AB}(1) + \tfrac{1}{2}S_{BA}(1) + C_{AB}S_{AB}(1) = \tfrac{1}{2}S_{AB}(1) + \tfrac{1}{2}S_{BA}(1) + C_{BA}S_{BA}$$

which simplifies to

$$C_{AB}S_{AB} = C_{BA}S_{BA} \tag{13.8}$$

13.6 FIXED END MOMENTS FROM JOINT TRANSLATION

Assume that nonprismatic member AB deforms as illustrated.

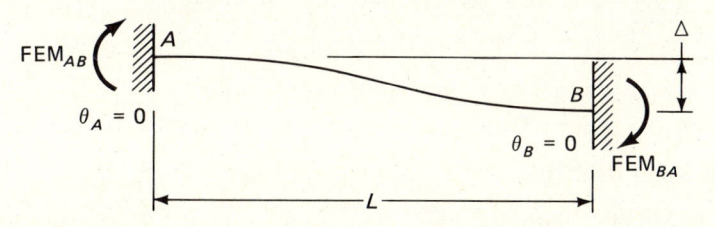

Figure 13.21

The deformed beam may be considered the superposition of the following three cases.

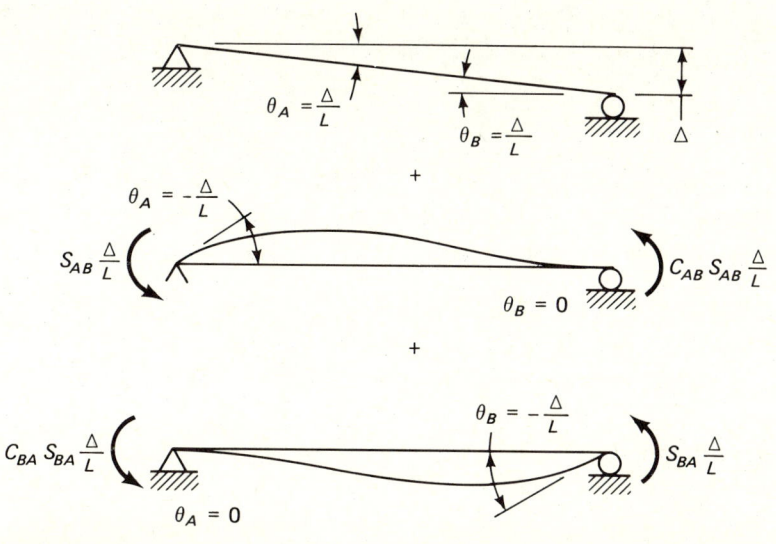

Figure 13.22

310 BEAMS AND RIGID FRAMES WITH NONPRISMATIC MEMBERS

Examination of the three cases shows that the total relative translation of end B with respect to A is Δ, and that the rotations of each end are zero. Further, the total end moments are

$$\text{FEM}_{AB} = -S_{AB}\frac{\Delta}{L} - C_{BA}S_{BA}\frac{\Delta}{L}$$

$$\text{FEM}_{BA} = -S_{BA}\frac{\Delta}{L} - C_{AB}S_{AB}\frac{\Delta}{L}$$

Substituting Eq. (13.8), these are rewritten as

$$\text{FEM}_{AB} = -S_{AB}(1 + C_{AB})\frac{\Delta}{L} \tag{13.9a}$$

and

$$\text{FEM}_{BA} = -S_{BA}(1 + C_{BA})\frac{\Delta}{L} \tag{13.9b}$$

These are the fixed end moments due to joint translation Δ.

Notice that for a prismatic member where

$$S_{AB} = S_{BA} = \frac{4EI}{L} \quad \text{and} \quad C_{AB} = C_{BA} = 0.5$$

These two fixed end moments are $-(6EI/L^2)\Delta$, agreeing with Eq. (12.1).

13.7 MODIFIED STIFFNESSES

The same reductions in calculational effort that came from the use of modified stiffnesses for prismatic members will also come when used for nonprismatic members.

We will first examine the modifier for the far end being pinned.

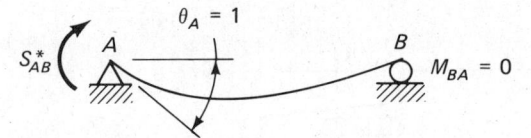

Figure 13.23

Moment S^*_{AB} is applied at A to cause a unit rotation there, while end B rotates freely, that is, $M_{BA} = 0$.

This can be considered the superposition of the two following cases:

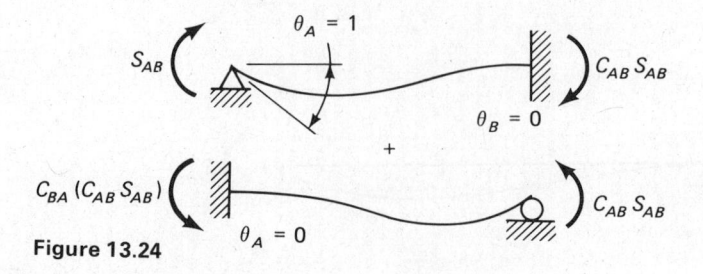

Figure 13.24

MODIFIED STIFFNESSES

In superposition the original conditions have been satisfied, namely, $M_{BA}=0$ and $\theta_A = 1$. Combining the end moments,

$$S^*_{AB} = S_{AB} - C_{AB}C_{BA}S_{AB}$$
$$= S_{AB}(1 - C_{AB}C_{BA}) \tag{13.10}$$

which is the modified stiffness for the far end pinned. Note that

$$S^*_{AB} = S_{AB}(1 - 0.5 \times 0.5)$$
$$= \tfrac{3}{4}S_{AB}$$

for a prismatic member, matching Eq. (11.5).

In cases where symmetry could apply, the member itself must be symmetric about its midpoint. Consider member AB with equal, opposite end moments S^*_{AB} and S^*_{BA} that together cause $\theta_A = 1$ and $\theta_B = -1$.

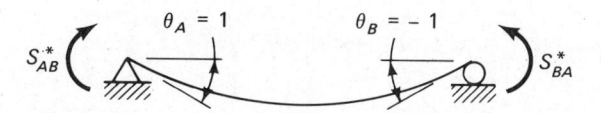

Figure 13.25

This can be considered the superposition of the two following cases:

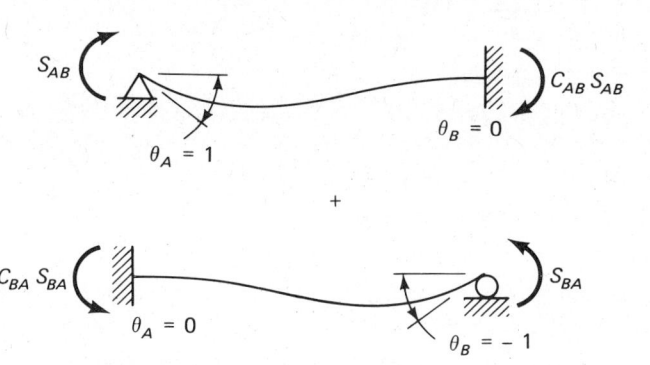

Figure 13.26

The end conditions required are satisfied by these two superposed cases. Thus

$$S^*_{AB} = S_{AB} - C_{BA}S_{BA}$$
$$= S_{AB} - C_{AB}S_{AB}$$
$$= S_{AB}(1 - C_{AB}) \tag{13.11}$$

which is the modified stiffness for the symmetric case. Note that it gives $S^*_{AB} = \tfrac{1}{2}S_{AB}$ for a prismatic member, agreeing with Eq. (11.6).

The derivation for the antisymmetric case is almost identical to that for the symmetric case. The sign of the rotation at B is reversed, as are the directions of S_{BA} and $C_{BA}S_{BA}$. The result is

$$S^*_{AB} = S_{AB}(1 + C_{AB}) \tag{13.12}$$

for antisymmetry, and gives $S^*_{AB} = \frac{3}{2} S_{AB}$ for a prismatic member, consistent with Eq. (11.7).

Finally, the fixed end moment for joint translation may be treated with a modifier if one end is pinned.

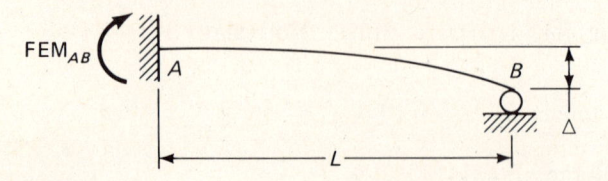

Figure 13.27

This may be considered the superposition of the following three cases:

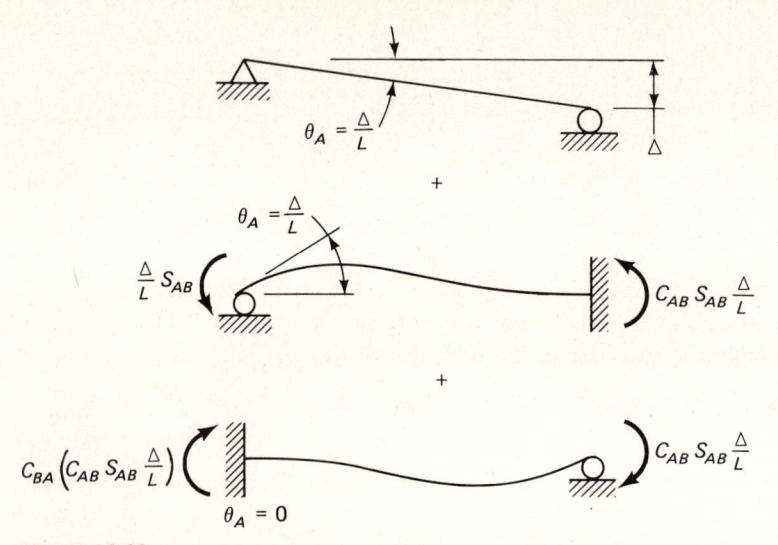

Figure 13.28

Thus

$$\text{FEM}_{AB} = -S_{AB}\frac{\Delta}{L} + S_{AB}\frac{\Delta}{L} C_{AB} C_{BA}$$

$$= -S_{AB}(1 - C_{AB}C_{BA})\frac{\Delta}{L} \qquad (13.13)$$

This is the fixed end moment due to joint translation for the far end pinned. It agrees with Eq. (12.2) when evaluated for a prismatic member.

13.8 MOMENT DISTRIBUTION FOR STRUCTURES WITH NONPRISMATIC MEMBERS, WITH SIDESWAY, AND USING MODIFIED STIFFNESSES

All of the methods used for moment distribution for prismatic members, using modified stiffnesses where applicable and including those with sidesway, may be used for structures with nonprismatic members. The values of the stiffnesses, carry-over factors, fixed end moments, and modifiers must be those for the specific member. Otherwise the procedures are unchanged from those for structures with only prismatic members.

13.9 SLOPE DEFLECTION FOR STRUCTURES WITH NONPRISMATIC MEMBERS

Recall the slope deflection equations for a prismatic member.

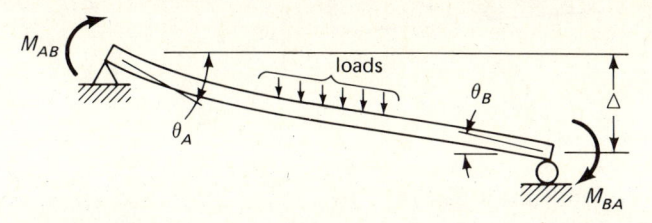

Figure 13.29

$$M_{AB} = \frac{4EI}{L}\theta_A + \frac{2EI}{L}\theta_B - \frac{6EI}{L^2}\Delta + \text{FEM}_{AB}$$

$$M_{BA} = \frac{2EI}{L}\theta_A + \frac{4EI}{L}\theta_B - \frac{6EI}{L^2}\Delta + \text{FEM}_{BA}$$

From the definitions of rotational stiffness and carry-over factor for prismatic members given in Eqs. (11.1) and (11.2) we can rewrite the slope definition equations,

$$M_{AB} = S_{AB}\theta_A + C_{BA}S_{BA}\theta_B - S_{AB}(1 + C_{AB})\frac{\Delta}{L} + \text{FEM}_{AB}$$

$$M_{BA} = C_{AB}S_{AB}\theta_A + S_{BA}\theta_B - S_{BA}(1 + C_{BA})\frac{\Delta}{L} + \text{FEM}_{BA}$$

(13.14)

Slope deflection analysis of structures with nonprismatic members may be performed using procedures identical to those for prismatic members by using Eqs. (13.14) in lieu of Eqs. (10.1).

PROBLEMS

Sequenced Problem 5, parts (a), (b), (c), and (d).
Sequenced Problem 7, parts (a), (b), (c), (d), and (e).

13.1

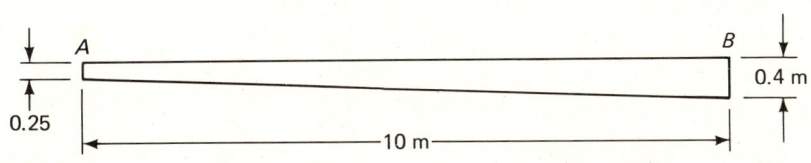

Figure P13.1

314 BEAMS AND RIGID FRAMES WITH NONPRISMATIC MEMBERS

Member AB has a constant width of 0.2 m, and has $E = 200 \times 10^6 \text{ kN/m}^2$. Determine its rotational stiffnesses and carry-over factors.

13.2 For the member AB in Problem 13.1, determine the fixed end moments resulting from a downward 12-kN force located 3 m in from end A.

13.3

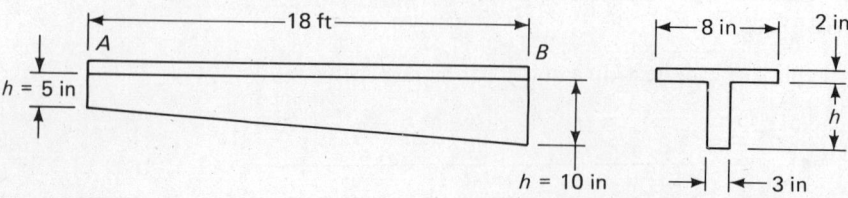

Figure P13.3

Member AB is the T section fabricated from a top flange 8 in × 2 in and a leg 3 in thick varying in depth from 5 in at A to 10 in at B. $E = 3 \times 10^4 \text{ kips/in}^2$. The flange and leg are welded together in a continuous weld from one end to the other. Determine its rotational stiffnesses and carry-over factors.

13.4 If member AB of Problem 13.3 is loaded with the distributed load increasing from 0 at A to 3.6 kips/ft at B as shown, determine the fixed end moments.

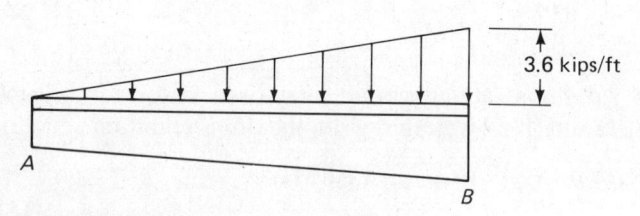

Figure P13.4

13.5 Develop a computer program that can be given expressions for I and M_{SS} as functions of position along the beam, as well as other necessary geometric and material property data, and which produces the rotational stiffnesses, carry-over factors, and fixed end moments. Use it to solve Problems 13.1, 2, 3, or 4, as well as Sequenced Problems 5 (a) and (b) and 7 (a) and (b).

13.6

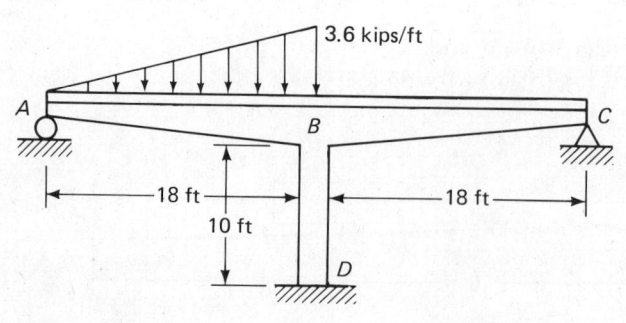

Figure P13.6

Members AB and BC are the members described in Problem 13.3. The loading of AB is as described in Problem 13.4. Member BD is a W 14 × 53 steel beam with $I = 542$ in^4 and $E = 3 \times 10^4$ kips/in^2. Use moment distribution to determine the end moments.

13.7 The pin support at C in Problem 13.6 is replaced with a roller. Thus sidesway is not prevented. Use moment distribution to determine the end moments. Note that the sidesway-prevented part of the solution may be taken from Problem 13.6.

13.8 Use slope deflection to solve Problem 13.7.

13.9

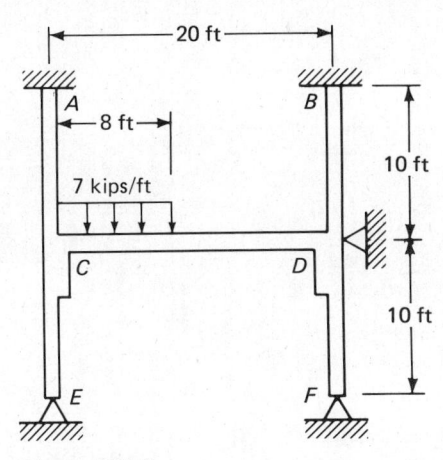

Figure P13.9

The following member properties are given:

$E = 3 \times 10^4$ kips/in^2

$I_{AB} = I_{BD} = 500$ in^4

$I_{CD} = 1500$ in^4

$S_{CE} = S_{DF} = 1.730 \times 10^5$ kip-in/rad

$S_{EC} = S_{FD} = 1.363 \times 10^5$ kip-in/rad

$C_{CE} = C_{DF} = 0.4682$

$C_{EC} = C_{FD} = 0.5941$

Use moment distribution to determine the end moments.

13.10 Use moment distribution to determine the end moments for Problem 13.9, assuming that the brace against sidesway at D has been removed.

13.11 Use slope deflection to solve Problem 13.10.

Matrix Formulation For Structural Analysis

Reprinted by permission: Tribune Company Syndicate, Inc.

14.0 INTRODUCTION

In practice, hand calculations using the classical methods of the earlier chapters are limited by the calculational capacity and endurance of the analyst. The effort needed to analyze a large, complex structure may be overwhelming, and thus possible only with the help of a computer.

Contemporary structural analysis software uses matrix methods, built from the classical methods, but applied in a more systematic format. There are two general classifications of methods, the force method and the displacement method, characterized by the use of forces vis-à-vis displacements as the unknowns in the systems of equations to be solved. These classifications apply to the classical as well as to the matrix approaches, so their features should be familiar to you already. Consistent deformations and least work are force methods, while slope deflection and moment distribution are displacement methods.

We will limit our study of matrix analysis to displacement methods since, for a variety of reasons, displacement methods are almost universally used in computer software. Many structures have a characteristic in which the number of unknown displacements decreases as the static indeterminacy increases. In such structures, especially those of high static indeterminacy, the unknown displacements may be fewer in number than the degrees of indeterminacy. Then the displacement method, with fewer unknowns, would seem to be the better choice. In addition, use of the force methods requires the selection of a primary structure, a critical step that may affect the accuracy of the solution. That selection requires use of judgment and is, therefore, not easily assigned to a computer. No similar judgmental step is associated with the displacement methods, unless it is that of assigning a suitable numbering system to the joint supports and displacements. While that may seem trivial, it is important, as will be seen later. Fortunately that numbering is done by the analyst prior to data entry on the computer and rarely would be a step assigned to the computer.[1]

We begin with structures composed of axial force members, that is, trusses. Since several unknown displacements are possible for each unsupported joint, the displacement method applied to a truss of more than a few joints involves a large number of unknowns. Thus for trusses the displacement method is not often used for hand calculations and, for that reason, was not included in our earlier studies. This first venture into matrix methods, therefore, not only introduces matrix methods themselves, but it also introduces the analysis of trusses using the displacement method.

14.1 SYSTEMS OF AXIAL FORCE MEMBERS; LOCAL AND GLOBAL STIFFNESSES

Consider axial force member AB with uniform AE for its entire length L.

---

[1] In some software systems for finite-element analysis, nodal point numbering is done by the computer.

SYSTEMS OF AXIAL FORCE MEMBERS; LOCAL AND GLOBAL STIFFNESSES

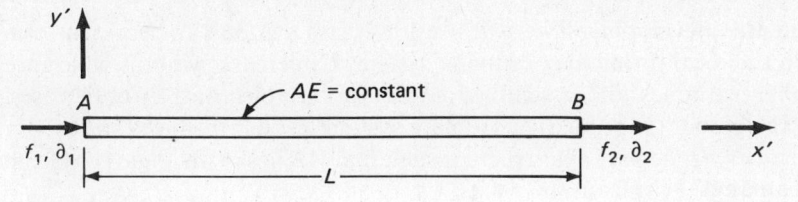

Figure 14.1

Axial forces f_1 and f_2, acting at the member ends, are directed in the positive x' direction, where x' and y' are a set of "local" coordinates oriented so that x' lies along the member.[2] Accompanying the end forces are axial end displacements ∂_1 and ∂_2, which also have positive values in the positive x' direction and subscripts that match the axial forces.

We will need to distinguish between the two ends. Since subscripting may change, it is convenient to refer to the end at the origin as the near end and to the other as the far end. Thus for the case illustrated, f_1 acts at the near end and f_2 at the far end.

From basic elastic theory, if we hold $\partial_1 = 0$ and induce a displacement ∂_2, the resulting end forces are

$$f_2 = -f_1 = \frac{AE}{L}\partial_2$$

and, similarly, for displacement ∂_1 with $\partial_2 = 0$,

$$f_1 = -f_2 = \frac{AE}{L}\partial_1$$

By superposition, if both ∂_1 and ∂_2 are induced,

$$f_1 = \frac{AE}{L}(\partial_1 - \partial_2) \quad \text{and} \quad f_2 = \frac{AE}{L}(\partial_2 - \partial_1)$$

which may be written in matrix form[3] as

$$\begin{Bmatrix} f_1 \\ f_2 \end{Bmatrix} = \frac{AE}{L} \begin{bmatrix} 1 & -1 \\ -1 & 1 \end{bmatrix} \begin{Bmatrix} \partial_1 \\ \partial_2 \end{Bmatrix} \tag{14.1}$$

or symbolically as

$$\{f\} = [k]\{\partial\} \tag{14.2}$$

In Eq. (14.2) $\{f\}$ is called the element force vector, $[k]$ the element stiffness matrix, and $\{\partial\}$ the element displacement vector. All are in local coordinates. It should be evident that

$$[k] = \frac{AE}{L}\begin{bmatrix} 1 & -1 \\ -1 & 1 \end{bmatrix} \tag{14.3}$$

---

[2] Several coordinate systems will be used at the same time in most matrix solutions, a separate local set for each member and a global coordinate set that is related to the assembled structure and its supports.
[3] Readers unfamiliar with matrix algebra are referred to Appendix A.

320 MATRIX FORMULATION FOR STRUCTURAL ANALYSIS

Throughout this development we will use lowercase symbols to represent element variables in the local (primed) coordinate system. Uppercase symbols will represent both member variables and assembled structure variables but, in both cases, expressed in the global (unprimed) coordinate system to be introduced next.

Let member AB be part of a structure such that AB lies at an angle $(x'x)$ from the x axis and an angle $(x'y)$ from the y axis.

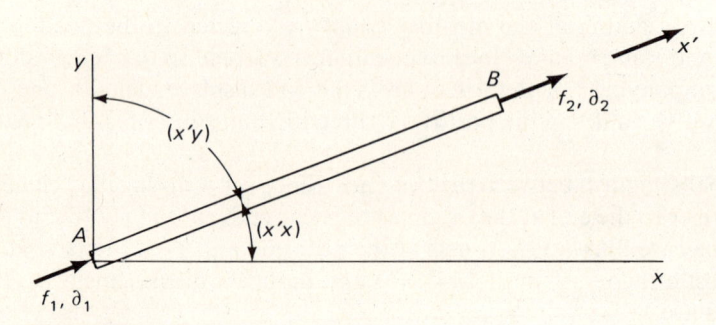

Figure 14.2

The x, y axes (unprimed), prescribed in relation to the geometry of the assembled structure and its supports, are called global coordinates.

With several coordinate systems in use, transformations between systems are necessary. First, we relate *displacements* in the two coordinate systems. Consider a displacement Δ_1 of end A that is solely in the $+x$ direction, with no displacements of end B.

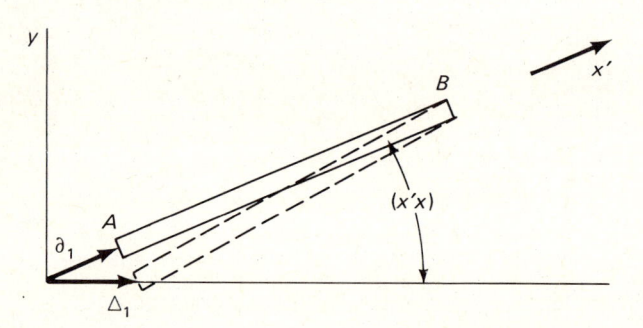

Figure 14.3

For a small displacement such that the change of angular orientation of AB is negligible,

$$\partial_1 = \Delta_1 \cos(x'x)$$

If end A were given a small displacement Δ_2 solely in the y direction, while end B remained undisplaced,

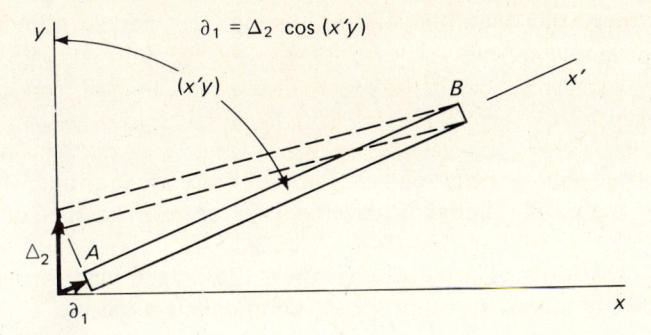

SYSTEMS OF AXIAL FORCE MEMBERS; LOCAL AND GLOBAL STIFFNESSES

Figure 14.4

If both Δ_1 and Δ_2 occur, displacement ∂_1 would be the superposition of the two parts, that is,

$$\partial_1 = \Delta_1 \cos(x'x) + \Delta_2 \cos(x'y) \tag{14.4}$$

A similar development relating small displacements of end B in global coordinates to local coordinate displacements gives

$$\partial_2 = \Delta_3 \cos(x'x) + \Delta_4 \cos(x'y) \tag{14.5}$$

where Δ_3 and Δ_4 are displacements of B in the positive x and y (global) directions, respectively.

Eqs. (14.4) and (14.5) may be written in matrix form as

$$\begin{Bmatrix} \partial_1 \\ \partial_2 \end{Bmatrix} = \begin{bmatrix} \cos(x'x) & \cos(x'y) & 0 & 0 \\ 0 & 0 & \cos(x'x) & \cos(x'y) \end{bmatrix} \begin{Bmatrix} \Delta_1 \\ \Delta_2 \\ \Delta_3 \\ \Delta_4 \end{Bmatrix} \tag{14.6}$$

or

$$\{\partial\} = [T]\{\Delta\} \tag{14.7}$$

Matrix $[T]$, called the global to local transformation matrix, is

$$[T] = \begin{bmatrix} \cos(x'x) & \cos(x'y) & 0 & 0 \\ 0 & 0 & \cos(x'x) & \cos(x'y) \end{bmatrix} \tag{14.8}$$

Signs for the elements in $[T]$ are often troublesome. That difficulty disappears if we define $\cos(x'x)$ and $\cos(x'y)$ in terms of the global coordinate locations of the near and far ends of the member according to the following equations:

$$\cos(x'x) = \frac{x_{\text{far}} - x_{\text{near}}}{L} \tag{14.9a}$$

$$\cos(x'y) = \frac{y_{\text{far}} - y_{\text{near}}}{L} \tag{14.9b}$$

where L is the member length.

There will also be occasions when we will need the element forces $\{f\}$, written in terms of the global displacements $\{\Delta\}$. Substituting Eqs. (14.7) in (14.1) gives

$$\{f\} = [k][T]\{\Delta\} \tag{14.10}$$

322 MATRIX FORMULATION FOR STRUCTURAL ANALYSIS

We will consistently list near end quantities first in our matrices, and then far end quantities. Thus in

$$\{f\} = \begin{Bmatrix} f_1 \\ f_2 \end{Bmatrix}$$

f_1 is the near end axial force and f_2 the far end axial force. There are four quantities in $\{\Delta\}$. They are, in order, the x and y global displacements of the near end, then of the far end.

We will have many occasions requiring us to express the element end forces in global coordinates. These are merely the appropriate components as shown.

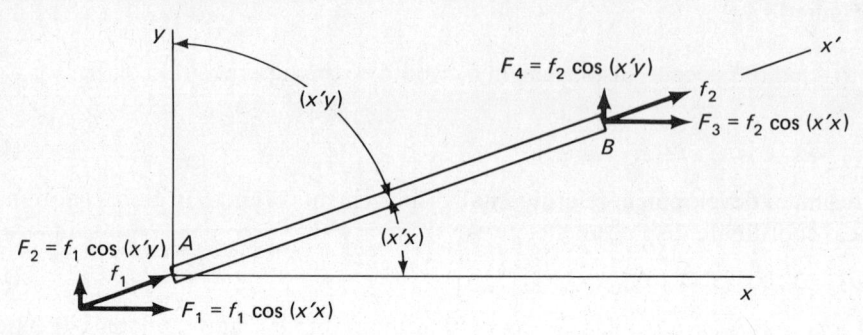

Figure 14.5

These are, in matrix form,

$$\begin{Bmatrix} F_1 \\ F_2 \\ F_3 \\ F_4 \end{Bmatrix} = \begin{bmatrix} \cos(x'x) & 0 \\ \cos(x'y) & 0 \\ 0 & \cos(x'x) \\ 0 & \cos(x'y) \end{bmatrix} \begin{Bmatrix} f_1 \\ f_2 \end{Bmatrix} \qquad (14.11)$$

or

$$\{F\} = [T]^T \{f\} \qquad (14.12)$$

Finally, we will need to write the global forces $\{F\}$ in terms of global displacements $\{\Delta\}$. Substituting Eq. (14.10) in Eq. (14.12) gives

$$\{F\} = [T]^T[k][T]\{\Delta\} \qquad (14.13)$$

or

$$\{F\} = [K]\{\Delta\} \qquad (14.14)$$

where $[K]$ is called the member global stiffness matrix. Evidently

$$[K] = [T]^T[k][T] \qquad (14.15)$$

If we carry out the matrix multiplications, we get

$$\begin{Bmatrix} F_1 \\ F_2 \\ F_3 \\ F_4 \end{Bmatrix} = \frac{AE}{L} \begin{bmatrix} \lambda_x^2 & \lambda_x\lambda_y & -\lambda_x^2 & -\lambda_x\lambda_y \\ \lambda_x\lambda_y & \lambda_y^2 & -\lambda_x\lambda_y & -\lambda_y^2 \\ -\lambda_x^2 & -\lambda_x\lambda_y & \lambda_x^2 & \lambda_x\lambda_y \\ -\lambda_x\lambda_y & \lambda_y^2 & \lambda_x\lambda_y & \lambda_y^2 \end{bmatrix} \begin{Bmatrix} \Delta_1 \\ \Delta_2 \\ \Delta_3 \\ \Delta_4 \end{Bmatrix} \qquad (14.16)$$

where $\lambda_x = \cos(x'x)$ and $\lambda_y = \cos(x'y)$.

This is the member global force/displacement relation, and the 4×4 matrix, with its coefficient AE/L, is the member global stiffness matrix $[K]$. As previously established, the elements in $\{\Delta\}$ and $\{F\}$ are for x, then y (global) of the near end, followed by those for the far end.

14.2 PLANE TRUSS ANALYSIS

Example 14.2.1 will illustrate the displacement method applied to a plane truss. In this our first example, our use of the matrix procedures will be informal, aimed at showing underlying principles. As a result, the method may seem awkward. In later examples we will streamline things.

EXAMPLE 14.2.1

Analyze truss $ABCD$ using the displacement method.

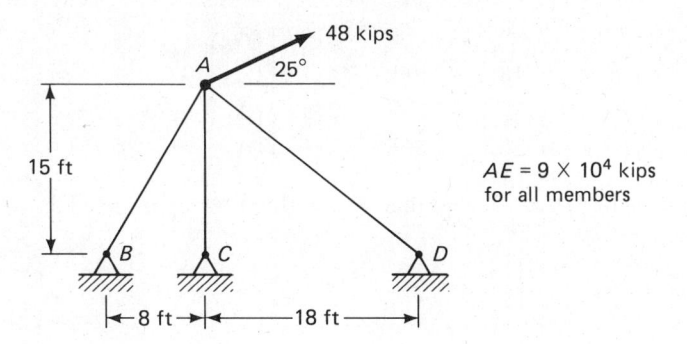

$AE = 9 \times 10^4$ kips for all members

Figure 14.6

First we establish a global coordinate system. For this structure we will use traditional x, y coordinates with origin at B. Then we assign subscripts serially from 1. The first ones are assigned to the *unknown* displacement components, after which subscripts are assigned to those x and y joint or support components that are constrained against motion. Thus the subscripting is as illustrated.

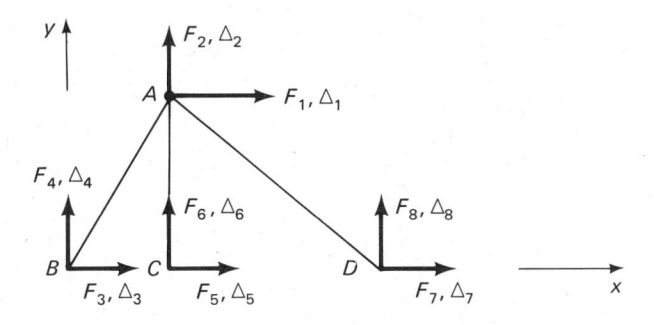

Figure 14.7

Since only joint A can displace, only subscripts 1 and 2 pertain to displacements.

Now we develop the global force/displacement relations for each member, using Eq. (14.16). For member AB, letting A be the far end, we use Eqs. (14.9) to get the direction cosines,

$$\cos(x'x) = \frac{8-0}{17} = 0.471$$

$$\cos(x'y) = \frac{15-0}{17} = 0.882$$

and combine them with

$$\frac{AE}{L} = \frac{9 \times 10^4}{17} = 5294 \text{ kips/ft}$$

in Eq. (14.16). The order of the subscripts that follow may seem peculiar. It is the result of our listing the near end displacements first. Thus

$$\begin{Bmatrix} F_3 \\ F_4 \\ F_1 \\ F_2 \end{Bmatrix}_{AB} = \begin{bmatrix} 1174 & 2199 & -1174 & -2199 \\ 2199 & 4118 & -2199 & -4118 \\ -1174 & -2199 & 1174 & 2199 \\ -2199 & -4118 & 2199 & 4118 \end{bmatrix} \begin{Bmatrix} \Delta_3 \\ \Delta_4 \\ \Delta_1 \\ \Delta_2 \end{Bmatrix}$$

A similar sequence is followed for member AC, with end A again used as the far end,

$$\cos(x'x) = \frac{8-8}{15} = 0$$

$$\cos(x'y) = \frac{15-0}{15} = 1$$

$$\frac{AE}{L} = \frac{9 \times 10^4 \text{ kips}}{15 \text{ ft}} = 6000 \text{ kips/ft}$$

$$\begin{Bmatrix} F_5 \\ F_6 \\ F_1 \\ F_2 \end{Bmatrix}_{AC} = \begin{bmatrix} 0 & 0 & 0 & 0 \\ 0 & 6000 & 0 & -6000 \\ 0 & 0 & 0 & 0 \\ 0 & -6000 & 0 & 6000 \end{bmatrix} \begin{Bmatrix} \Delta_5 \\ \Delta_6 \\ \Delta_1 \\ \Delta_2 \end{Bmatrix}$$

Again the peculiar sequence of subscripts arises from listing the near end first and then the far end for member AC.

Finally for AD, with A again the far end,

$$\cos(x'x) = -0.769$$

$$\cos(x'y) = 0.640$$

$$\frac{AE}{L} = 3841 \text{ kips/ft}$$

PLANE TRUSS ANALYSIS

$$\begin{Bmatrix} F_7 \\ F_8 \\ F_1 \\ F_2 \end{Bmatrix}_{AD} = \begin{bmatrix} 2271 & -1890 & -2271 & 1890 \\ -1890 & 1573 & 1890 & -1573 \\ -2271 & 1890 & 2271 & -1890 \\ 1890 & -1573 & -1890 & 1573 \end{bmatrix} \begin{Bmatrix} \Delta_7 \\ \Delta_8 \\ \Delta_1 \\ \Delta_2 \end{Bmatrix}$$

You are urged to write the expressions for the direction cosines for AD and satisfy yourself about the signs and values that evolved.

Although this is not usually a part of the matrix procedure, we will examine these member force/displacement relations pictorially. For AB,

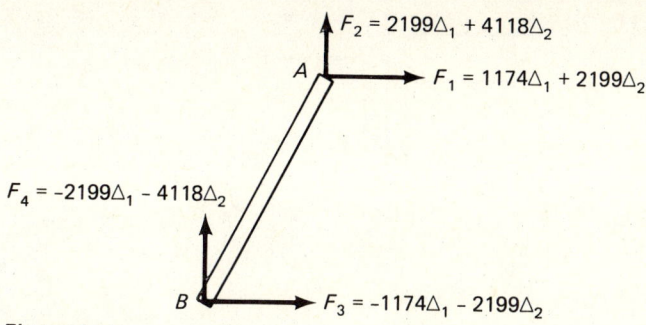

$F_2 = 2199\Delta_1 + 4118\Delta_2$
$F_1 = 1174\Delta_1 + 2199\Delta_2$
$F_4 = -2199\Delta_1 - 4118\Delta_2$
$F_3 = -1174\Delta_1 - 2199\Delta_2$

Figure 14.8

For AC,

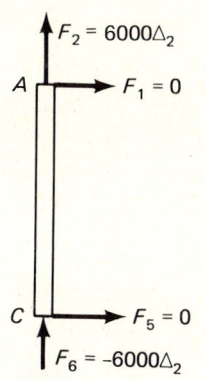

$F_2 = 6000\Delta_2$
$F_1 = 0$
$F_5 = 0$
$F_6 = -6000\Delta_2$

Figure 14.9

For AD,

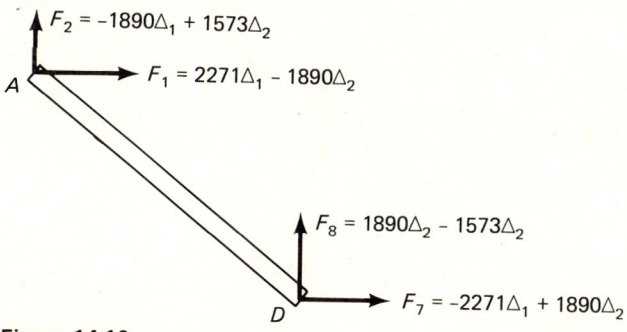

$F_2 = -1890\Delta_1 + 1573\Delta_2$
$F_1 = 2271\Delta_1 - 1890\Delta_2$
$F_8 = 1890\Delta_2 - 1573\Delta_2$
$F_7 = -2271\Delta_1 + 1890\Delta_2$

Figure 14.10

In all three, contributions to the forces from Δ_3 through Δ_8 were omitted since each of those displacements is zero.

Now we draw the free-body diagram of joint A, including the equal, opposite of each of the forces at end A of each member, as well as the original externally applied 48-kip force that acts on joint A but is shown in its x and y components of 43.5 kips and 20.3 kips, respectively.

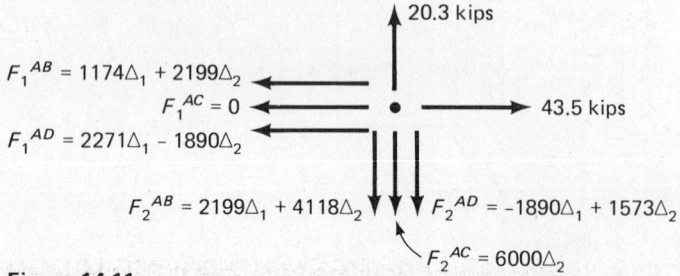

$$F_1^{AB} = 1174\Delta_1 + 2199\Delta_2$$
$$F_1^{AC} = 0$$
$$F_1^{AD} = 2271\Delta_1 - 1890\Delta_2$$
$$F_2^{AB} = 2199\Delta_1 + 4118\Delta_2$$
$$F_2^{AD} = -1890\Delta_1 + 1573\Delta_2$$
$$F_2^{AC} = 6000\Delta_2$$

Figure 14.11

Force equilibrium in x and y gives

$$(1174 + 2271)\Delta_1 + (2199 - 1890)\Delta_2 = 43.5$$

and

$$(2199 - 1890)\Delta_1 + (4118 + 6000 + 1573)]\Delta_2 = 20.3$$

for which $\Delta_1 = 0.01250$ ft and $\Delta_2 = 0.001406$ ft.

Notice that *the coefficients* for Δ_1 and Δ_2 found in these equilibrium equations *are sums of elements in the global stiffness matrices of the individual members*. Specifically, $1174 + 2271$ is the sum of the K_{11} elements of AB and AC (there is no K_{11} for AC); $2199 - 1890$ is the sum of the K_{12} elements and also of the K_{21} elements for AB and AD (the K_{12} and K_{21} elements for AC are zero); and $4118 + 6000 + 1573$ is the sum of the K_{22} elements for all three members. That the coefficients for the unknown Δ's can be generated directly by summing stiffness matrix elements will be used to advantage in solutions henceforth.

Now that the unknown displacements have been determined, the support reactions can be calculated. They are forces F_3 through F_8, each having been expressed in terms of Δ_1 and Δ_2. Thus

$$F_3 = -1174(0.0125) - 2199(0.001406) = -17.8 \text{ kips}$$
$$F_4 = -2199(0.0125) - 4118(0.001406) = -33.3 \text{ kips}$$
$$F_5 = 0$$
$$F_6 = -6000(0.001406) = -8.44 \text{ kips}$$
$$F_7 = -2271(0.01250) + 1890(0.001406) = -25.7 \text{ kips}$$
$$F_8 = 1890(0.01250) - 1573(0.001406) = 21.4 \text{ kips}$$

The structure free-body diagram with all the known external forces and support reactions is shown.

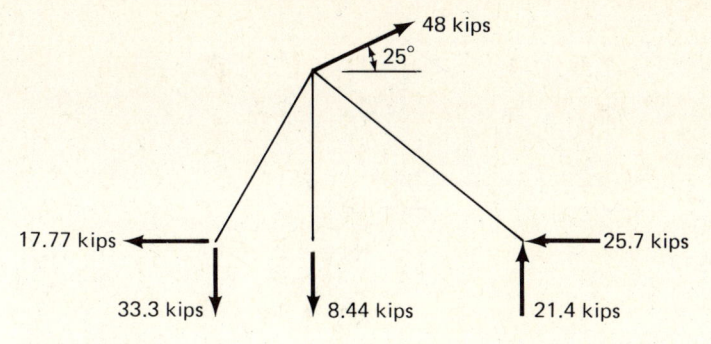

Figure 14.12

Finally the member forces may be found using Eq. (14.10), applied for each member. Only the second force listed in $\{f\}$ need be calculated. That is for the far end, the one that is nominally in tension and, hence, adheres to our usual tension $(+)$ and compression $(-)$ conventions. For AB, with $\cos(x'x) = 0.471$, $\cos(x'y) = 0.882$, and $AE/L = 5294$ kips/ft from before, we substitute into Eq. (14.10), listing the near end, then the far end displacements in $\{\Delta\}$.

$$\begin{Bmatrix} f_1 \\ f_2 \end{Bmatrix} = 5294 \begin{bmatrix} 1 & -1 \\ -1 & 1 \end{bmatrix} \begin{bmatrix} 0.471 & 0.882 & 0 & 0 \\ 0 & 0 & 0.471 & 0.882 \end{bmatrix} \begin{Bmatrix} 0 \\ 0 \\ 0.0125 \\ 0.00146 \end{Bmatrix}$$

Evaluated, we find

$f_2 = 37.7$ kips

Similar calculations for AC and AD give $+8.44$ kips and -33.5 kips, respectively. Thus

$f_{AB} = 37.7$ kips
$f_{AC} = 8.44$ kips
$f_{AD} = -33.5$ kips

completing the analysis.

Now let us work through another example, following more formal matrix procedures.

EXAMPLE 14.2.2

Analyze truss $ABCD$.

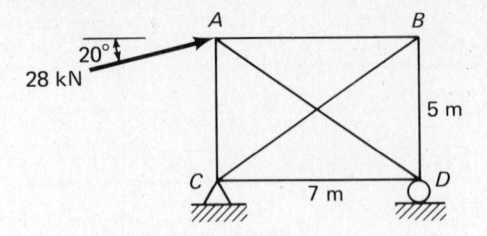

328 MATRIX FORMULATION FOR STRUCTURAL ANALYSIS

Figure 14.13

Member stiffnesses are

$$AE_{AC} = AE_{AB} = AE_{BD} = AE_{CD} = 1.28 \times 10^6 \text{ kN}$$
$$AE_{AD} = AE_{BC} = 1.55 \times 10^6 \text{ kN}$$

We select the global coordinate system using the usual x and y directions with the origin at C. Then we assign subscripts serially from 1, first to all the "freedoms," then to all the "supports." Thus

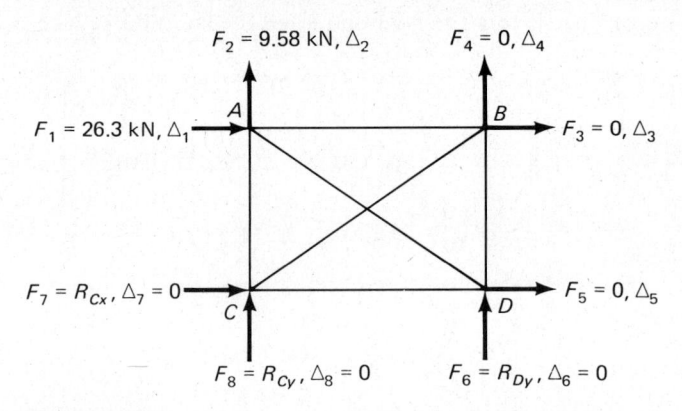

Figure 14.14

The subscripts relate to both force and displacement in an indicated direction at an indicated joint. Where values of force or displacement are known, they are shown on the diagram. For clarity, the unknown support reactions are related to the subscripts in the same way.

Next we write global force/displacement relations in global coordinates for each member, using Eq. (14.16) with $\cos(x'x)$ and $\cos(x'y)$ determined according to Eq. (14.9). This may be done readily on a computer by entering the near end and far end coordinates and computing $\cos(x'x)$ and $\cos(x'y)$. Then we establish a $[T]$ matrix.

$$[T] = \begin{bmatrix} \cos(x'x) & \cos(x'y) & 0 & 0 \\ 0 & 0 & \cos(x'x) & \cos(x'y) \end{bmatrix} \quad (14.8)$$

and carry out the matrix multiplication

$$[K] = [T]^T[k][T] \quad (14.15)$$

where

$$[k]' = \frac{AE}{L}\begin{bmatrix} 1 & -1 \\ -1 & 1 \end{bmatrix} \qquad (14.3)$$

Thus for member AB, with end B as the far end,

$$\cos(x'x) = \frac{7-0}{7} = 1$$

$$\cos(x'y) = \frac{0-0}{7} = 0$$

$$\frac{AE}{L} = \frac{1.28 \times 10^6}{7} = 1.829 \times 10^5 \text{ kN/m}$$

$$[k][T] = 1.829 \times 10^5 \begin{bmatrix} 1 & -1 \\ -1 & 1 \end{bmatrix}\begin{bmatrix} 1 & 0 & 0 & 0 \\ 0 & 0 & 1 & 0 \end{bmatrix}$$

$$= 1.829 \times 10^5 \begin{bmatrix} 1 & 0 & -1 & 0 \\ -1 & 0 & 1 & 0 \end{bmatrix}$$

$$[K] = [T]^T[k][T] = 1.829 \times 10^5 \begin{bmatrix} 1 & 0 \\ 0 & 0 \\ 0 & 1 \\ 0 & 0 \end{bmatrix}\begin{bmatrix} 1 & 0 & -1 & 0 \\ -1 & 0 & 1 & 0 \end{bmatrix}$$

$$[K]_{AB} = 1.829 \times 10^5 \begin{bmatrix} 1 & 0 & -1 & 0 \\ 0 & 0 & 0 & 0 \\ -1 & 0 & 1 & 0 \\ 0 & 0 & 0 & 0 \end{bmatrix}$$

and thus

$$\begin{Bmatrix} F_1 \\ F_2 \\ F_3 \\ F_4 \end{Bmatrix}_{AB} = 10^5 \begin{bmatrix} 1.829 & 0 & -1.829 & 0 \\ 0 & 0 & 0 & 0 \\ -1.829 & 0 & 1.829 & 0 \\ 0 & 0 & 0 & 0 \end{bmatrix}\begin{Bmatrix} \Delta_1 \\ \Delta_2 \\ \Delta_3 \\ \Delta_4 \end{Bmatrix}$$

That the subscripts are in the order 1, 2, 3, 4 is not from listing them in numerical order. Rather, subscripts 1 and 2, which relate to x and y for the near end, precede subscripts 3 and 4, which relate to x and y for the far end. That sequence of subscripts establishes the subscripting in the rows and columns of the associated $[K]$ matrix.

Similar steps, taken for member AC with A as the far end, gives

$$\begin{Bmatrix} F_7 \\ F_8 \\ F_1 \\ F_2 \end{Bmatrix}_{AC} = 10^5 \begin{bmatrix} 0 & 0 & 0 & 0 \\ 0 & 2.56 & 0 & -2.56 \\ 0 & 0 & 0 & 0 \\ 0 & -2.56 & 0 & 2.56 \end{bmatrix}\begin{Bmatrix} \Delta_7 \\ \Delta_8 \\ \Delta_1 \\ \Delta_2 \end{Bmatrix}$$

For *AD*, using *A* as the far end,

$$\begin{Bmatrix} F_5 \\ F_6 \\ F_1 \\ F_2 \end{Bmatrix}_{AD} = 10^5 \begin{bmatrix} 1.193 & -0.851 & -1.193 & 0.851 \\ -0.851 & 0.609 & 0.851 & -0.609 \\ -1.193 & 0.851 & 1.193 & -0.851 \\ 0.851 & -0.609 & -0.851 & 0.609 \end{bmatrix} \begin{Bmatrix} \Delta_5 \\ \Delta_6 \\ \Delta_1 \\ \Delta_2 \end{Bmatrix}$$

For *CB*, with far end *B*,

$$\begin{Bmatrix} F_7 \\ F_8 \\ F_3 \\ F_4 \end{Bmatrix}_{CB} = 10^5 \begin{bmatrix} 1.193 & 0.851 & -1.193 & -0.851 \\ 0.851 & 0.609 & -0.851 & -0.609 \\ -1.193 & -0.851 & 1.193 & 0.851 \\ -0.851 & -0.609 & 0.851 & 0.609 \end{bmatrix} \begin{Bmatrix} \Delta_7 \\ \Delta_8 \\ \Delta_3 \\ \Delta_4 \end{Bmatrix}$$

For *CD*, with far end *D*,

$$\begin{Bmatrix} F_7 \\ F_8 \\ F_5 \\ F_6 \end{Bmatrix}_{CD} = 10^5 \begin{bmatrix} 1.829 & 0 & -1.829 & 0 \\ 0 & 0 & 0 & 0 \\ -1.829 & 0 & 1.829 & 0 \\ 0 & 0 & 0 & 0 \end{bmatrix} \begin{Bmatrix} \Delta_7 \\ \Delta_8 \\ \Delta_5 \\ \Delta_6 \end{Bmatrix}$$

And finally, for *BD*, with *B* as the far end,

$$\begin{Bmatrix} F_5 \\ F_6 \\ F_3 \\ F_4 \end{Bmatrix}_{BD} = 10^5 \begin{bmatrix} 0 & 0 & 0 & 0 \\ 0 & 2.56 & 0 & -2.56 \\ 0 & 0 & 0 & 0 \\ 0 & -2.56 & 0 & 2.56 \end{bmatrix} \begin{Bmatrix} \Delta_5 \\ \Delta_6 \\ \Delta_3 \\ \Delta_4 \end{Bmatrix}$$

Having written all the *member* global stiffness equations, we write the structure global stiffness equations in the form

$$\{F\} = [K]\{\Delta\} \tag{14.14}$$

The vector $\{F\}$ is composed of all the forces applied to the structure externally, including the support reactions. Thus

$$\{F\} = \begin{Bmatrix} 26.3 \\ 9.58 \\ 0 \\ 0 \\ 0 \\ R_{Dy} \\ R_{Cx} \\ R_{Cy} \end{Bmatrix}$$

The vector Δ contains all the joint displacements, with the unknowns first. Thus

$$\{\Delta\} = \begin{Bmatrix} \Delta_1 \\ \Delta_2 \\ \Delta_3 \\ \Delta_4 \\ \Delta_5 \\ 0 \\ 0 \\ 0 \end{Bmatrix}$$

The structure global stiffness matrix consists of elements K_{ij}, each with values equal to the sum of the values for the corresponding K_{ij} for each member, as found in each member stiffness matrix. Thus for structure stiffness element K_{11} we added $(K_{11})_{AB} = 1.829 \times 10^5$, $(K_{11})_{AC} = 0$, and $(K_{11})_{AD} = 1.193 \times 10^5$ to get $K_{11} = 3.02 \times 10^5$ kN/m for the structure. Similarly, by summing,

$(K_{78})_{AC} = 0$

$(K_{78})_{CB} = 0.851 \times 10^5$

$(K_{78})_{CD} = 0$

we get $K_{78} = 0.851 \times 10^5$ kN/m for the structure.

After summing the member stiffness elements for each subscript pair, we write the structure stiffness equations in the form of Eq. (14.14),

$$\{F\} = [K]\{\Delta\} \tag{14.14}$$

$$\begin{Bmatrix} 26.3 \\ 9.58 \\ 0 \\ 0 \\ 0 \\ \hline R_{Dy} \\ R_{Cx} \\ R_{Cy} \end{Bmatrix} = 10^5 \begin{bmatrix} 3.02 & -0.851 & -1.829 & 0 & -1.193 & \vdots & 0.851 & 0 & 0 \\ -0.851 & 3.169 & 0 & 0 & 0.851 & \vdots & -0.609 & 0 & -2.56 \\ -1.829 & 0 & 3.02 & 0.851 & 0 & \vdots & 0 & -1.193 & -0.851 \\ 0 & 0 & 0.851 & 3.169 & 0 & \vdots & -2.56 & -0.851 & -0.609 \\ -1.193 & 0.851 & 0 & 0 & 3.02 & \vdots & -0.851 & 1.829 & 0 \\ \hline 0.851 & -0.609 & 0 & -2.56 & -0.851 & \vdots & 3.169 & 0 & 0 \\ 0 & 0 & -1.193 & -0.851 & -1.829 & \vdots & 0 & 3.02 & 0.851 \\ 0 & -2.56 & -0.851 & -0.609 & 0 & \vdots & 0 & 0.851 & 3.169 \end{bmatrix} \begin{Bmatrix} \Delta_1 \\ \Delta_2 \\ \Delta_3 \\ \Delta_4 \\ \Delta_5 \\ \hline 0 \\ 0 \\ 0 \end{Bmatrix}$$

For convenience, the matrix equation has been partitioned. We give labels to the submatrices. It is easy to see that our matrix equation is of the form

$$\begin{Bmatrix} F_F \\ \hline F_S \end{Bmatrix} = \begin{bmatrix} K_{FF} & \vdots & K_{FS} \\ \hline K_{SF} & \vdots & K_{SS} \end{bmatrix} \begin{Bmatrix} \Delta_F \\ \hline \Delta_S \end{Bmatrix} \tag{14.17}$$

Subscripts F and S refer to freedom and support. Through partitioning we can work conveniently with the freedom related equations or the support related equations without having to write each set out in detail.

The freedom related equations are

$$\{F_F\} = [K_{FF}]\{\Delta_F\} + [K_{FS}]\{\Delta_S\} \tag{14.18}$$

which, because $\{\Delta_S\} = 0$, reduce to

$$\{F_F\} = [K_{FF}]\{\Delta_F\} \tag{14.19}$$

Thus

$$\begin{Bmatrix} 26.3 \\ 9.58 \\ 0 \\ 0 \\ 0 \end{Bmatrix} = 10^5 \begin{bmatrix} 3.02 & -0.851 & -1.829 & 0 & -1.193 \\ -0.851 & 3.169 & 0 & 0 & 0.851 \\ -1.829 & 0 & 3.02 & 0.851 & 0 \\ 0 & 0 & 0.851 & 3.169 & 0 \\ -1.193 & 0.851 & 0 & 0 & 3.02 \end{bmatrix} \begin{Bmatrix} \Delta_1 \\ \Delta_2 \\ \Delta_3 \\ \Delta_4 \\ \Delta_5 \end{Bmatrix}$$

which may be solved by the computer to give

$$\{\Delta_F\} = \begin{Bmatrix} \Delta_1 \\ \Delta_2 \\ \Delta_3 \\ \Delta_4 \\ \Delta_5 \end{Bmatrix} = \begin{Bmatrix} 2.22 \\ 0.718 \\ 1.456 \\ -0.391 \\ 0.675 \end{Bmatrix} \times 10^{-4} \text{ m}$$

Now with the displacements known, we are ready to solve for the support reactions and member forces. For the support reactions we use the support related forces of the partitioned matrix equation (14.17),

$$\{F_S\} = [K_{SF}]\{\Delta_F\} + [K_{SS}]\{\Delta_S\} \tag{14.20}$$

which, because $\{\Delta_S\} = 0$, reduces to

$$\{F_S\} = [K_{SF}]\{\Delta_F\} \tag{14.21}$$

$$= \begin{bmatrix} 0.851 & -0.609 & 0 & -2.56 & -0.851 \\ 0 & 0 & -1.193 & -0.851 & -1.829 \\ 0 & -2.56 & -0.851 & -0.609 & 0 \end{bmatrix} \begin{Bmatrix} 2.22 \\ 0.718 \\ 1.456 \\ -0.391 \\ 0.675 \end{Bmatrix} \times 10^{-4}$$

Expanded by the computer we get

$$\{F_S\} = \begin{Bmatrix} R_{Dy} \\ R_{Cx} \\ R_{Cy} \end{Bmatrix} = \begin{Bmatrix} 18.82 \\ -26.4 \\ -28.4 \end{Bmatrix} \text{ kN}$$

Next we use Eq. (14.10) to determine the member forces,

$$\{f\} = [k][T]\{\Delta\} \tag{14.10}$$

If $[k]$ and $[T]$ have been saved for each member, we proceed with the matrix multiplication. If storage space is limited, $[k]$ and $[T]$ may not have been stored, and it will be necessary to regenerate them. In using Eq. (14.10) it is important that subscripts used for $\{\Delta\}$ be those related to the *member* in the usual near end, far end sequence. Finally, only the second element (the assumed tension one) of $\{f\}$ need be calculated.

For member AB, applying Eq. (14.10),

$$\{f\} = \begin{Bmatrix} f_1 \\ f_2 \end{Bmatrix}_{AB} = 1.829 \times 10^5 \begin{bmatrix} 1 & -1 \\ -1 & 1 \end{bmatrix} \begin{bmatrix} 1 & 0 & 0 & 0 \\ 0 & 0 & 1 & 0 \end{bmatrix} \begin{Bmatrix} 1.456 \\ -0.391 \\ 2.22 \\ 0.718 \end{Bmatrix} \times 10^{-4}$$

$$f_2 = f_{AB} = -14.0 \text{ kN}$$

Similar calculations for other members give

$f_{AC} = 18.38$ kN
$f_{AD} = -15.17$ kN
$f_{CB} = 17.13$ kN
$f_{BD} = -10.0$ kN
$f_{CD} = 12.34$ kN

14.3 COMPUTER CONSIDERATIONS

In Example 14.2.2 we made frequent references to solutions using a computer. We will look more closely at computer methods in this section, but in general terms. It is not our plan to overwhelm you with programming details, nor to give you an all-purpose program. Rather, we wish to bring out a few ideas that may form the base for preparing your own programs and may, in addition, help you become a more wary user of packaged software.

In Example 14.2.2 we solved a set of simultaneous equations that came from a partitioned matrix equation. All the required coefficients were in submatrix $[K_{FF}]$. Assembling all those coefficients in that corner of $[K]$ so that the partitioning was possible was the direct result of our assigning subscripts for *all* the *unknown* displacements before assigning any to the known displacements. A solution can be reached if that scheme is not used, but it requires the computer to search through all the rows and columns to select and assemble those it needs before solving the equations.

So far nothing has been said about how the computer solves the set of simultaneous equations once they have been established. Our use of matrix methods might suggest a solution by matrix inversion. In matrix inversion, a solution of

$$[K_{FF}]\{\Delta_F\} = \{F_F\} \tag{14.19}$$

for $\{\Delta_F\}$ is accomplished by applying

$$\{\Delta_F\} = [K_{FF}]^{-1}\{F_F\} \tag{14.22}$$

where $[K_{FF}]^{-1}$ is the direct inverse[4] of $[K_{FF}]$, found using

$$[K_{FF}]^{-1} = \frac{[\text{adjoint } K_{FF}]}{[K_{FF}]} \tag{14.23}$$

Direct inversion using Eq. (14.23) is done easily for 2×2 or 3×3 matrices. But the matrices for larger, more complex structures usually are much bigger than 3×3. Their direct inversion is much more difficult, requiring the evaluation of a large number of higher order determinants. For very large systems direct inversion becomes impractical even for a computer. Several more practical methods for solving simultaneous equations are in common use today. Two of these are Gauss elimination, described in Appendix A, and Cholesky's method.[5]

---
[4] Readers not familiar with Eq. (14.23) are referred to any textbook on matrix algebra.
[5] Cholesky's method is an offshoot of Gauss elimination that requires a symmetric matrix, a property of all our stiffness matrices. It is described in most textbooks on matrix algebra.

Many structure stiffness matrices $[K]$ are sparse matrices, which means that a high proportion of their elements are zero-valued. Through judicious assignment of subscripts, these zero-valued elements can be positioned within the matrix to make it a *banded* matrix, preferably one with a *low bandwidth*. A banded matrix has all of its nonzero elements inside a strip centered on the main diagonal. The closer the nonzero elements are clustered around the diagonal, the smaller is the bandwidth. Formally, the largest numerical difference between the two subscripts associated with any of the nonzero elements in the matrix is the bandwidth.

A structure stiffness matrix bandwidth can be minimized by minimizing the numerical difference between subscripts assigned to the unknown global displacements related to each member. Two possible subscripting systems are shown for the same structure in the following diagrams.

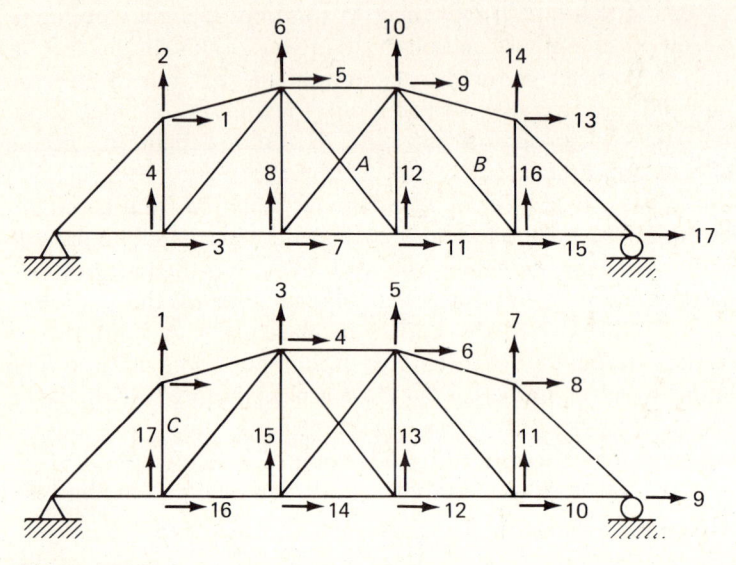

Figure 14.15

In the first the bandwidth is 7, as determined by the subscript differences in either member A or member B. In the second system the bandwidth is 16, found by the subscript differences for member C.

Two advantages accrue from low bandwidth matrices. First, special methods, including some adaptations of Gauss elimination and Cholesky's method, have been devised that are very efficient for use with banded matrices.

Second, low bandwidth matrices can produce large savings in computer storage, an important consideration if a large, complex structure is to be analyzed on a relatively small computer. Coupled with savings from the symmetry of the stiffness matrices, a banded matrix can reduce the required storage space to that needed to save the diagonal elements plus the elements within the bandwidth on just *one side* of the diagonal.

Perhaps of greatest importance in our matrix formulation is the idea that the method used to analyze the plane truss with its axial force elements is exactly the same when applied to other kinds of structures, including those composed of more

U. S. Customs Court (foreground) and Federal Office Building. The 41-story office building is a typical steel-framed tower, with columns supporting beams. The 8-story courthouse hangs from two parallel trusses above roof level to provided courtrooms and a pedestrian concourse at street level. (Courtesy of the Bethlehem Steel Corporation).

complicated members. The summing of the member K_{ij}'s to form the structure $[K]$ is the same for any structure. The element stiffness matrices $[k]$ and the transformation matrices $[T]$ will be different in size and makeup. But for all structures it will involve $[T]^T[k][T]$ to produce $[K]$ for each member. The required matrix multiplication may be done on the computer with equal ease for simple or complex $[k]$ and $[T]$.

Let us expand our analysis to space trusses and see how little the method changes.

14.4 SPACE TRUSS ANALYSIS

Aside from including the third dimension, analysis of space trusses is the same as for plane trusses. The members are axial force members with the same stiffness matrices $[k]$ in local coordinates. Thus Eq. (14.1) applies without change,

$$\begin{Bmatrix} f_1 \\ f_2 \end{Bmatrix} = \frac{AE}{L} \begin{bmatrix} 1 & -1 \\ -1 & 1 \end{bmatrix} \begin{Bmatrix} \partial_1 \\ \partial_2 \end{Bmatrix} \qquad (14.1)$$

Equation (14.7), transforming from global to local displacements, is unchanged,

$$\{\partial\} = [T]\{\Delta\} \qquad (14.7)$$

although both $[T]$ and $\{\Delta\}$ are changed from those for a plane truss to provide for the third dimension,

$$[T] = \begin{bmatrix} \cos(x'x) & \cos(x'y) & \cos(x'z) & 0 & 0 & 0 \\ 0 & 0 & 0 & \cos(x'x) & \cos(x'y) & \cos(x'z) \end{bmatrix} \quad (14.24)$$

$$\{\Delta\} = \begin{Bmatrix} \Delta_1 \\ \Delta_2 \\ \Delta_3 \\ \Delta_4 \\ \Delta_5 \\ \Delta_6 \end{Bmatrix} \quad (14.25)$$

The elements of $\{\Delta\}$ are, in order, the three global displacement components (in order x, y, and z) of the near end, followed by the x, y, and z global displacement components for the far end. In Eq. (14.24) the direction cosines are calculated based on differences in the coordinate values of the near and far ends. Thus comparable to Eqs. (14.9),

$$\cos(x'x) = \frac{x_{far} - x_{near}}{L} \quad (14.26a)$$

$$\cos(x'y) = \frac{y_{far} - y_{near}}{L} \quad (14.26b)$$

$$\cos(x'z) = \frac{z_{far} - z_{near}}{L} \quad (14.26c)$$

where L is the member length.

Member global stiffness matrices $[K]$ are generated exactly as for plane trusses by applying Eq. (14.15),

$$[K] = [T]^T[k][T] \quad (14.15)$$

Done on a computer, the matrix multiplication of Eq. (14.15) is no more taxing than that for a plane truss. The resulting $[K]$ is the member stiffness matrix in global coordinates, and thus is part of the member force/displacement relation of Eq. (14.14),

$$\{F\} = [K]\{\Delta\} \quad (14.14)$$

Once all the *member* stiffness matrices have been developed and stored, the *structure* stiffness matrix $[K]$ may be developed. As in plane trusses, it is the sum of the member stiffness matrices, taken by adding separately, for each subscript combination ij, all of the member K_{ij}'s.

As structures increase in size and complexity, computer storage requirements expand rapidly. Thus conserving storage space may be necessary. Large savings are possible if we avoid storing of each of the individual member stiffness matrices. We establish a *structure* stiffness matrix with all of its elements initialized at zero, and add to it the elements of each of the member stiffness matrices developed one member at a time.

We will illustrate some of the preceding with Example 14.4.1.

EXAMPLE 14.4.1

Analyze the space truss.

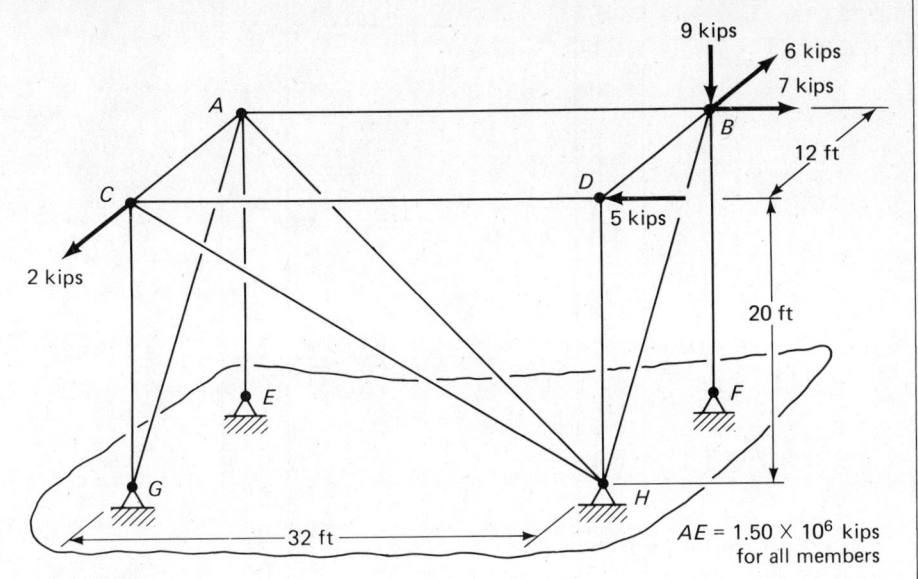

Figure 14.16

$AE = 1.50 \times 10^6$ kips for all members

We define a global coordinate system and assign the subscripting.

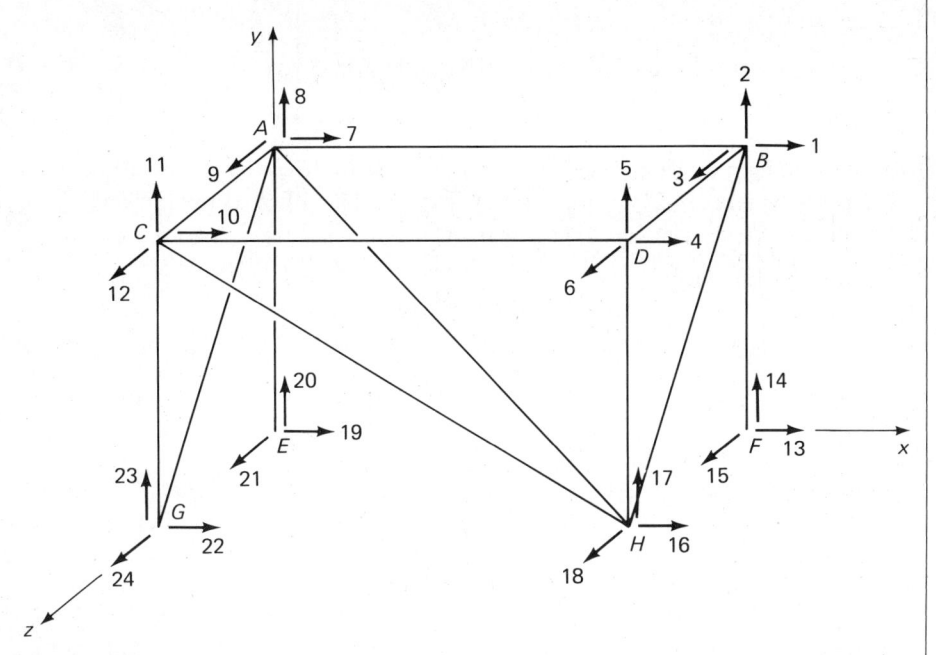

Figure 14.17

338 MATRIX FORMULATION FOR STRUCTURAL ANALYSIS

Thus $\{F\}$ and $\{\Delta\}$ become

$$\{F\} = \begin{Bmatrix} F_F \\ \hline F_S \end{Bmatrix} = \begin{Bmatrix} 7 \\ -9 \\ -6 \\ -5 \\ 0 \\ 0 \\ 0 \\ 0 \\ 0 \\ 0 \\ 0 \\ 2 \\ \hline F_x \\ F_y \\ F_z \\ H_x \\ H_y \\ H_z \\ E_x \\ E_y \\ E_z \\ G_x \\ G_y \\ G_z \end{Bmatrix} \text{kips} \qquad \{\Delta\} = \begin{Bmatrix} \Delta_F \\ \hline \Delta_S \end{Bmatrix} = \begin{Bmatrix} \Delta_1 \\ \Delta_2 \\ \Delta_3 \\ \Delta_4 \\ \Delta_5 \\ \Delta_6 \\ \Delta_7 \\ \Delta_8 \\ \Delta_9 \\ \Delta_{10} \\ \Delta_{11} \\ \Delta_{12} \\ \hline 0 \\ 0 \\ 0 \\ 0 \\ 0 \\ 0 \\ 0 \\ 0 \\ 0 \\ 0 \\ 0 \\ 0 \end{Bmatrix}$$

We begin by calculating member stiffness matrices by substituting Eqs. (14.3) and (14.24) in Eq. (14.15), first for AB with B as the far end, $\cos(x'x) = 1$, $\cos(x'y) = \cos(x'z) = 0$, and

$$\frac{AE}{L} = \frac{1.5 \times 10^6}{32} = 0.469 \times 10^5 \text{ kips/ft}$$

$$[K]_{AB} = 0.469 \times 10^5 \begin{bmatrix} 7 & 8 & 9 & 1 & 2 & 3 & \\ 1 & 0 & 0 & -1 & 0 & 0 & 7 \\ 0 & 0 & 0 & 0 & 0 & 0 & 8 \\ 0 & 0 & 0 & 0 & 0 & 0 & 9 \\ -1 & 0 & 0 & 1 & 0 & 0 & 1 \\ 0 & 0 & 0 & 0 & 0 & 0 & 2 \\ 0 & 0 & 0 & 0 & 0 & 0 & 3 \end{bmatrix}$$

The matrix row and column subscripts have been labeled outside of the brackets. They are the applicable subscripts for the global displacements of AB, listing x, y, and z of the near end first, followed by x, y, and z of the far end. The nonzero values stored in the *structure* stiffness matrix for this member are

$$K_{11} = K_{77} = 0.469 \times 10^5 \text{ kips/ft}$$
$$K_{17} = K_{71} = -0.469 \times 10^5 \text{ kips/ft}$$

Member CD is parallel to AB. With far end at D, its stiffness matrix is identical to AB's, except for the subscripts that represent x, y, and z of the near and far ends, which are 10, 11, 12, 4, 5, and 6.

$$[K]_{CD} = 0.469 \times 10^5 \begin{array}{c} \begin{matrix} 10 & 11 & 12 & 4 & 5 & 6 \end{matrix} \\ \begin{bmatrix} 1 & 0 & 0 & -1 & 0 & 0 \\ 0 & 0 & 0 & 0 & 0 & 0 \\ 0 & 0 & 0 & 0 & 0 & 0 \\ -1 & 0 & 0 & 1 & 0 & 0 \\ 0 & 0 & 0 & 0 & 0 & 0 \\ 0 & 0 & 0 & 0 & 0 & 0 \end{bmatrix} \end{array} \begin{matrix} 10 \\ 11 \\ 12 \\ 4 \\ 5 \\ 6 \end{matrix}$$

Nonzero values stored in the structure stiffness matrix are

$$K_{4,4} = K_{10,10} = 0.469 \times 10^5 \text{ kips/ft}$$
$$K_{4,10} = K_{10,4} = -0.469 \times 10^5 \text{ kips/ft}$$

Member AC, with far end at C, has $\cos(x'z) = 1$, $\cos(x'x) = \cos(x'y) = 0$, and

$$\frac{AE}{L} = \frac{1.5 \times 10^6}{12} = 1.25 \times 10^5 \text{ kips/ft}$$

Thus

$$[K]_{AC} = 1.25 \times 10^5 \begin{array}{c} \begin{matrix} 7 & 8 & 9 & 10 & 11 & 12 \end{matrix} \\ \begin{bmatrix} 0 & 0 & 0 & 0 & 0 & 0 \\ 0 & 0 & 0 & 0 & 0 & 0 \\ 0 & 0 & 1 & 0 & 0 & -1 \\ 0 & 0 & 0 & 0 & 0 & 0 \\ 0 & 0 & 0 & 0 & 0 & 0 \\ 0 & 0 & -1 & 0 & 0 & 1 \end{bmatrix} \end{array} \begin{matrix} 7 \\ 8 \\ 9 \\ 10 \\ 11 \\ 12 \end{matrix}$$

Member BD, parallel to AC, with far end at D, has a member stiffness matrix identical in form, except that its subscripts are 1, 2, 3, 4, 5, and 6.

The four verticals have member stiffness matrices that are identical in form when each is assigned its top joint as the far end. One of these, member FB, with far end at B, has

$$[K]_{FB} = 0.75 \times 10^5 \begin{array}{c} \begin{matrix} 13 & 14 & 15 & 1 & 2 & 3 \end{matrix} \\ \begin{bmatrix} 0 & 0 & 0 & 0 & 0 & 0 \\ 0 & 1 & 0 & 0 & -1 & 0 \\ 0 & 0 & 0 & 0 & 0 & 0 \\ 0 & 0 & 0 & 0 & 0 & 0 \\ 0 & -1 & 0 & 0 & 1 & 0 \\ 0 & 0 & 0 & 0 & 0 & 0 \end{bmatrix} \end{array} \begin{matrix} 13 \\ 14 \\ 15 \\ 1 \\ 2 \\ 3 \end{matrix}$$

Member *HD*, with far end at *D* uses subscripts 16, 17, 18, 4, 5, and 6. Member *CG*, with far end at *C*, has subscripts 22, 23, 24, 10, 11, and 12, while member *AE*, with far end at *A*, uses 19, 20, 21, 7, 8, and 9.

Diagonal member *HB*, with far end at *B*, is slightly more complicated; $\cos(x'x) = 0$, $\cos(x'y) = 20/23.0 = 0.857$, $\cos(x'z) = -12/23.3 = -0.514$, and

$$\frac{AE}{L} = \frac{1.5 \times 10^6}{23.3} = 0.643 \text{ kips/ft}$$

Thus

$$[K]_{HB} = 0.643 \times 10^5 \begin{bmatrix} 0 & 0 & 0 & 0 & 0 & 0 \\ 0 & 0.735 & -0.441 & 0 & -0.735 & 0.441 \\ 0 & -0.441 & 0.265 & 0 & 0.441 & -0.265 \\ 0 & 0 & 0 & 0 & 0 & 0 \\ 0 & -0.735 & 0.441 & 0 & 0.735 & -0.441 \\ 0 & 0.441 & -0.265 & 0 & -0.441 & 0.265 \end{bmatrix} \begin{matrix} 16 \\ 17 \\ 18 \\ 1 \\ 2 \\ 3 \end{matrix}$$

$$\begin{matrix} 16 & 17 & 18 & 1 & 2 & 3 \end{matrix}$$

Diagonal member *GA*, with far end at *A*, is identical to and parallel to *HB*. Its $[K]$ matches *HB*'s, but uses subscripts 22, 23, 24, 7, 8, and 9.

For diagonal *HC*, with far end at *C*, $\cos(x'x) = -32/37.74 = -0.848$, $\cos(x'y) = 20/37.74 = 0.530$, $\cos(x'z) = 0$, and

$$\frac{AE}{L} = \frac{1.5 \times 10^6}{37.74} = 0.3975 \times 10^5 \text{ kips/ft}$$

Thus

$$[K]_{HC} = 0.3975 \times 10^5 \begin{bmatrix} 0.719 & -0.449 & 0 & -0.719 & 0.449 & 0 \\ -0.449 & 0.281 & 0 & 0.449 & -0.281 & 0 \\ 0 & 0 & 0 & 0 & 0 & 0 \\ -0.719 & 0.449 & 0 & 0.719 & -0.449 & 0 \\ 0.449 & -0.281 & 0 & -0.449 & 0.281 & 0 \\ 0 & 0 & 0 & 0 & 0 & 0 \end{bmatrix} \begin{matrix} 16 \\ 17 \\ 18 \\ 10 \\ 11 \\ 12 \end{matrix}$$

$$\begin{matrix} 16 & 17 & 18 & 10 & 11 & 12 \end{matrix}$$

Finally, for diagonal *HA*, with far end at *A*, $\cos(x'x) = -32/39.6 = -0.808$, $\cos(x'y) = 12/39.6 = 0.303$, $\cos(x'z) = 20/39.6 = 0.505$, and

$$\frac{AE}{L} = \frac{1.5 \times 10^6}{39.6} = 0.379 \times 10^5 \text{ kips/ft}$$

Thus

$$[K]_{HA} = 0.379 \times 10^5 \begin{bmatrix} 0.653 & -0.245 & -0.408 & -0.653 & 0.245 & 0.408 \\ -0.245 & 0.0918 & 0.153 & 0.245 & -0.0918 & -0.153 \\ -0.408 & 0.153 & 0.255 & 0.408 & -0.153 & -0.255 \\ -0.653 & 0.245 & 0.408 & 0.653 & -0.245 & -0.408 \\ 0.245 & -0.0918 & -0.153 & -0.245 & 0.0918 & 0.153 \\ 0.408 & -0.153 & -0.255 & -0.408 & 0.153 & 0.255 \end{bmatrix} \begin{matrix} 16 \\ 17 \\ 18 \\ 7 \\ 8 \\ 9 \end{matrix}$$

$$\begin{matrix} 16 & 17 & 18 & 7 & 8 & 9 \end{matrix}$$

SPACE TRUSS ANALYSIS   341

Having added the elements of each of the member stiffness matrices as they were completed, the structure stiffness matrix has been assembled. At this stage we will show only submatrix $[K_{FF}]$, needed to solve for $\{\Delta_F\}$. Matrix Eq. (14.19), structure force/displacement, follows:

$$\begin{Bmatrix} 7 \\ -9 \\ -6 \\ -5 \\ 0 \\ 0 \\ 0 \\ 0 \\ 0 \\ 0 \\ 0 \\ 2 \end{Bmatrix} = 10^5 \begin{bmatrix} 0.469 & 0 & 0 & 0 & 0 & 0 & -0.469 & 0 & 0 & 0. & 0 & 0 \\ 0 & 1.22 & -0.284 & 0 & 0 & 0 & 0 & 0 & 0 & 0 & 0 & 0 \\ 0 & -0.284 & 1.42 & 0 & 0 & -1.25 & 0 & 0 & 0 & 0 & 0 & 0 \\ 0 & 0 & 0. & 0.469 & 0 & 0 & 0 & 0 & 0 & -0.469 & 0 & 0 \\ 0 & 0 & 0 & 0 & 0.75 & 0 & 0 & 0 & 0 & 0 & 0 & 0 \\ 0 & 0 & 1.25 & 0 & 0 & 1.25 & 0 & 0 & 0 & 0 & 0 & 0 \\ -0.469 & 0 & 0 & 0 & 0 & 0 & 0.717 & -0.093 & -0.155 & 0 & 0 & 0 \\ 0 & 0 & 0 & 0 & 0 & 0 & -0.093 & 1.258 & -0.226 & 0 & 0 & 0 \\ 0 & 0 & 0 & 0 & 0 & 0 & -0.155 & -0.226 & 1.517 & 0 & 0 & -1.25 \\ 0 & 0 & 0 & -0.469 & 0 & 0 & 0 & 0 & 0 & 0.755 & -0.1785 & 0 \\ 0 & 0 & 0 & 0 & 0 & 0 & 0 & 0 & 0 & -0.1785 & 0.862 & 0 \\ 0 & 0 & 0 & 0 & 0 & 0 & 0 & 0 & -1.25 & 0 & 0 & 1.25 \end{bmatrix} \begin{Bmatrix} \Delta_1 \\ \Delta_2 \\ \Delta_3 \\ \Delta_4 \\ \Delta_5 \\ \Delta_6 \\ \Delta_7 \\ \Delta_8 \\ \Delta_9 \\ \Delta_{10} \\ \Delta_{11} \\ \Delta_{12} \end{Bmatrix}$$

Solution by Gauss elimination gives

$$\{\Delta_F\} = \begin{Bmatrix} 9.173 \\ -2.55 \\ -7.79 \\ -3.07 \\ 0.00 \\ -7.79 \\ 7.68 \\ 1.773 \\ 6.71 \\ -2.01 \\ -0.416 \\ 6.87 \end{Bmatrix} \times 10^{-4}$$

Now, with $\{\Delta_F\}$ known, the support reactions may be found using Eq. (14.21)

$$\{F_S\} = [K_{SF}]\{\Delta_F\} \tag{14.21}$$

Submatrix $[K_{SF}]$, was developed as part of $[K]$. Thus

$$\begin{Bmatrix} F_x \\ F_y \\ F_z \\ H_x \\ H_y \\ H_z \\ E_x \\ E_y \\ E_z \\ G_x \\ G_y \\ G_z \end{Bmatrix} = 10^5 \begin{bmatrix} 0 & 0 & 0 & 0 & 0 & 0 & 0 & 0 & 0 & 0 & 0 & 0 \\ 0 & -0.75 & 0 & 0 & 0 & 0 & 0 & 0 & 0 & 0 & 0 & 0 \\ 0 & 0 & 0 & 0 & 0 & 0 & 0 & 0 & 0 & 0 & 0 & 0 \\ 0 & 0 & 0 & 0 & 0 & 0 & -0.248 & 0.093 & 0.155 & -0.286 & 0.1785 & 0 \\ 0 & -0.473 & -0.284 & 0 & -0.75 & 0 & 0.093 & -0.035 & -0.058 & 0.1785 & -0.1117 & 0 \\ 0 & 0.284 & -0.170 & 0 & 0 & 0 & 0.155 & -0.058 & -0.097 & 0 & 0 & 0 \\ 0 & 0 & 0 & 0 & 0 & 0 & 0 & 0 & 0 & 0 & 0 & 0 \\ 0 & 0 & 0 & 0 & 0 & 0 & -0.75 & 0 & 0 & 0 & 0 & 0 \\ 0 & 0 & 0 & 0 & 0 & 0 & 0 & 0 & 0 & 0 & 0 & 0 \\ 0 & 0 & 0 & 0 & 0 & 0 & 0 & 0 & 0 & 0 & 0 & 0 \\ 0 & 0 & 0 & 0 & 0 & 0 & -0.473 & -0.284 & 0 & -0.75 & 0 & 0 \\ 0 & 0 & 0 & 0 & 0 & 0 & 0.284 & -0.170 & 0 & 0 & 0 & 0 \end{bmatrix} \begin{Bmatrix} 9.173 \\ -2.55 \\ -7.79 \\ -3.07 \\ 0 \\ -7.79 \\ 7.68 \\ 1.773 \\ 6.71 \\ -2.01 \\ -0.416 \\ 6.87 \end{Bmatrix} 10^{-4}$$

Solution of the matrix equations give

$$\begin{Bmatrix} F_x \\ F_y \\ F_z \\ H_x \\ H_y \\ H_z \\ E_x \\ E_y \\ E_z \\ G_x \\ G_y \\ G_z \end{Bmatrix} = \begin{Bmatrix} 0 \\ 19.1 \\ 0 \\ -2.0 \\ 33.7 \\ 10.37 \\ 0 \\ -13.3 \\ 0 \\ 0 \\ -24.3 \\ -6.4 \end{Bmatrix} \text{ kips}$$

Only the member forces remain to be calculated. Again, Eq. (14.10) may be used, applied for each member, taking care that appropriate $\{\Delta\}$ are used for each member.

For member AB,

$$\begin{Bmatrix} f_1 \\ f_2 \end{Bmatrix} = 0.469 \times 10^5 \begin{bmatrix} 1 & -1 \\ -1 & 1 \end{bmatrix} \begin{bmatrix} 1 & 0 & 0 & 0 & 0 & 0 \\ 0 & 0 & 0 & 1 & 0 & 0 \end{bmatrix} \begin{Bmatrix} 7.68 \\ 1.773 \\ 6.71 \\ 9.173 \\ -2.55 \\ -7.79 \end{Bmatrix} \times 10^{-4}$$

The $[k]$ and $[T]$ used are for AB, while $\{\Delta\}$ consists of $\Delta_7, \Delta_8, \Delta_9, \Delta_1, \Delta_2$ and Δ_3, AB's near and far end displacements.

$$\begin{Bmatrix} f_1 \\ f_2 \end{Bmatrix} = \begin{Bmatrix} -7.00 \\ 7.00 \end{Bmatrix}$$

or

$$f_{AB} = 7 \text{ kips}$$

an obviously correct value if joint B is examined.

Similar calculations for the other members give

$f_{CD} - 4.97$ kips

$f_{AC} = 2.00$ kips

$f_{BD} = 0$

$f_{BF} = -19.1$ kips

$f_{DH} = 0$

$f_{GC} = -3.12$ kips

$f_{AE} = 13.3$ kips

$$f_{GA} = -12.4 \text{ kips}$$
$$f_{CH} = 5.90 \text{ kips}$$
$$f_{AH} = -8.46 \text{ kips}$$
$$f_{HB} = 11.69 \text{ kips}$$

14.5 RIGID PLANE FRAME ANALYSES

Let us now consider member bending in addition to axial deformation. We will limit our studies to plane frames with in-plane loads and deformations. The displacement method parallels our previous approaches for plane and space trusses exactly.

First we define the member end displacements in a local coordinate system $(x'y')$.

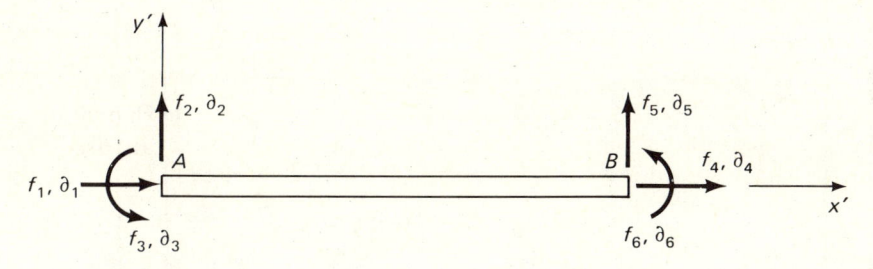

Figure 14.18

This time there are six end forces and displacements identified by subscripts. All are shown in their positive sense, with f and ∂, subscripted identically, referring to a force or displacement at a specific location. The word force here may mean couple, as in f_3 and f_6, as well as a linear force.

From elastic theory, and particularly from the slope deflection equations of Chapter 10, we note the following relations.

(1) For displacements ∂_1 and ∂_4,

$$f_1 = -f_4 = \frac{AE}{L}\partial_1 - \frac{AE}{L}\partial_4$$

$$f_2 = f_3 = f_5 = f_6 = 0$$

(2) For displacements ∂_2 and ∂_5,

$$f_2 = -f_5 = \frac{12EI}{L^3}\partial_2 - \frac{12EI}{L^3}\partial_5$$

$$f_3 = f_6 = \frac{6EI}{L^2}\partial_2 - \frac{6EI}{L^2}\partial_5$$

$$f_1 = f_4 = 0$$

(3) For displacements ∂_3 and ∂_6,

$$f_3 = \frac{4EI}{L}\partial_3 + \frac{2EI}{L}\partial_6$$

$$f_6 = \frac{4EI}{L}\partial_6 + \frac{2EI}{L}\partial_3$$

$$f_2 = -f_5 = \frac{6EI}{L^2}\partial_3 + \frac{6EI}{L^2}\partial_6$$

$$f_1 = f_4 = 0$$

In matrix form these are

$$\begin{Bmatrix} f_1 \\ f_2 \\ f_3 \\ f_4 \\ f_5 \\ f_6 \end{Bmatrix} = \begin{bmatrix} \frac{AE}{L} & 0 & 0 & -\frac{AE}{L} & 0 & 0 \\ 0 & \frac{12EI}{L^3} & \frac{6EI}{L^2} & 0 & -\frac{12EI}{L^3} & \frac{6EI}{L^2} \\ 0 & \frac{6EI}{L^2} & \frac{4EI}{L^2} & 0 & -\frac{6EI}{L^2} & \frac{2EI}{L} \\ -\frac{AE}{L} & 0 & 0 & \frac{AE}{L} & 0 & 0 \\ 0 & -\frac{12EI}{L^3} & -\frac{6EI}{L^2} & 0 & \frac{12EI}{L^3} & -\frac{6EI}{L^3} \\ 0 & \frac{6EI}{L^2} & \frac{2EI}{L} & 0 & -\frac{6EI}{L^2} & \frac{4EI}{L} \end{bmatrix} \begin{Bmatrix} \partial_1 \\ \partial_2 \\ \partial_3 \\ \partial_4 \\ \partial_5 \\ \partial_6 \end{Bmatrix} \quad (14.27)$$

or

$$\{f\} = [k]\{\partial\} \quad (14.2)$$

The composition of $[k]$ may be inferred from Eqs. (14.2) and (14.27).

We will need to transform displacements in global coordinates to local coordinates. Consider displacement Δ_1, a displacement of *near* end A in the global x direction.

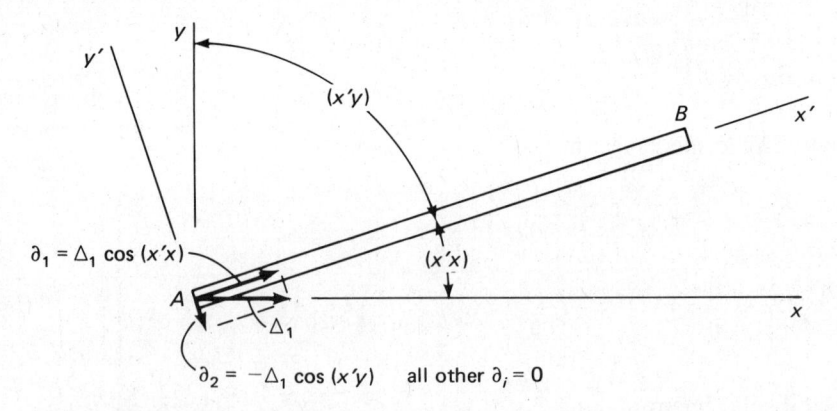

Figure 14.19

Next consider displacement Δ_2, a displacement of *near* end A in the global y direction.

RIGID PLANE FRAME ANALYSIS 345

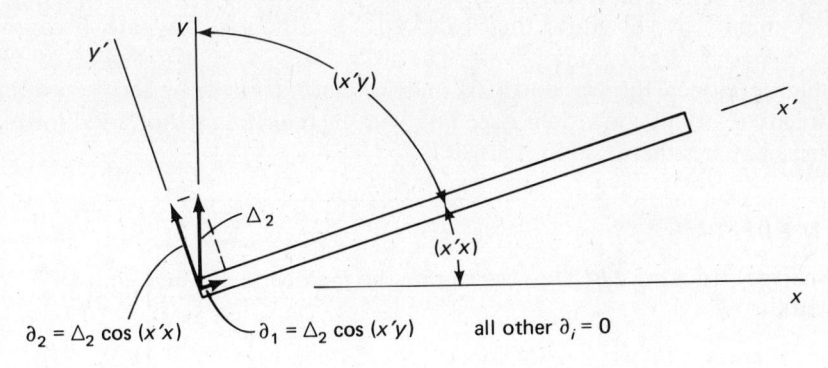

$\partial_2 = \Delta_2 \cos(x'x)$ $\partial_1 = \Delta_2 \cos(x'y)$ all other $\partial_i = 0$

Figure 14.20

Rotation Δ_3 of near end A is duplicated by an equal rotation ∂_3. Similar displacements at far end B of Δ_4 in the global x direction, Δ_5 in the global y direction, and rotation Δ_6 give

$$\partial_4 = \Delta_4 \cos(x'x) \quad \text{and} \quad \partial_5 = -\Delta_4 \cos(x'y)$$
$$\partial_4 = \Delta_5 \cos(x'y) \quad \text{and} \quad \partial_5 = \Delta_5 \cos(x'x)$$
$$\partial_6 = \Delta_6$$

These are expressed in matrix form,

$$\begin{Bmatrix} \partial_1 \\ \partial_2 \\ \partial_3 \\ \partial_4 \\ \partial_5 \\ \partial_6 \end{Bmatrix} = \begin{bmatrix} \cos(x'x) & \cos(x'y) & 0 & 0 & 0 & 0 \\ -\cos(x'y) & \cos(x'x) & 0 & 0 & 0 & 0 \\ 0 & 0 & 1 & 0 & 0 & 0 \\ 0 & 0 & 0 & \cos(x'x) & \cos(x'y) & 0 \\ 0 & 0 & 0 & -\cos(x'y) & \cos(x'x) & 0 \\ 0 & 0 & 0 & 0 & 0 & 1 \end{bmatrix} \begin{Bmatrix} \Delta_1 \\ \Delta_2 \\ \Delta_3 \\ \Delta_4 \\ \Delta_5 \\ \Delta_6 \end{Bmatrix} \quad (14.28)$$

or

$$\{\partial\} = [T]\{\Delta\} \quad (14.7)$$

That the 6×6 square matrix (14.28) is $[T]$ should be obvious. Its cosines are as defined in Eqs. (14.9).

Analogous to Eqs. (14.11) and (14.12) we may express the end forces in global coordinates from those in local coordinates by taking appropriate components. Without showing the details,

$$\begin{Bmatrix} F_1 \\ F_2 \\ F_3 \\ F_4 \\ F_5 \\ F_6 \end{Bmatrix} = \begin{bmatrix} \cos(x'x) & -\cos(x'y) & 0 & 0 & 0 & 0 \\ \cos(x'y) & \cos(x'x) & 0 & 0 & 0 & 0 \\ 0 & 0 & 1 & 0 & 0 & 0 \\ 0 & 0 & 0 & \cos(x'x) & -\cos(x'y) & 0 \\ 0 & 0 & 0 & \cos(x'y) & \cos(x'x) & 0 \\ 0 & 0 & 0 & 0 & 0 & 1 \end{bmatrix} \begin{Bmatrix} f_1 \\ f_2 \\ f_3 \\ f_4 \\ f_5 \\ f_6 \end{Bmatrix} \quad (14.29)$$

or

$$\{F\} = [T]^T\{f\} \quad (14.12)$$

Careful examination will show that $[T]^T \neq [T]$, although they are deceptively similar.

Having developed the $[k]$ and $[T]$ matrices and defined the local coordinate system, we are ready to show their use. You will see that the method used for plane rigid frames matches that used for trusses.

EXAMPLE 14.5.1

Analyze structure $ABCD$ to determine the member end forces and moments.

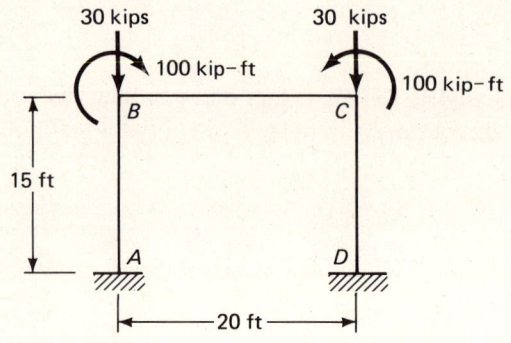

Figure 14.21

Member properties are for AB and CD,

$$I = 310 \text{ in}^4 \qquad A = 11.8 \text{ in}^2$$

and for BC,

$$I = 238 \text{ in}^4 \qquad A = 8.79 \text{ in}^2$$

$E = 3 \times 10^4$ kips/in² for all members

We start by prescribing the global coordinate system. The usual horizontal x and vertical y axes with origin at A are convenient. Then we first assign subscripts with the freedoms assigned serially from 1, followed by the supports.

Figure 14.22

Now $\{F\}$ and $\{\Delta\}$ are written,

$$\{F\} = \begin{Bmatrix} F_F \\ \hline F_S \end{Bmatrix} = \begin{Bmatrix} 0 \\ -30 \text{ kips} \\ -1200 \text{ kip-in} \\ 0 \\ -30 \text{ kips} \\ 1200 \text{ kip-in} \\ \hline R_{Ax} \\ R_{Ay} \\ M_A \\ R_{Dx} \\ R_{Dy} \\ M_D \end{Bmatrix} \quad \text{and} \quad \{\Delta\} = \begin{Bmatrix} \Delta_F \\ \hline \Delta_S \end{Bmatrix} = \begin{Bmatrix} \Delta_1 \\ \Delta_2 \\ \Delta_3 \\ \Delta_4 \\ \Delta_5 \\ \Delta_6 \\ \hline 0 \\ 0 \\ 0 \\ 0 \\ 0 \\ 0 \end{Bmatrix}$$

Next we calculate member $[k]$ matrices and $[T]$ matrices from Eqs. (14.27) and (14.28). Since members AB and CD are identical in properties and orientation, their matrices will be identical, although the applicable subscripts will change. Using their top ends B and C as the far ends,

$$\frac{AE}{L} = \frac{11.8(3 \times 10^4)}{15(12)} = 1967 \text{ kips/in}$$

$$\frac{12EI}{L^3} = \frac{12(3 \times 10^4)(310)}{(15 \times 12)^3} = 19.14 \text{ kips/in}$$

$$\frac{6EI}{L^2} = \frac{6(3 \times 10^4)(310)}{(15 \times 12)^2} = 1722 \text{ kips}$$

$$\frac{4EI}{L} = \frac{4(3 \times 10^4)(310)}{15 \times 12} = 2.067 \times 10^5 \text{ kip-in}$$

$$\frac{2EI}{L} = \frac{1}{2}\left(\frac{4EI}{L}\right) = 1.033 \times 10^5 \text{ kip-in}$$

$$\cos(x'x) = 0 \qquad \cos(x'y) = 1$$

Thus

$$[k]_{AB,CD} = \begin{bmatrix} 1967 & 0 & 0 & -1967 & 0 & 0 \\ 0 & 19.14 & 1722 & 0 & -19.14 & 1722 \\ 0 & 1722 & 2.067 \times 10^5 & 0 & -1722 & 1.033 \times 10^5 \\ -1967 & 0 & 0 & 1967 & 0 & 0 \\ 0 & -19.14 & -1722 & 0 & 19.14 & -1722 \\ 0 & 1722 & 1.033 \times 10^5 & 0 & -1722 & 2.067 \times 10^5 \end{bmatrix}$$

and

$$[T] = \begin{bmatrix} 0 & 1 & 0 & 0 & 0 & 0 \\ -1 & 0 & 0 & 0 & 0 & 0 \\ 0 & 0 & 1 & 0 & 0 & 0 \\ 0 & 0 & 0 & 0 & 1 & 0 \\ 0 & 0 & 0 & -1 & 0 & 0 \\ 0 & 0 & 0 & 0 & 0 & 1 \end{bmatrix}$$

Applying

$$[K] = [T]^T[k][T] \tag{14.15}$$

we get

$$[K]_{AB,CD} = \begin{bmatrix} 19.14 & 0 & -1722 & -19.14 & 0 & -1722 \\ 0 & 1967 & 0 & 0 & -1967 & 0 \\ -1722 & 0 & 2.067 \times 10^5 & 1722 & 0 & 1.033 \times 10^5 \\ -19.14 & 0 & 1722 & 19.14 & 0 & 1722 \\ 0 & -1967 & 0 & 0 & 1967 & 0 \\ -1722 & 0 & 1.033 \times 10^5 & 1722 & 0 & 2.067 \times 10^5 \end{bmatrix}$$

In order, the rows and columns are 7, 8, 9, 1, 2, and 3 for AB and 10, 11, 12, 4, 5, and 6 for CD.

A similar sequence, followed for BC, with far end at C, gives

$$[K]_{BC} = \begin{bmatrix} & 1 & 2 & 3 & 4 & 5 & 6 & \\ 1099 & 0 & 0 & -1099 & 0 & 0 & \end{bmatrix} \begin{matrix} 1 \\ 0 & 6.2 & 744 & 0 & -6.2 & 744 & \end{matrix} \begin{matrix} 2 \\ 0 & 744 & 1.195 \times 10^5 & 0 & -744 & 5.95 \times 10^4 & \end{matrix} \begin{matrix} 3 \\ -1099 & 0 & 0 & 1099 & 0 & 0 & \end{matrix} \begin{matrix} 4 \\ 0 & -6.2 & -744 & 0 & 6.2 & -744 & \end{matrix} \begin{matrix} 5 \\ 0 & 744 & 5.95 \times 10^4 & 0 & -744 & 1.19 \times 10^5 & \end{matrix} \begin{matrix} 6 \end{matrix}$$

The applicable subscripts are 1 through 6 consecutively for the rows and columns.

As each member $[K]$ was developed, its elements were combined in a structure stiffness matrix $[K]$. We will use only submatrix $[K_{FF}]$ of that $[K]$ at first. It is

$$[K_{FF}] = \begin{bmatrix} & 1 & 2 & 3 & 4 & 5 & 6 & \\ 1118 & 0 & 1722 & -1099 & 0 & 0 & \\ 0 & 1973 & 744 & 0 & -6.2 & 744 & \\ 1722 & 744 & 3.26 \times 10^5 & 0 & -744 & 5.95 \times 10^4 & \\ -1099 & 0 & 0 & 1118 & 0 & 1722 & \\ 0 & -6.2 & -744 & 0 & 1973 & -744 & \\ 0 & 744 & 5.95 \times 10^4 & 1722 & -744 & 3.26 \times 10^5 & \end{bmatrix} \begin{matrix} 1 \\ 2 \\ 3 \\ 4 \\ 5 \\ 6 \end{matrix}$$

With $\{F_F\}$ and $\{\Delta_F\}$ as defined earlier, we use a computer to solve $\{F_F\} = [K_{FF}]\{\Delta_F\}$ for $\{\Delta_F\}$,

$$\{\Delta_F\} = \begin{Bmatrix} 3.52 \times 10^{-3} \text{ in} \\ -1.525 \times 10^{-2} \text{ in} \\ -4.53 \times 10^{-3} \text{ rad} \\ -3.52 \times 10^{-3} \text{ in} \\ -1.525 \times 10^{-2} \text{ in} \\ 4.53 \times 10^{-3} \text{ rad} \end{Bmatrix}$$

Now we can calculate the support reactions, using Eq. (14.21),

$$\{F_S\} = [K_{SF}]\{\Delta_F\} \tag{14.21}$$

using the $[K_{SF}]$ submatrix from the already assembled structure stiffness matrix $[K]$.

$$\{F_S\} = \begin{bmatrix} -19.14 & 0 & -1722 & 0 & 0 & 0 \\ 0 & -1967 & 0 & 0 & 0 & 0 \\ 1722 & 0 & 1.033 \times 10^5 & 0 & 0 & 0 \\ 0 & 0 & 0 & -19.14 & 0 & -1722 \\ 0 & 0 & 0 & 0 & -1967 & 0 \\ 0 & 0 & 0 & 1722 & 0 & 1.033 \times 10^5 \end{bmatrix} \times$$

$$\begin{Bmatrix} 3.52 \times 10^{-3} \\ -1.525 \times 10^{-2} \\ -4.53 \times 10^{-3} \\ -3.52 \times 10^{-3} \\ -1.525 \times 10^{-2} \\ 4.53 \times 10^{-3} \end{Bmatrix}$$

$$\begin{Bmatrix} R_{Ax} \\ R_{Ay} \\ M_A \\ R_{Dx} \\ R_{Dy} \\ M_D \end{Bmatrix} = \begin{Bmatrix} 7.73 \text{ kips} \\ 30.0 \text{ kips} \\ -462 \text{ kip-in} \\ -7.73 \text{ kips} \\ 30.0 \text{ kips} \\ 462 \text{ kip-in} \end{Bmatrix}$$

Finally for the individual member forces we apply Eq. (14.10). For member $[AB]$, using its $[k]$ and $[T]$ previously written,

$$\{f\}_{AB} = [k]_{AB}[T]_{AB}\{\Delta\}_{AB}$$

$$= \begin{bmatrix} 1967 & 0 & 0 & -1967 & 0 & 0 \\ 0 & 19.14 & 1722 & 0 & -19.14 & 1722 \\ 0 & 1722 & 206700 & 0 & -1722 & 103300 \\ -1967 & 0 & 0 & 1967 & 0 & 0 \\ 0 & -19.14 & -1722 & 0 & 19.14 & -1722 \\ 0 & 1722 & 103300 & 0 & -1722 & 206700 \end{bmatrix} \times$$

$$\begin{bmatrix} 0 & 1 & 0 & 0 & 0 & 0 \\ -1 & 0 & 0 & 0 & 0 & 0 \\ 0 & 0 & 1 & 0 & 0 & 0 \\ 0 & 0 & 0 & 0 & 1 & 0 \\ 0 & 0 & 0 & -1 & 0 & 0 \\ 0 & 0 & 0 & 0 & 0 & 1 \end{bmatrix} \begin{Bmatrix} 0 \\ 0 \\ 0 \\ 0.00352 \\ -0.01525 \\ -0.00453 \end{Bmatrix}$$

The elements of displacement $\{\Delta_{AB}\}$ used are, in order, the global displacements in x, y and rotation at AB's near end and far end. Thus

$$\{f\}_{AB} = \begin{Bmatrix} f_1 \\ f_2 \\ f_3 \\ f_4 \\ f_5 \\ f_6 \end{Bmatrix} = \begin{Bmatrix} 30 \text{ kips} \\ -7.73 \text{ kips} \\ -462 \text{ kip-in} \\ -30 \text{ kips} \\ 7.73 \text{ kips} \\ -930 \text{ kip-in} \end{Bmatrix}$$

Thus for member AB, sketched in its actual orientation,

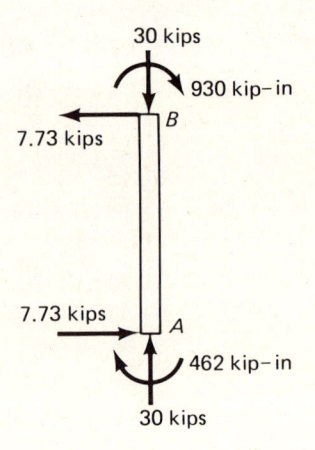

Figure 14.23

Similar calculation for BC gives

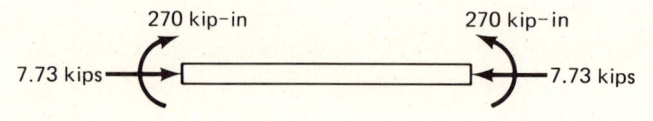

Figure 14.24

and for CD,

Figure 14.25

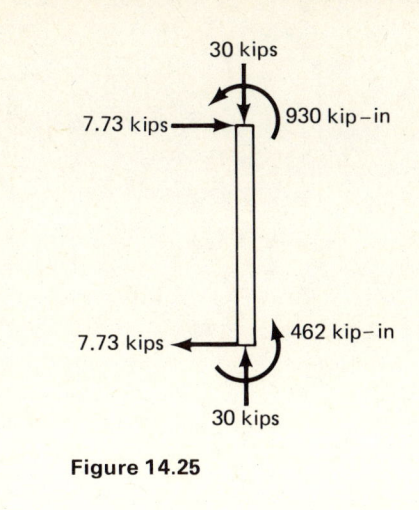

[Figure 14.25: vertical member with 30 kips down and 7.73 kips right at top, 930 kip-in moment at top; 7.73 kips left, 30 kips up, and 462 kip-in moment at bottom]

At this point, having worked your way through this example, it would be valuable for you to do it again, but this time using the slope deflection method. Comparison of the equations for the slope deflection solution with those of the matrix solution should help you understand the matrix method. At least it should show you that the matrix methods produce solutions by a well organized application of the displacement method and not by magic or voodoo.

Now let us modify the structure of Example 14.5.1 to add some new elements to the solution.

EXAMPLE 14.5.2

Analyze structure $ABCD$ to determine the member end forces and moments.

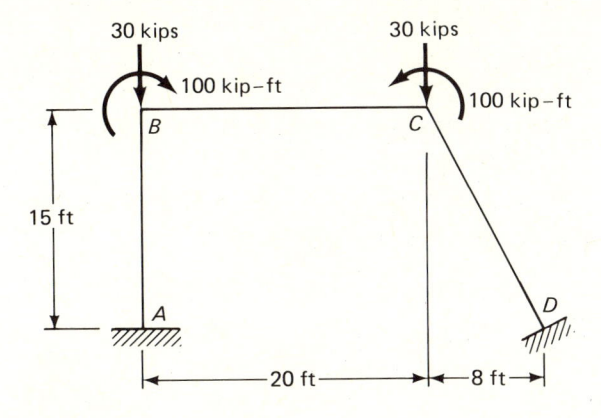

[Figure 14.26: frame ABCD with fixed support at A, column AB of 15 ft, horizontal beam BC of 20 ft, inclined member CD with D fixed, 8 ft horizontal distance from C to D; 30 kips downward and 100 kip-ft moment at B; 30 kips downward and 100 kip-ft moment at C]

Figure 14.26

352 MATRIX FORMULATION FOR STRUCTURAL ANALYSIS

Member properties are

$AB:$ $I = 310$ in^4 $\quad A = 11.8$ in^2
$BC:$ $I = 238$ in^4 $\quad A = 8.79$ in^2
$CD:$ $I = 394$ in^4 $\quad A = 14.7$ in^2
$E = 3 \times 10^4$ kips/in^2

Subscripting and global coordinates are chosen.

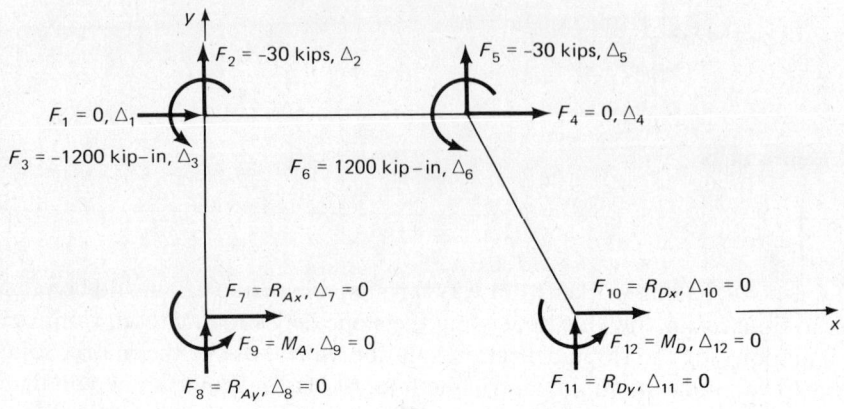

Figure 14.27

$\{F\}$ and $\{\Delta\}$ duplicate those for Example 14.5.1 exactly, as do $[K]_{AB}$ and $[K]_{BC}$. Member CD, different from that of Example 14.5.1, has a new $[K]$ to be developed here.

$$\frac{AE}{L} = 2162 \text{ kips/in} \qquad \frac{2EI}{L} = 1.159 \times 10^5 \text{ kip-in}$$

$$\frac{12EI}{L^3} = 16.71 \text{ kips/in} \qquad \cos(x'x) = -0.4706$$

$$\frac{6EI}{L^2} = 1704 \text{ kips} \qquad \cos(x'y) = +0.882$$

$$\frac{4EI}{L} = 2.32 \times 10^5 \text{ kip-in}$$

The direction cosines are for C as the far end. These values are used to assemble $[k]_{CD}$ and $[T]_{CD}$, which are substituted in Eq. (14.15).

The resulting member stiffness matrix is

$$[K_{CD}] = \begin{bmatrix} 491.6 & -890 & -1503 & -491.6 & 890 & -1503 \\ -890 & 1686 & -802 & 890 & -1686 & -802 \\ -1503 & -802 & 2.32 \times 10^5 & 1503 & 802 & 1.159 \times 10^5 \\ -491.6 & 890 & 1503 & 491.6 & -890 & 1503 \\ 890 & -1686 & 802 & -890 & 1686 & 802 \\ -1503 & -802 & 1.159 \times 10^5 & 1503 & 802 & 2.32 \times 10^5 \end{bmatrix} \begin{matrix} 10 \\ 11 \\ 12 \\ 4 \\ 5 \\ 6 \end{matrix}$$

$\qquad\qquad\qquad 10 \qquad 11 \qquad 12 \qquad\quad 4 \qquad\quad 5 \qquad\quad 6$

RIGID PLANE FRAME ANALYSIS 353

Its elements are added to the values stored in the structure stiffness matrix for members AB and BC to get $[K]$ for the structure.

Taking submatrix $[K_{FF}]$ from $[K]$ and $\{F_F\}$ from Example 14.5.1, we set up the matrix equation to solve for the displacements $\{\Delta_F\}$,

$$\begin{bmatrix} 1118 & 0 & 1722 & -1099 & 0 & 0 \\ 0 & 1973 & 744 & 0 & -6.2 & 744 \\ 1722 & 744 & 3.26 \times 10^5 & 0 & -744 & 5.95 \times 10^4 \\ -1099 & 0 & 0 & 1590.6 & -890 & 1503 \\ 0 & -6.2 & -744 & -890 & 1692 & 58 \\ 0 & 744 & 5.95 \times 10^4 & 1503 & 58 & 3.51 \times 10^5 \end{bmatrix} \begin{Bmatrix} \Delta_1 \\ \Delta_2 \\ \Delta_3 \\ \Delta_4 \\ \Delta_5 \\ \Delta_6 \end{Bmatrix}$$

$$= \begin{Bmatrix} 0 \\ -30 \\ -1200 \\ 0 \\ -30 \\ 1200 \end{Bmatrix}$$

$$\{\Delta_F\} = \begin{Bmatrix} \Delta_1 \\ \Delta_2 \\ \Delta_3 \\ \Delta_4 \\ \Delta_5 \\ \Delta_6 \end{Bmatrix} = \begin{Bmatrix} -5.052 \times 10^{-1} \text{ in} \\ -1.742 \times 10^{-2} \text{ in} \\ -2.77 \times 10^{-3} \text{ rad} \\ -5.18 \times 10^{-1} \text{ in} \\ -2.92 \times 10^{-1} \text{ in} \\ 6.20 \times 10^{-3} \text{ rad} \end{Bmatrix}$$

Support reactions are found using Eq. (14.21), for which submatrix $[K_{SF}]$ is taken from $[K]$.

$$\{F_S\} = \begin{bmatrix} -19.14 & 0 & -1722 & 0 & 0 & 0 \\ 0 & -1967 & 0 & 0 & 0 & 0 \\ 1722 & 0 & 1.033 \times 10^5 & 0 & 0 & 0 \\ 0 & 0 & 0 & -491.7 & 891 & -1504 \\ 0 & 0 & 0 & 891 & -1687 & -802 \\ 0 & 0 & 0 & 1504 & 802 & 1.159 \times 10^5 \end{bmatrix} \times \begin{Bmatrix} -5.052 \times 10^{-1} \\ -1.742 \times 10^{-2} \\ -2.77 \times 10^{-3} \\ -5.18 \times 10^{-1} \\ -2.92 \times 10^{-1} \\ 6.20 \times 10^{-3} \end{Bmatrix}$$

$$\begin{Bmatrix} R_{Ax} \\ R_{Ay} \\ M_A \\ R_{Dx} \\ R_{Dy} \\ M_D \end{Bmatrix} = \begin{Bmatrix} 14.44 \text{ kips} \\ 34.25 \text{ kips} \\ -1156.7 \text{ kip-in} \\ -14.44 \text{ kips} \\ 25.75 \text{ kips} \\ -295.3 \text{ kip-in} \end{Bmatrix}$$

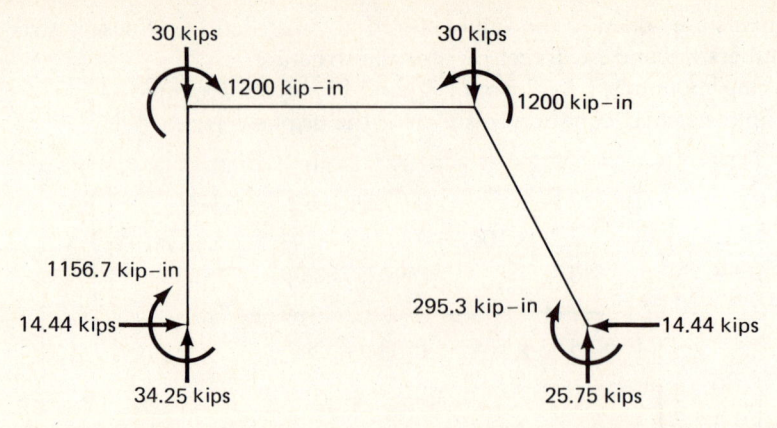

354 MATRIX FORMULATION FOR STRUCTURAL ANALYSIS

Figure 14.28

Finally we calculate the member end forces and moments by applying Eq. (14.10). We will use the $[k]_{AB}$ and $[T]_{AB}$ established in Example 14.5.1, but with the $\{\Delta\}$ for the global end displacements of member AB for this example, listing the three near end displacements followed by the three far end displacements.

$$\{f\}_{AB} = [k]_{AB}[T]_{AB}\{\Delta\}_{AB}$$

$$= \begin{bmatrix} 1967 & 0 & 0 & -1967 & 0 & 0 \\ 0 & 19.14 & 1722 & 0 & -19.14 & 1722 \\ 0 & 1722 & 206700 & 0 & -1722 & 103300 \\ -1967 & 0 & 0 & 1967 & 0 & 0 \\ 0 & -19.14 & -1722 & 0 & 19.14 & -1722 \\ 0 & 1722 & 103300 & 0 & -1722 & 206700 \end{bmatrix} \times$$

$$\begin{bmatrix} 0 & 1 & 0 & 0 & 0 & 0 \\ -1 & 0 & 0 & 0 & 0 & 0 \\ 0 & 0 & 1 & 0 & 0 & 0 \\ 0 & 0 & 0 & 0 & 1 & 0 \\ 0 & 0 & 0 & -1 & 0 & 0 \\ 0 & 0 & 0 & 0 & 0 & 1 \end{bmatrix} \begin{Bmatrix} 0 \\ 0 \\ 0 \\ -0.5052 \\ -0.01742 \\ -0.00277 \end{Bmatrix}$$

$$= \begin{Bmatrix} 34.25 \text{ kips} \\ -14.44 \text{ kips} \\ -1156.7 \text{ kip-in} \\ -34.25 \text{ kips} \\ 14.44 \text{ kips} \\ -1443.1 \text{ kip-in} \end{Bmatrix}$$

The three members with end reactions are shown.

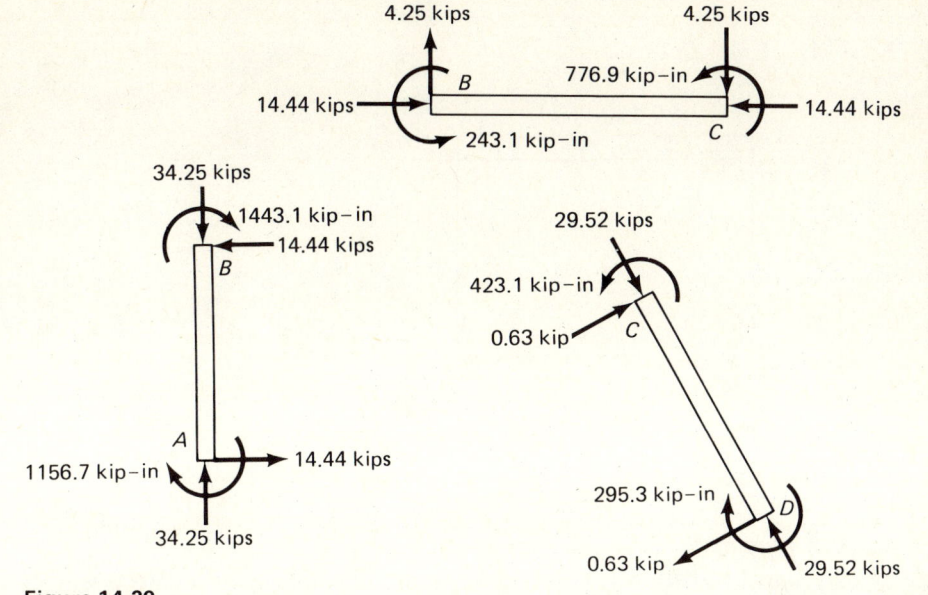

Figure 14.29

Matrix calculations are sensitive to round off errors. Readers may note some minor discrepancies in the third or fourth significant digits in some of the end forces. The end forces were actually calculated on a computer. They are a close, but not exact, match with those found from hand calculations using the rounded off stiffness elements of this example.

14.6 LOADS BETWEEN JOINTS

In most rigid frames loads are applied between the joints, not just at the joints. The loading may consist of concentrated loads or distributed loads, or both, in a limitless number of possible combinations.

One way to include this loading in the analysis is to introduce extra nodes (joints) at various locations along the loaded member. Each added node is treated as a rigid joint with unknown displacements. If the loading consists only of concentrated forces or couples, an extra node is added at each load point with the load applied to the joint.

For distributed forces a number of extra nodes may be added, each carrying a concentrated force equivalent to the total distributed load that is applied to the portion of the member centered at the node and extending halfway to the adjacent node on either side. Accuracy increases as the number of these extra nodes increases, but at the expense of adding more unknown displacements and requiring that added node subscripts, coordinates, and so on, be entered.

An alternate method, one especially useful when added nodes would tax the capacity of the computer, uses fixed end forces and moments. Consider Example 14.6.1 in which the structure of Example 14.5.1 is loaded with a distributed force.

EXAMPLE 14.6.1

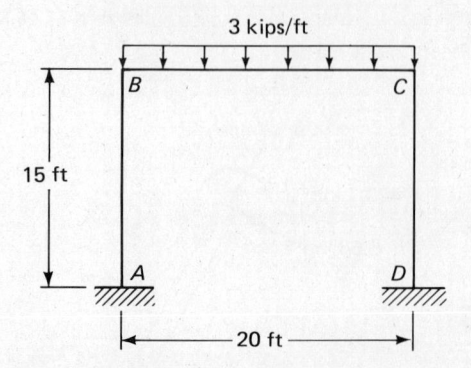

Figure 14.30

Member properties are as given in Example 14.5.1

We start by artificially constraining joints B and C against displacement, applying fixed end forces and moments for member BC based on Table 10.1.

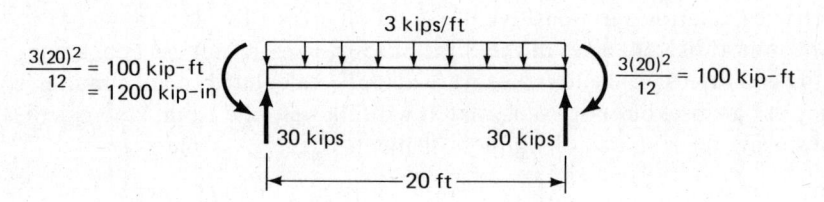

$\dfrac{3(20)^2}{12} = 100$ kip-ft $= 1200$ kip-in

$\dfrac{3(20)^2}{12} = 100$ kip-ft

30 kips   30 kips

Figure 14.31

Thus if structure $ABCD$ had been loaded as shown, there would be no rotational displacements at B or C.

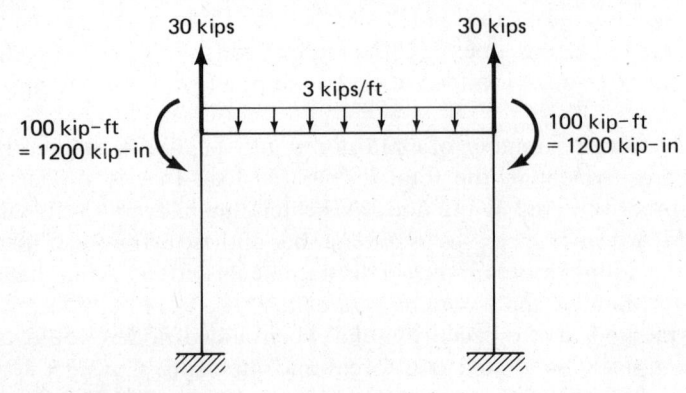

30 kips   30 kips

100 kip-ft = 1200 kip-in

100 kip-ft = 1200 kip-in

3 kips/ft

Figure 14.32

This is the original loading combined with two extra 30-kip forces and two extra 100 kip-ft couples at B and C. Because there are no joint displacements, only member BC would have end forces and moments.

We restore our original loading by superposing the loading of Fig. 14.32 with that of Fig. 14.33.

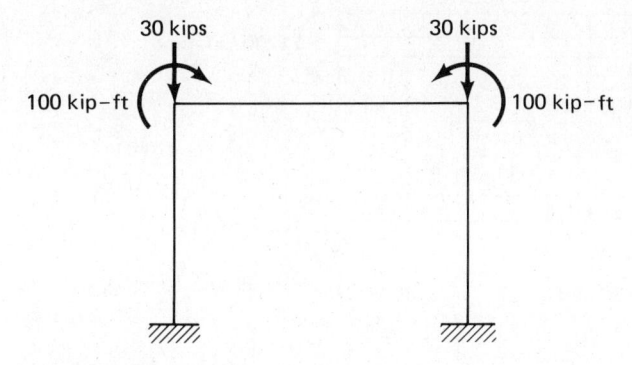

Figure 14.33

which is exactly the loading of Example 14.5.1. In essence, we superposed as follows.

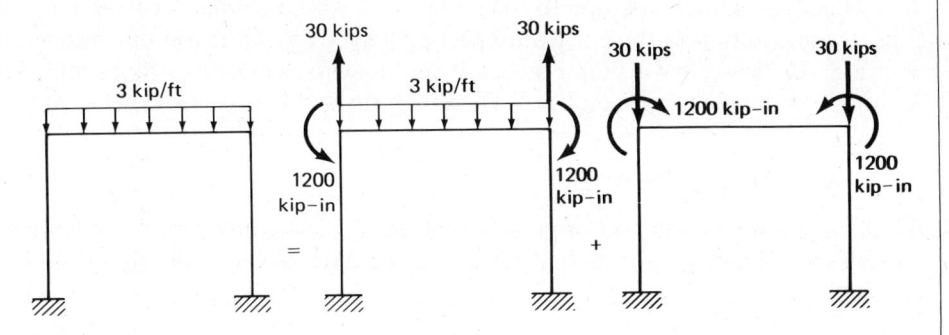

Figure 14.34

If we were analyzing Example 14.6.1 without having done Example 14.5.1 in advance, the sequence would be the reverse of that used here. Fixed end forces and moments found for the loaded members would be reversed and applied at the member end joints and an analysis completed such as for Example 14.5.1. Then the fixed end moments of the loaded members would be superposed.

For this example members AB and CD have end forces and moments exactly as found in Example 14.5.1. Member BC is the superposition of the end forces and moments from Example 14.5.1 and the fixed end forces and moments of the loaded members. Thus for BC,

358 MATRIX FORMULATION FOR STRUCTURAL ANALYSIS

Figure 14.35

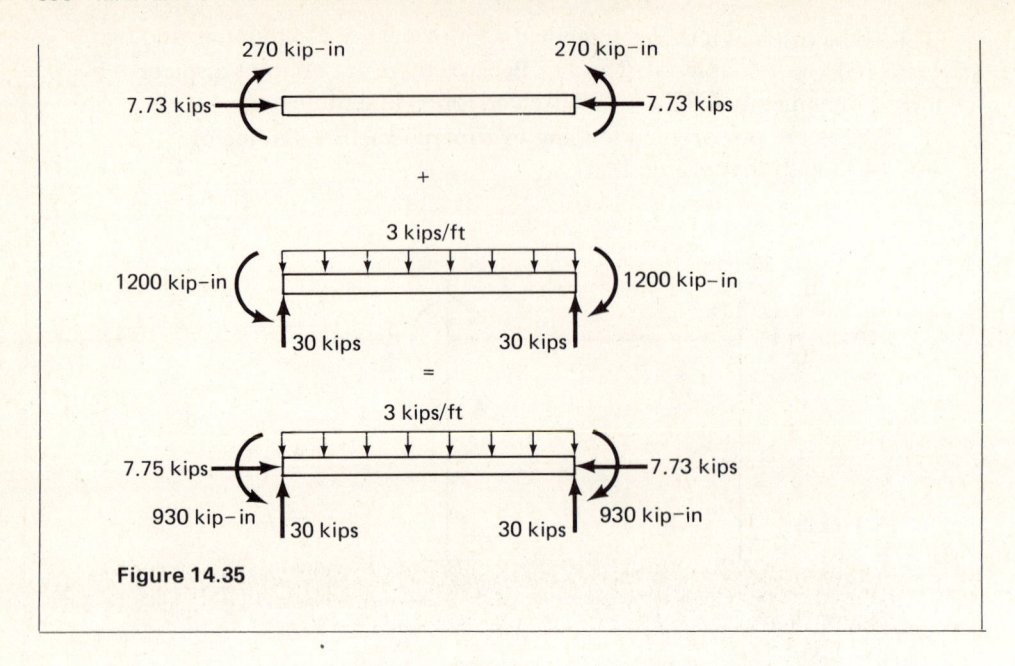

There are many commercially available software packages available that may be used to treat cases such as that of Example 14.6.1. Several of these systems reduce the "between joint" loading to that for which the forces and couples are applied to the joints and then neglect to add in the fixed end moments and forces of the loaded members, that is, their solution to Example 14.6.1 would be that for Example 14.5.1. The documentation of these software packages may not warn you of this omission. It is a good idea to try a simple structure and load system, comparing simple hand solutions to the computer solutions. Then such surprise omissions won't betray you.

14.7 SUPPORT DISPLACEMENTS

Until now our matrix formulations have been for structures with unmoving supports, that is, $\{\Delta_S\} = 0$. This permitted us to discard the second term from Eqs. (14.18) and (14.20),

$$\{F_F\} = [K_{FF}]\{\Delta_F\} + [K_{FS}]\{\Delta_S\} \tag{14.18}$$

$$\{F_S\} = [K_{SF}]\{\Delta_F\} + [K_{SS}]\{\Delta_S\} \tag{14.20}$$

If there are support displacements, the second terms must be retained and used. In Eq. (14.18) with known displacements $\{\Delta_S\}$, $[K_{FS}]\{\Delta_S\}$ becomes a set of numbers that are combined with $\{F_F\}$ algebraically and Eq. (14.18) is solved for $\{\Delta_F\}$. Support reactions are then found from Eq. (14.20), also involving just one added set of values.

EXAMPLE 14.7.1

Determine the member end forces of the structure of Example 14.5.1 that would be the result of a 1.0-in downward settlement of support A.

Using the subscripting of Example 14.5.1, the given settlements are

$$\{\Delta_S\} = \begin{Bmatrix} \Delta_7 \\ \Delta_8 \\ \Delta_9 \\ \Delta_{10} \\ \Delta_{11} \\ \Delta_{12} \end{Bmatrix} = \begin{Bmatrix} 0 \\ -1 \text{ in} \\ 0 \\ 0 \\ 0 \\ 0 \end{Bmatrix}$$

For this example, involving only support motion, we will assume that there are no externally applied forces. Thus

$$\{F_F\} = \{0\}$$

We take submatrix $[K_{FS}]$ from the structure stiffness matrix $[K]$ that was generated in Example 14.5.1.

$$[K_{FS}] = \begin{bmatrix} -19.14 & 0 & 1722 & 0 & 0 & 0 \\ 0 & -1967 & 0 & 0 & 0 & 0 \\ -1722 & 0 & 1.033 \times 10^5 & 0 & 0 & 0 \\ 0 & 0 & 0 & -19.14 & 0 & 1722 \\ 0 & 0 & 0 & 0 & -1967 & 0 \\ 0 & 0 & 0 & -1722 & 0 & 1.033 \times 10^5 \end{bmatrix} \begin{matrix} 1 \\ 2 \\ 3 \\ 4 \\ 5 \\ 6 \end{matrix}$$

$$\begin{matrix} 7 & 8 & 9 & 10 & 11 & 12 \end{matrix}$$

so that

$$[K_{FS}]\{\Delta_S\} = \begin{Bmatrix} 0 \\ 1967 \\ 0 \\ 0 \\ 0 \\ 0 \end{Bmatrix}$$

Substituting in Eq. (14.18) and rearranging with submatrix $[K_{FF}]$ taken from Example 14.5.1,

$$\begin{bmatrix} 1118 & 0 & 1722 & -1099 & 0 & 0 \\ 0 & 1973 & 744 & 0 & -6.2 & 744 \\ 1722 & 744 & 3.26 \times 10^5 & 0 & -744 & 5.95 \times 10^4 \\ -1099 & 0 & 0 & 1118 & 0 & 1722 \\ 0 & -6.2 & -744 & 0 & 1973 & -744 \\ 0 & 744 & 5.95 \times 10^4 & 1722 & -744 & 3.26 \times 10^5 \end{bmatrix} \begin{Bmatrix} \Delta_1 \\ \Delta_2 \\ \Delta_3 \\ \Delta_4 \\ \Delta_5 \\ \Delta_6 \end{Bmatrix}$$

$$= \begin{Bmatrix} 0 \\ -1967 \\ 0 \\ 0 \\ 0 \\ 0 \end{Bmatrix}$$

$$\{\Delta_F\} = \begin{Bmatrix} \Delta_1 \\ \Delta_2 \\ \Delta_3 \\ \Delta_4 \\ \Delta_5 \\ \Delta_6 \end{Bmatrix} = \begin{Bmatrix} -0.2904 \text{ in} \\ -0.9995 \text{ in} \\ 3.227 \times 10^{-3} \text{ rad} \\ -0.2904 \text{ in} \\ -7.016 \times 10^{-4} \text{ in} \\ 3.227 \times 10^{-3} \text{ rad} \end{Bmatrix}$$

Support reactions are found using Eq. (14.20) involving submatrices $[K_{SF}]$ and $[K_{SS}]$, both taken from the previously assembled $[K]$ in Example 14.5.1,

$$\{F_S\} = [K_{SF}]\{\Delta_F\} + [K_{SS}]\{\Delta_S\} \tag{14.20}$$

$$= \begin{bmatrix} -19.14 & 0 & -1722 & 0 & 0 & 0 \\ 0 & -1967 & 0 & 0 & 0 & 0 \\ 1722 & 0 & 1.033 \times 10^5 & 0 & 0 & 0 \\ 0 & 0 & 0 & -19.14 & 0 & -1722 \\ 0 & 0 & 0 & 0 & -1967 & 0 \\ 0 & 0 & 0 & 1722 & 0 & 1.033 \times 10^5 \end{bmatrix} \times \begin{Bmatrix} -0.2904 \\ -0.9993 \\ 3.227 \times 10^{-3} \\ -0.2904 \\ -7.016 \times 10^{-4} \\ 3.227 \times 10^{-3} \end{Bmatrix}$$

$$+ \begin{bmatrix} 19.14 & 0 & -1722 & 0 & 0 & 0 \\ 0 & 1967 & 0 & 0 & 0 & 0 \\ -1722 & 0 & 2.067 \times 10^5 & 0 & 0 & 0 \\ 0 & 0 & 0 & 19.14 & 0 & -1722 \\ 0 & 0 & 0 & 0 & 1967 & 0 \\ 0 & 0 & 0 & -1722 & 0 & 2.067 \times 10^5 \end{bmatrix} \begin{Bmatrix} 0 \\ -1 \\ 0 \\ 0 \\ 0 \\ 0 \end{Bmatrix}$$

$$= \begin{Bmatrix} 0 \\ 1965.61 \\ -166.8 \\ 0 \\ 1.380 \\ -166.8 \end{Bmatrix} + \begin{Bmatrix} 0 \\ -1967 \\ 0 \\ 0 \\ 0 \\ 0 \end{Bmatrix} = \begin{Bmatrix} 0 \\ -1.38 \text{ kips} \\ -166.8 \text{ kip-in} \\ 0 \\ 1.38 \text{ kips} \\ -166.8 \text{ kip-in} \end{Bmatrix}$$

Thus

166.8 kip-in   166.8 kip-in

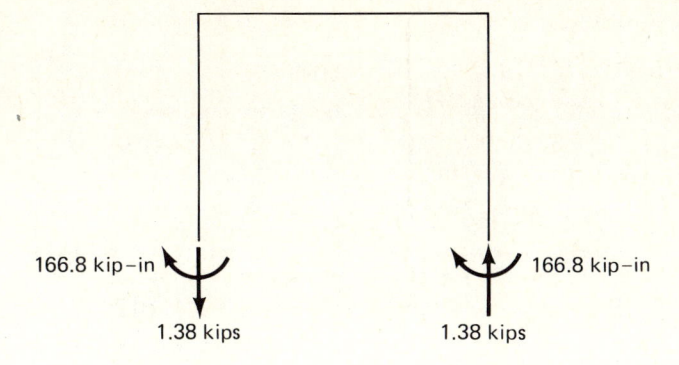

1.38 kips   1.38 kips

Figure 14.36

Member end forces are calculated from Eq. (14.10). First for member AB,

$$\{f\}_{AB} = [k]_{AB}[T]_{AB}\{\Delta\}_{AB}$$

$$= \begin{bmatrix} 1967 & 0 & 0 & -1967 & 0 & 0 \\ 0 & 19.14 & 1722 & 0 & -19.14 & 1722 \\ 0 & 1722 & 2.067 \times 10^5 & 0 & -1722 & 1.033 \times 10^5 \\ -1967 & 0 & 0 & 1967 & 0 & 0 \\ 0 & -19.14 & -1722 & 0 & 19.14 & -1722 \\ 0 & 1722 & 1.033 \times 10^5 & 0 & -1722 & 2.067 \times 10^5 \end{bmatrix}$$

$$\times \begin{bmatrix} 0 & 1 & 0 & 0 & 0 & 0 \\ -1 & 0 & 0 & 0 & 0 & 0 \\ 0 & 0 & 1 & 0 & 0 & 0 \\ 0 & 0 & 0 & 0 & 1 & 0 \\ 0 & 0 & 0 & -1 & 0 & 0 \\ 0 & 0 & 0 & 0 & 0 & 1 \end{bmatrix} \begin{Bmatrix} 0 \\ -1 \\ 0 \\ -0.2904 \\ -0.9993 \\ 3.227 \times 10^{-3} \end{Bmatrix}$$

As before, the values used for $\{\Delta\}_{AB}$ are the x, y, and rotational (global) displacements for the near end and far end of AB, that is, Δ_7, Δ_8, Δ_9, Δ_1, Δ_2, and Δ_3.

$$\{f\}_{AB} = \begin{Bmatrix} -1.38 \\ 0 \\ -166.8 \\ 1.38 \\ 0 \\ 166.8 \end{Bmatrix}$$

For member *BC*,

$$\{f\}_{BC} = \begin{bmatrix} 1099 & 0 & 0 & -1099 & 0 & 0 \\ 0 & 6.2 & 744 & 0 & -6.2 & 744 \\ 0 & 744 & 1.195 \times 10^5 & 0 & -744 & 5.95 \times 10^4 \\ -1099 & 0 & 0 & 1099 & 0 & 0 \\ 0 & -6.2 & -744 & 0 & 6.2 & -744 \\ 0 & 744 & 5.95 \times 10^4 & 0 & -744 & 1.195 \times 10^5 \end{bmatrix}$$

$$\times \begin{Bmatrix} -0.2904 \\ -0.9993 \\ 3.227 \times 10^{-3} \\ -0.2904 \\ -7.016 \times 10^{-4} \\ 3.227 \times 10^{-3} \end{Bmatrix}$$

$$= \begin{Bmatrix} 0 \\ -1.38 \text{ kips} \\ -166.8 \text{ kip-in} \\ 0 \\ 1.38 \text{ kips} \\ -166.8 \text{ kip-in} \end{Bmatrix}$$

For member *CD*,

$$\{f\}_{CD} = \begin{bmatrix} 1967 & 0 & 0 & -1967 & 0 & 0 \\ 0 & 19.14 & 1722 & 0 & -19.14 & 1722 \\ 0 & 1722 & 2.067 \times 10^5 & 0 & -1722 & 1.033 \times 10^5 \\ -1967 & 0 & 0 & 1967 & 0 & 0 \\ 0 & -19.14 & -1722 & 0 & 19.14 & -1722 \\ 0 & 1722 & 1.033 \times 10^5 & 0 & -1722 & 2.067 \times 10^5 \end{bmatrix}$$

$$\times \begin{bmatrix} 0 & 1 & 0 & 0 & 0 & 0 \\ -1 & 0 & 0 & 0 & 0 & 0 \\ 0 & 0 & 1 & 0 & 0 & 0 \\ 0 & 0 & 0 & 0 & 1 & 0 \\ 0 & 0 & 0 & -1 & 0 & 0 \\ 0 & 0 & 0 & 0 & 0 & 1 \end{bmatrix} \begin{Bmatrix} 0 \\ 0 \\ 0 \\ -0.2904 \\ -7.016 \times 10^{-4} \\ 3.27 \times 10^{-3} \end{Bmatrix}$$

$$= \begin{Bmatrix} 1.38 \text{ kips} \\ 0 \\ -166.8 \text{ kip-in} \\ -1.38 \text{ kips} \\ 0 \\ 166.8 \text{ kip-in} \end{Bmatrix}$$

Figure 14.37

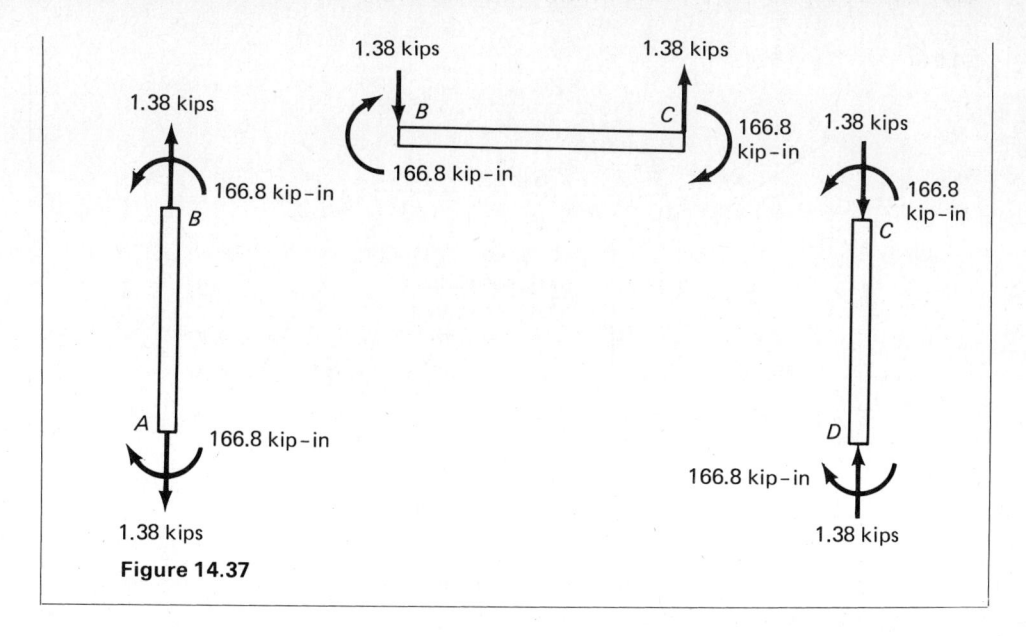

14.8 CLOSURE

You have reached the end of this book. I hope it has been rewarding for you to have worked your way through it. Before you close its cover, let me remind you that this has only been a beginning. You will find that structural analysis is central in design using a variety of materials: steel, concrete, timber and so on. Thus the topics you have studied will be used extensively in courses to follow. They should serve you well.

But don't get over confident. This book is *ELEMENTARY Structural Analysis*. Thus there are many topics and concepts that were not presented: cable structures, arches, shells and so on. There were assumptions made such as those that limited us to linear, elastic materials, and such as that of small deflections to permit our ignoring changes in the lines of action of forces. But there are cases where these assumptions do not apply. Lines of action of forces do move in some cases. Many materials are nonlinear, for example concrete or steel when the steel is loaded beyond its yield point. Such loadings would lead us to topics of plastic hinges and limit loads. These topics often are considered in courses in steel design but could as well be considered in structural analysis.

So while I hope you have confidence in your abilities as structural analysts, I also hope you remain aware that you don't know it all. I will end all this a bit whimsically by quoting a baseball manager: "Its what you learn after you know it all that counts."—Earl Weaver.

PROBLEMS

Sequenced Problem 9.

14.1

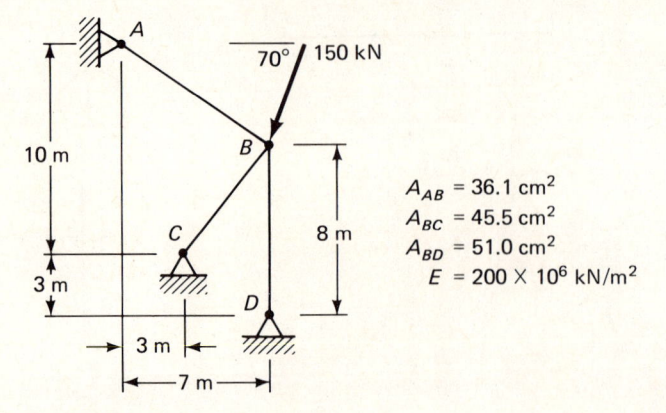

Figure P14.1

$A_{AB} = 36.1 \text{ cm}^2$
$A_{BC} = 45.5 \text{ cm}^2$
$A_{BD} = 51.0 \text{ cm}^2$
$E = 200 \times 10^6 \text{ kN/m}^2$

Use the matrix formulation of the displacement method to determine the support reactions and the member forces.

14.2

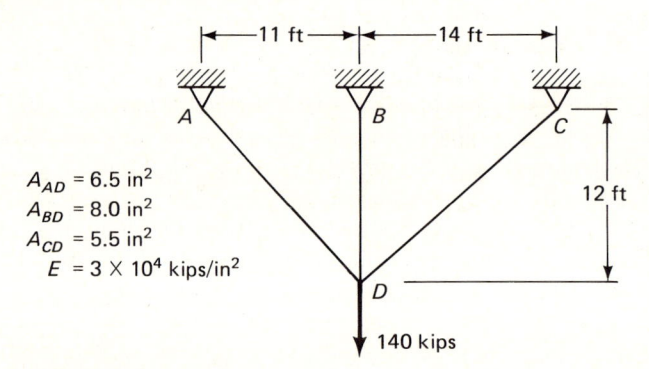

Figure P14.2

$A_{AD} = 6.5 \text{ in}^2$
$A_{BD} = 8.0 \text{ in}^2$
$A_{CD} = 5.5 \text{ in}^2$
$E = 3 \times 10^4 \text{ kips/in}^2$

Use the matrix formulation of the displacement method to determine the support reactions and the member forces.

14.3

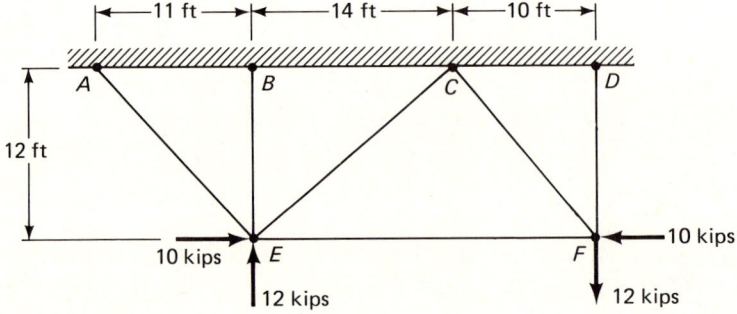

Figure P14.3

14.4 All members have $A = 7.5$ in^2, $E = 3 \times 10^4$ kips/in^2. Use the matrix formulation of the displacement method to determine the support reactions and member forces.

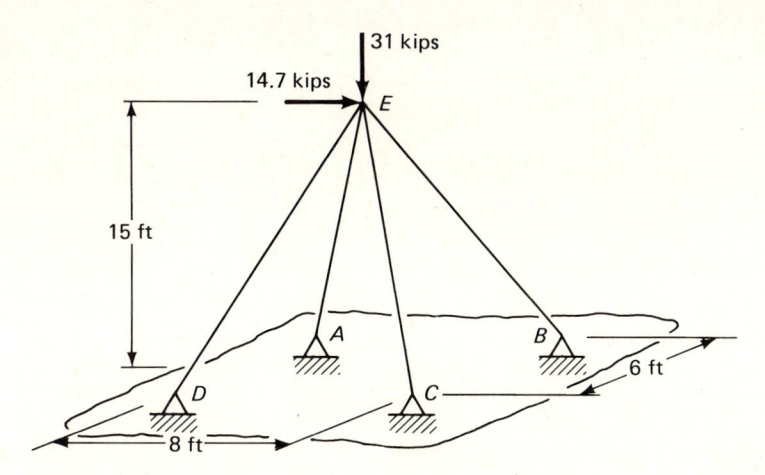

Figure P14.4

Joint E is directly above a point midway between joints A, B, C, and D. The 31-kip force is vertical, and the 14.7-kip force acts along a line parallel to AB and DC. Each member has area $A = 11.5$ in^2 and $E = 3 \times 10^4$ kips/in^2. Use the matrix formulation of the displacement method to determine the support reactions and the member forces.

14.5 Solve Problem 14.4 but with the two given forces replaced by a force of 29.5 kips acting at joint E and directed toward joint A.

14.6

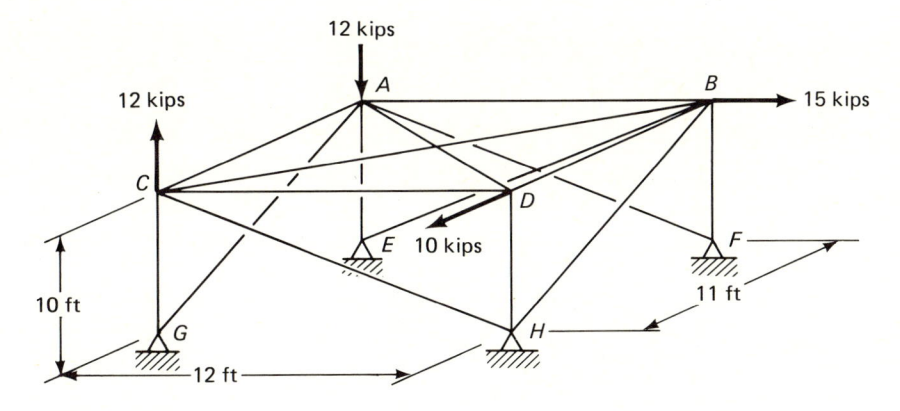

Figure P14.6

All members have $A = 8.1$ in^2, $E = 3 \times 10^4$ kips/in^2. Use the matrix formulation of the displacement method to determine the support reactions and member forces.

14.7 Repeat Problem 14.6, but with 18 kips acting at D along line DA, toward A, instead of the four forces shown.

14.8 Determine the member forces and support reactions if support C of the structure of Problem 14.4 settles downward 1.5 in while the other supports remain in position. Do not include the externally applied forces at joint E.

14.9

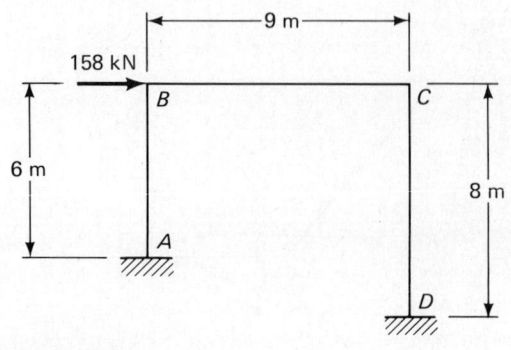

Figure P14.9

$I_{AB} = 2.91 \times 10^{-4} \text{ m}^4$ $\quad A_{AB} = 0.0194 \text{ m}^2$
$I_{BC} = 3.63 \times 10^{-4} \text{ m}^4$ $\quad A_{BC} = 0.0248 \text{ m}^2$
$I_{CD} = 4.12 \times 10^{-4} \text{ m}^4$ $\quad A_{CD} = 0.0298 \text{ in}^2$
$E = 200 \times 10^6 \text{ kN/m}^2$

Use the matrix formulation of the displacement method to determine support reactions and member end forces and moments.

14.10 Solve Problem 14.9, but loaded with a uniformly distributed force of 25 kN/m acting downward on BC instead of the 158-kN force at B.

14.11 Determine the support reactions and member end forces for the structure of Problem 14.9 if the support at D settles vertically downward 3 cm and rotates clockwise 1.5°. Do not consider any externally applied forces. Use the matrix formulation of the displacement method.

14.12

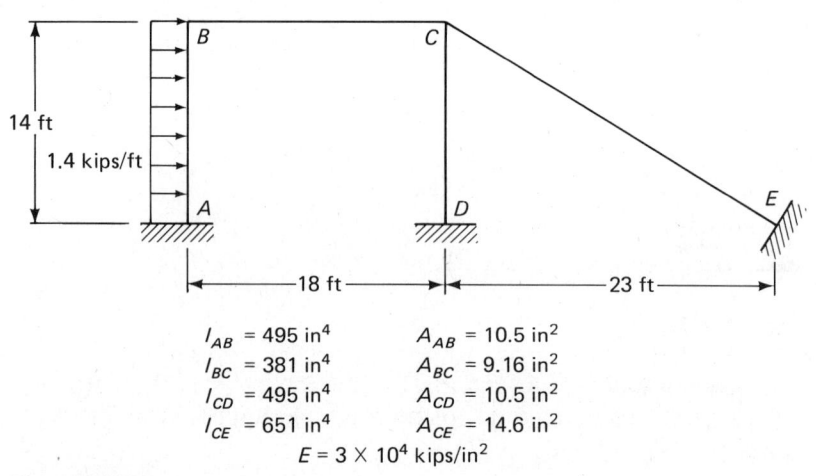

$I_{AB} = 495 \text{ in}^4$ $\quad A_{AB} = 10.5 \text{ in}^2$
$I_{BC} = 381 \text{ in}^4$ $\quad A_{BC} = 9.16 \text{ in}^2$
$I_{CD} = 495 \text{ in}^4$ $\quad A_{CD} = 10.5 \text{ in}^2$
$I_{CE} = 651 \text{ in}^4$ $\quad A_{CE} = 14.6 \text{ in}^2$
$E = 3 \times 10^4 \text{ kips/in}^2$

Figure P14.12

Use the matrix formulation of the displacement method to determine the support reactions and member end forces and moments.

14.13 Support D of the structure of Problem 14.12 settles downward 1.5 in. Use the matrix formulation of the displacement method to determine the support reactions and member end forces and moments. Ignore any external loading.

14.14 Using available packaged structure software, solve for the end forces and moments.

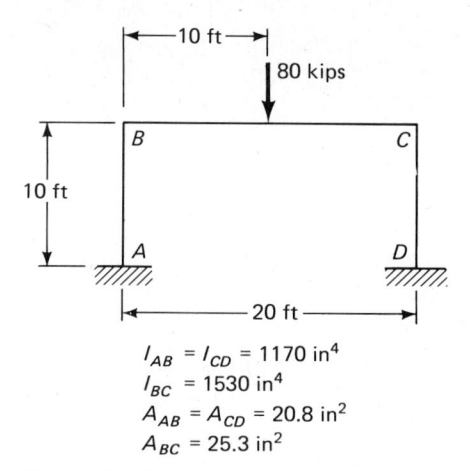

$I_{AB} = I_{CD} = 1170$ in^4
$I_{BC} = 1530$ in^4
$A_{AB} = A_{CD} = 20.8$ in^2
$A_{BC} = 25.3$ in^2

Figure P14.14

Using any classical technique (moment distribution, slope deflection, and so on), solve for the end forces and moments. Compare the results. Draw the shear and moment diagrams. Check to see whether the computer output indicates the maximum bending moments for each member or only the end moments. Is this important?

14.15 Use available packaged software, rather than the method prescribed with the problems, to solve the following problems. Supplementary member data needed for computer solutions are listed with the problem number. In addition, use $E = 2.9 \times 10^4$ kips/in$^2 = 2.0 \times 10^8$ kN/m^2.
  (a) Problem 7.2.
  (b) Problem 7.3.
  (c) Problem 7.5.
  (d) Problem 7.7.
  (e) Problem 7.9. Use $EA = 3.75 \times 10^6$ kN for all members.
  (f) Problem 10.1(a). Use $I = 4.21 \times 10^4$ cm^4, $A = 115$ cm^2.
  (g) Problem 10.1(b). Use $I = 130$ in^4, $A = 5.57$ in^2.
  (h) Problem 10.1(c). Use $I = 510$ in^4, $A = 10.3$ in^2.
  (i) Problem 10.1(d). Use $EA_{AB} = 2.60 \times 10^6$ kN, $EA_{BC} = 2.89 \times 10^6$ kN, $EA_{CD} = 4.05 \times 10^6$ kN.
  (j) Problem 10.1(e). Use $EI = 8.42 \times 10^4$ kN·m^2, $EA = 2.31 \times 10^6$ kN.
  (k) Problem 10.1(f). Use $I_{AB} = 1300$ in^4, $A_{AB} = 26.2$ in^2, $I_{BC} = 1950$ in^4, $A_{BC} = 31.0$ in^2.
  (l) Problem 10.1(g). Use $I = 1300$ in^4, $A = 26.2$ in^2.
  (m) Problem 10.2. Use $A = 29.0$ in^2.
  (n) Problem 10.3. Use $EA = 6.15 \times 10^5$ kips.

368 MATRIX FORMULATION FOR STRUCTURAL ANALYSIS

(o) Problem 10.5. Use $EA = 3.75 \times 10^6$ kN for all members.
(p) Problem 10.7. Use $I_{AD} = I_{BE} = I_{CF} = 1600$ in^4, $A_{AD} = A_{BE} = A_{CF} = 21.5$ in^2, $I_{AB} = I_{BC} = 2400$ in^2, $A_{AB} = A_{BC} = 56.8$ in^2.
(q) Problem 10.8. Use $I_{AB} = I_{DE} = 1600$ in^4, $A_{AB} = A_{DE} = 21.5$ in^2, $I_{BC} = I_{CD} = =2400$ in^4, $A_{BC} = A_{CD} = 56.8$ in^2.
(r) Problem 10.9. Use $EA = 8.73 \times 10^5$ kN.

Sequenced Problems

Nine problem sets are given here. Each set is composed of a variety of individual problems which are related to a structure and are to be solved as homework problems in conjunction with topics throughout this book. Scheduling of these individual problems to match the topics is indicated by listing them with the homework problems in the appropriate chapters. In completing all the problems of one set as they appear with their subjects, the reader will have progressed in reasonable steps through a major structural analysis.

Sequenced Problem 1

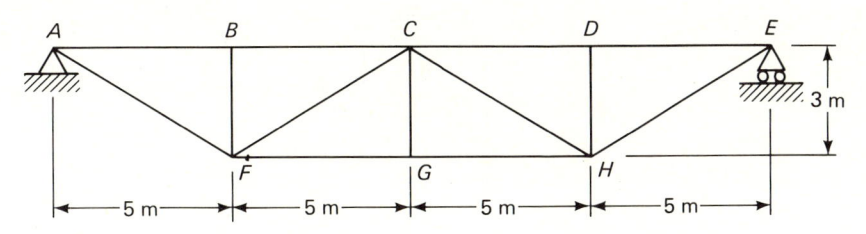

Figure SP-1

Cross-sectional areas A:

$AB, BC, CD, DE = 0.00650 \text{ m}^2$

$FG, GH = 0.00550 \text{ m}^2$

$AF, CF, CH, EH = 0.00700 \text{ m}^2$

$BF, CG, DH = 0.00500 \text{ m}^2$

All members steel, $E = 200 \times 10^6 \text{ kN/m}^2$

(a) Use the method of joints to determine the internal forces in each member if 200 kN is applied downward at joints B, C, and D (a total of 600 kN). Take advantage of symmetry to reduce your work. In each case state whether the force is tension of compression by using the sign convention, (+) for tension and (−) for compression.

(b) Use the method of sections to determine the internal forces in each member caused by a unit force acting upward at joint B. In each case indicate whether the force is tension or compression by using the sign convention, (+) for tension and (−) for compression.

(c) Use superposition of the internal forces found in part (a) and −75 kN times those of part (b) to determine the internal forces in each member caused by a 275 kN downward force at joint B and 200-kN forces downward at both C and D. Indicate tension (+) or compression (−) for each member.

(d) Use any method to determine the forces in each member caused by a unit force acting horizontally to the right at joint E. Indicate whether tension (+) or compression (−).

(e) Determine the internal forces in each member caused by a unit force acting upward at joint C. Indicate whether tension (+) or compression (−).

(f) An additional member, BG, is added to the basic structure. It has an area of 0.00700 m², is connected at joint G, and is positioned to lie along the line from B to G, but it is *not* connected at joint B. The structure is as shown.

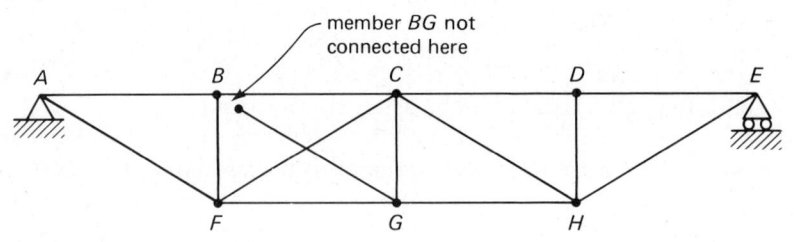

Figure SP1-F

With no other forces acting, two *unit* forces are applied as follows: One, applied at joint B, is directed toward joint G, and the other, applied at end B of the added member BG, is directed away from joint G. (Note that this puts member BG in tension.) Determine the internal forces in all members (including the added one). Indicate tension (+) or compression (−) in each.

(g) Determine the vertical deflections of joints B and C caused by three 200-kN forces acting at joints B, C, and D (600 kN total). Use the N values found in part (a) and the u values found in parts (b) and (e) in applying the method of virtual work, Eq. (6.7).

(h) Recommend changes from the nominal lengths of members AB, DE, FG, and GH to produce an initial camber of the structure which will offset the vertical deflections of joints B, C, and D caused by the three 200-kN forces acting downward at joints B, C, and D. Assume that the changes in the nominal length of AB and DE will be the same, and that the changes in FG and GH also will be the same. Use the internal forces u found in parts (b) and (e) and the real load deflections found in part (g), and apply Eq. (6.8).

(i) Determine the distance along line BG that joint B and end B of added member BG move relative to each other as a result of the two unit forces of part (f) acting simultaneously. *Hint:* Use the internal forces previously calculated in part (f).

(j) (1) Determine the horizontal displacement of joint *E* caused by the two unit forces of part (f) acting at the same time.

(2) Determine the horizontal displacement of joint *E* caused by the unit horizontal force of part (d).

(3) Determine the distance along line *BG* that joint *B* and end *B* of added member *BG* move relative to each other as a result of a unit horizontal force to the right acting at *E*.

Hint: In each of these parts use the internal forces previously found in parts (d) and (f).

(k) Define the following actions:

(1) Pertains to relative displacements along line *BG* of joint *B* and end *B* of added member *BG*, and to the pair of forces described in part (f).

(2) Pertains to the horizontal displacement of joint *E* and to horizontal forces at joint *E*.

Then calculate flexibility coefficients δ_{11}, δ_{12}, δ_{21}, and δ_{22}. *Hint:* Study the *definitions* of these flexibility coefficients and compare them to the displacements calculated in parts (i) and (j).

(l) Determine the horizontal displacement of joint *E* caused by the three forces, 275 kN at *B*, 200 kN at *C*, and 200 kN at *D*, acting downward at the same time. use the internal forces of parts (c) and (d) in applying the method of virtual work.

(m) Determine the relative displacement along line *BG* of joint *B* and end *B* of added member *BG* as a result of simultaneous downward loading of 275 kN, 200 kN, and 200 kN at *B*, *C*, and *D*, respectively. Use the internal forces found in parts (c) and (f) in applying the method of virtual work.

(n)

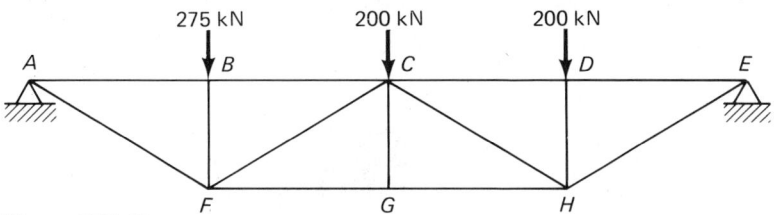

Figure SP1-N

Determine the internal forces for the structure loaded and supported as illustrated. Note that there is a pin support at *E*, not a roller as used previously. *Hint:* Release the horizontal constraint at *E* to form the primary structure and then use flexibility coefficients and the real load displacements of the redundant point, as found in parts (k) and (l).

(o)

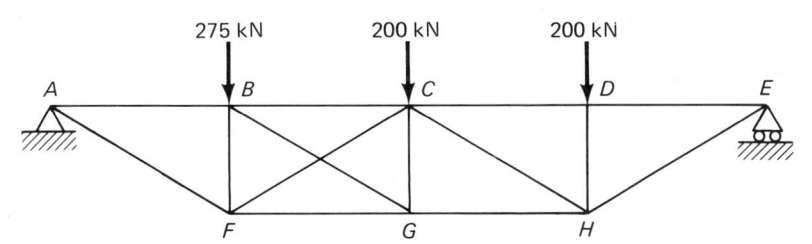

Figure SP1-O

Determine the internal forces for the structure illustrated. Note that member *BG* has been connected at *B*. *Hint:* Form the primary structure by releasing the connection of *BG* at *B* and then use flexibility coefficients and the real load displacement of the redundant points found in parts (k) and (m).

(p)

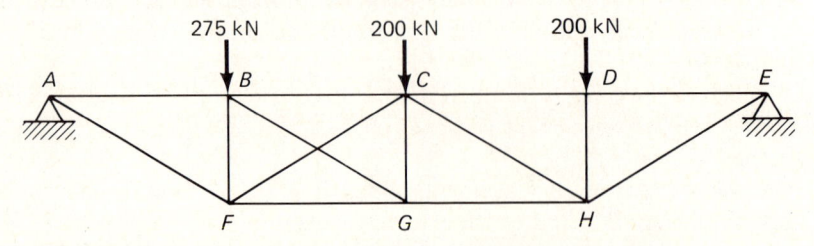

Figure SP1-P

Determine the internal forces for the structure illustrated. Member *BG* is connected at *B* and there is a pin support at *E*. *Hint:* Form the primary structure by disconnecting *BG* at *B* and releasing the horizontal constraint at *E*, and then use real load displacements and flexibility coefficients found in parts (k), (l), and (m).

Sequenced Problem 2

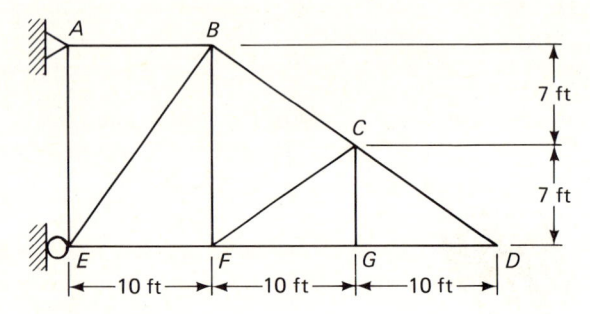

Figure SP-2

Cross-sectional areas A:

$AB, BE, EF, FG, GD, CF = 20$ in^2

$BC, CD, AE = 12$ in^2

$CG, BF = 8$ in^2

All members steel, $E = 3 \times 10^4$ kips/in^2

(a) Use the method of joints to determine the internal forces in each member if two 100-kip forces are applied downward, one at joint *G* and one at joint *D*. Indicate tension or compression for each member by using the sign convention, (+) for tension and (−) for compression.

(b) Use the method of sections to determine the internal force in each member caused by a unit force directed upward at *F*. In each indicate whether tension or

compression by using the sign convention, (+) for tension and (−) for compression.
(c) Use the results of parts (a) and (b) to determine the forces carried in each member if three downward directed 100-kip forces are applied, one each at joints F, G, and D.
(d) An extra member, BH, is added by connecting it at joint B. Its other end is not connected to anything. The member has a cross-sectional area of 12 in^2. A unit force is applied to it at H as shown. Determine the resulting internal force in each member, and indicate whether tension (+) or compression (−) in each. Assume that there are *no* other forces acting on the structure.

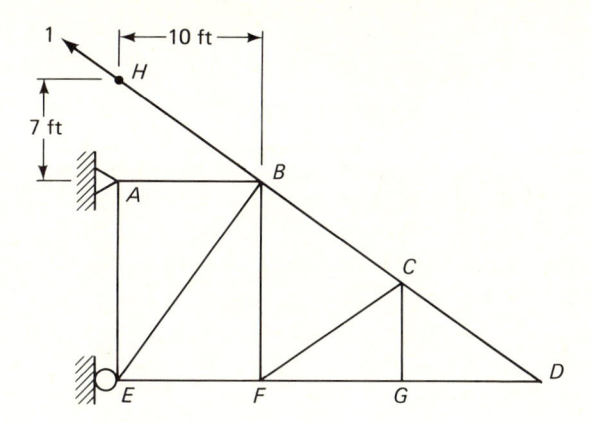

Figure SP2-D

(e) An extra member, BG, is added to the basic structure. It has an area of 12 in^2, is connected to the structure only at joint G, lies along line BG, but is not connected at joint B. A pair of unit forces is applied as follows: a unit force at joint B directed toward joint G, and a unit force at end B of member BG, directed away from joint G (so that member BG is placed in tension). Determine the internal forces in all members (including BG). Indicate tension (+) or compression (−) for each member. Assume that there are no other forces acting on the structure.
(f) Determine the axial displacement of end H of member BH [added to the structure in part (d)] caused by two 100-kip forces acting downward at joints G and D [per part (a)]. Use internal forces evaluated in parts (a) and (d) in applying the method of virtual work.
(g) Determine the relative movement along line BG by end B of member BG [added to the basic structure in part (e)] and joint B caused by two 100-kip forces acting downward at points G and D [per part (a)]. Use internal forces evaluated in parts (a) and (e) in applying the method of virtual work.
(h) Determine the displacements of end H of member BH [added to the structure in part (d)] caused by the two unit forces of part (e). Use internal forces found in parts (d) and (e) in applying the method of virtual work.
(i) Define the motion of end B of member BG relative to joint B along the line BG as action 1. Define the displacement along line BH of end H of member BH as action 2. Calculate flexibility coefficients δ_{11}, δ_{12}, δ_{21}, and δ_{22}. Use internal

forces found in parts (d) and (e) in applying Eq. (6.19). Compare your solution for part (h) to your solution for δ_{21}.

(j) Determine the internal forces in each member caused by the loading shown. Use real load displacements of the primary structure (*BH* not connected at *H*) from part (f) and appropriate flexibility coefficients from part (i).

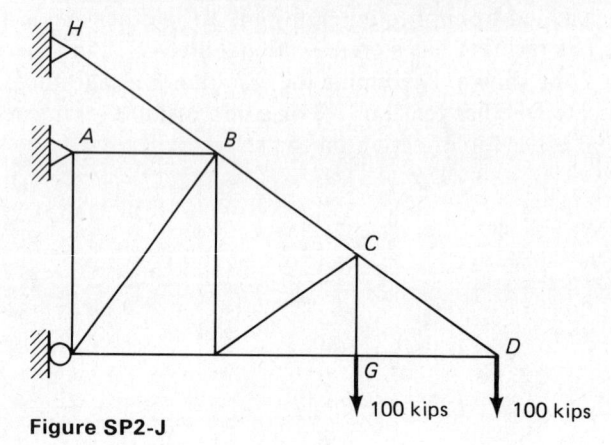

Figure SP2-J

(k) Determine the internal forces in each member caused by the loading shown. Use real load displacements of the primary structure (*H* not connected, *BG* not connected at *B*) from parts (f) and (g) and flexibility coefficients found in part (i).

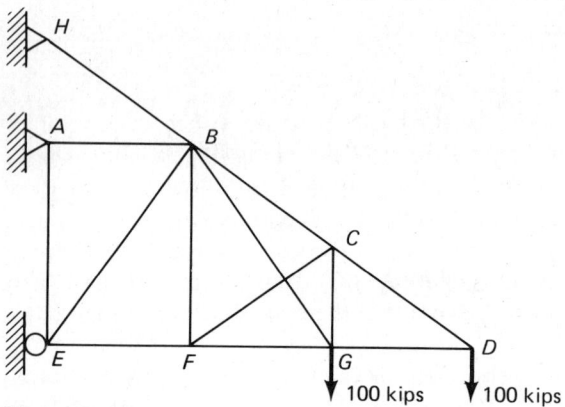

Figure SP2-K

Sequenced Problem 3

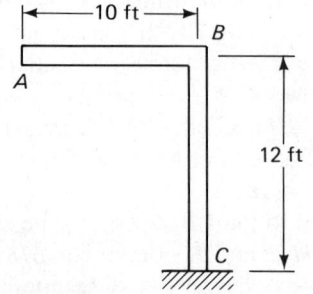

Figure SP-3

Member properties:

$A_{AB} = 30 \text{ in}^2 \qquad I_{AB} = 1200 \text{ in}^4$

$A_{BC} = 60 \text{ in}^2 \qquad I_{BC} = 2500 \text{ in}^4$

steel, $E = 3 \times 10^4 \text{ kips/in}^2$

(a) Determine algebraic expressions for the bending moment and draw the moment diagram for the structure loaded as shown.

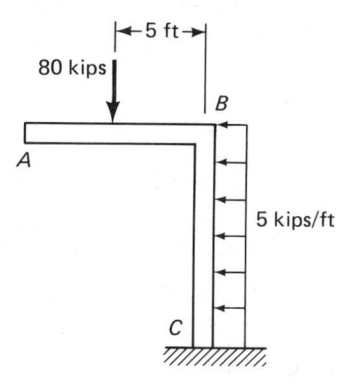

Use the following coordinates and sign conventions. For AB,

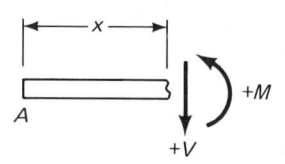

For BC,

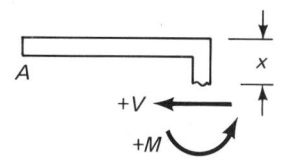

Figure SP3-A

(b) Using the sign conventions and coordinate systems specified in part (a), determine algebraic expressions for the bending moments for the following loading cases:

(1)

(2)

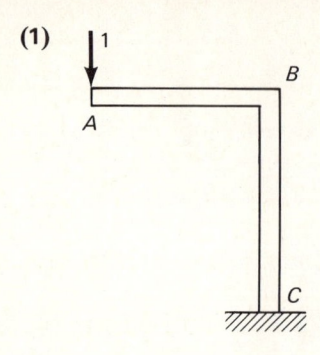

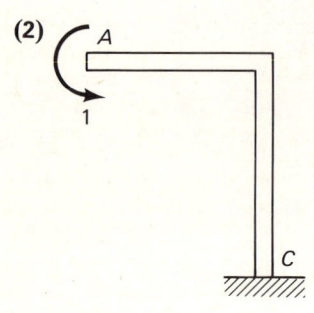

Figure SP3-B

Note that the loads are unit forces or couples, presumed to be dimensionless. In each case the unit force or the couple are the only loads applied.

(c) Use the method of virtual work to determine the vertical and rotational displacements of point A caused by the loading of part (a). Use internal bending moments found in parts (a) and (b).

(d) Using the coordinate systems specified in part (a), determine algebraic expressions for the bending moments for the structure with horizontal force H applied to the left at A.

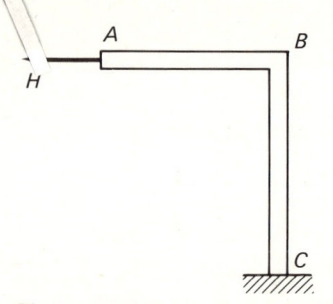

Figure SP3-D

(e) Use Castigliano's theorem to calculate the horizontal displacement of point A caused by the loading of part (a). Use moment expressions developed in parts (a) and (d).
(f) Use the method of virtual work to calculate:
 (1) The downward displacement of point A caused by a downward directed unit force at A.

(2) The downward displacement of point A caused by a counterclockwise unit couple at A.
(3) The rotation of end A caused by a downward directed unit force at A.
(4) The rotation of end A caused by a counterclockwise unit couple at A.
Use internal bending moments determined in part (b).

(g) Use Castigliano's theorem to determine the horizontal displacement of point A caused by a downward directed unit force at A. Use internal bending moments found in parts (b) and (d).

(h) Use Castigliano's theorem to determine the horizontal displacement of point A caused by a counterclockwise unit couple acting at A. Use internal bending moments found in parts (b) and (d).

(i) Use Castigliano's theorem to determine the horizontal displacement of point A caused by a horizontal unit force to the left, acting at A. Use internal bending moments found in part (d).

(j) For the following actions:
(1) A vertical downward force or displacement at point A
(2) A counterclockwise couple or rotation at point A
(3) A leftward horizontal force or displacement at point A
calculate the following flexibility coefficients:

$$\delta_{11}, \delta_{12}, \delta_{13}, \delta_{21}, \delta_{22}, \delta_{23}, \delta_{31}, \delta_{32}, \text{ and } \delta_{33}$$

Hint: Study the definitions of these flexibility coefficients and compare them with the loads and displacements found in parts (f), (g), (h), and (i). Be sure to indicate the units of each flexibility coefficient.

(k)

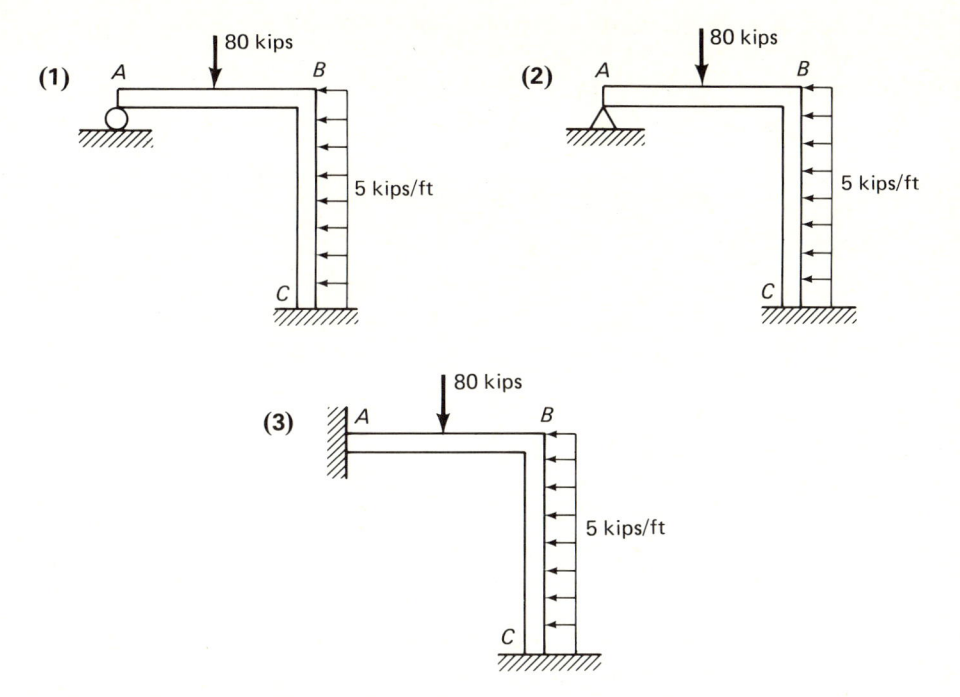

Figure SP3-K

378 SEQUENCED PROBLEMS

Using bent *ABC*, free at *A*, as the primary structure, use real load displacements and flexibility coefficients found in (c) and (j) to calculate the support reactions for cases (1)–(3). Draw the shear and moment diagrams for each case.

(l) Use the method of least work to find the support reactions. Member *AD* has cross-sectional area $A = 2.25 \text{ in}^2$.

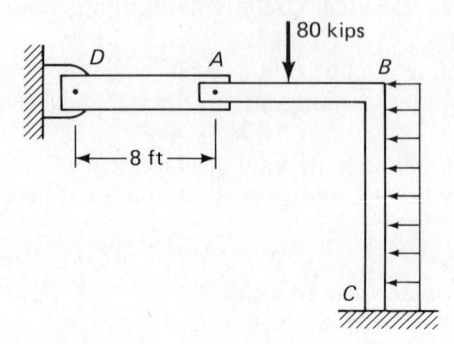

Figure SP3-L

(m) Use the slope deflection method to find the support reactions.

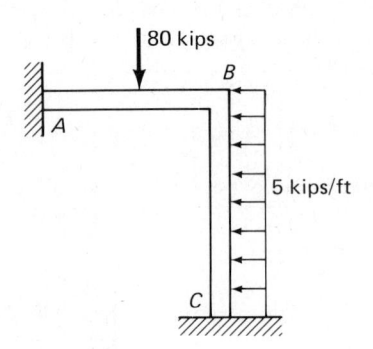

Figure SP3-M

(n) Use the slope deflection method to find the support reactions.

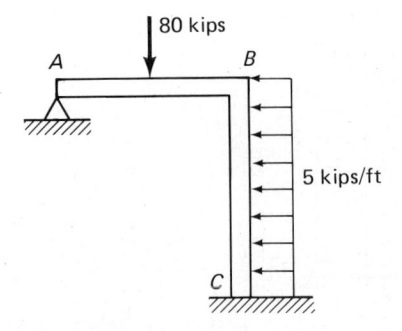

Figure SP3-N

(o) Use moment distribution to find the support reactions.

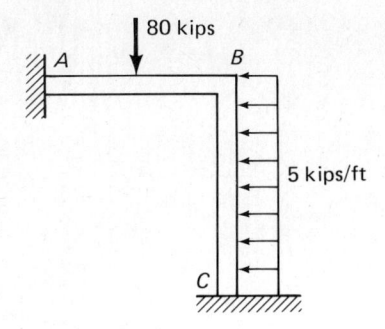

Figure SP3-O

(p) Use moment distribution to find the support reactions.

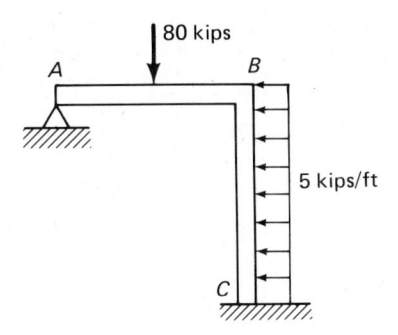

Figure SP3-P

(q) Use moment distribution to find the support reaction. *Hint:* The sidesway prevented part of this may be taken from part (p).

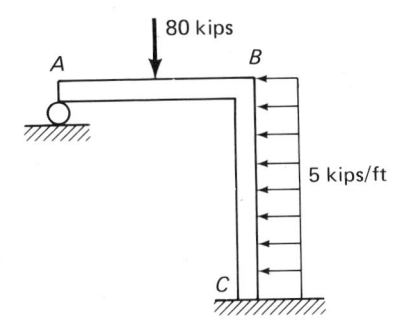

Figure SP3-Q

(r) Use slope deflection to find the support reactions for part (q).

Sequenced Problem 4

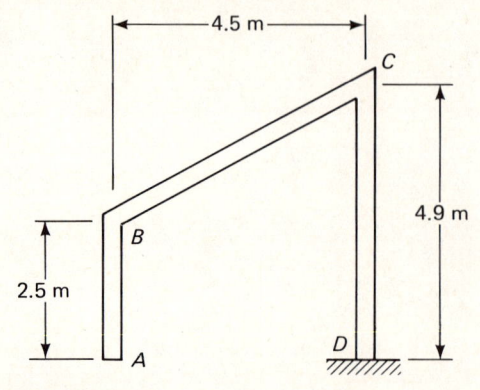

Figure SP-4

Member properties:

AB: $I = 129 \times 10^{-6}$ m^4 $A = 7.61 \times 10^{-3}$ m^2
BC: $I = 462 \times 10^{-6}$ m^4 $A = 14.58 \times 10^{-3}$ m^2
CD: $I = 315 \times 10^{-6}$ m^4 $A = 10.8 \times 10^{-3}$ m^2

All members are steel, $E = 2 \times 10^8$ kN/m^2

(a) Determine algebraic expressions for the bending moments and draw the moment diagram for the structure loaded as shown.

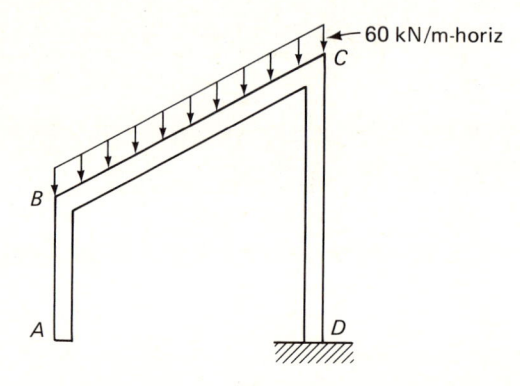

Use the following coordinates and sign conventions: For *AB*,

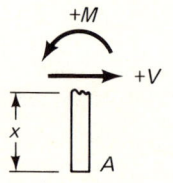

For *BC*,

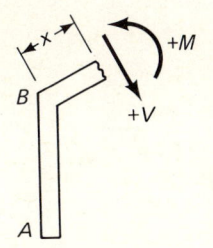

[Figure: segment BC with coordinate x from B, showing +M moment and +V shear]

For *CD*,

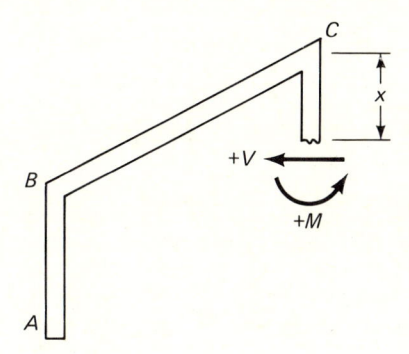

[Figure: frame with segment CD, coordinate x measured down from C, showing +V and +M]

Figure SP4-A

(b) Using the sign conventions and coordinate systems specified in part (a), determine algebraic expressions for the bending moments for the following loading cases:

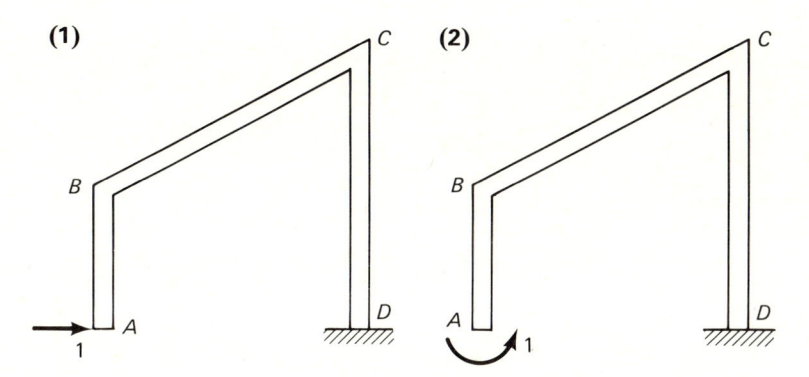

(1) [Frame ABCD with unit horizontal force at A]   (2) [Frame ABCD with unit couple at A]

Figure SP4-B

Note that these are unit forces and couples, and are considered dimensionless.

(c) Use the method of virtual work to determine the horizontal and rotational displacements of end *A* caused by the loading of part (a). Use bending moments found in parts (a) and (b).

(d) Using the coordinate systems and sign conventions specified in part (a), determine the algebraic expressions for the bending moments caused by the vertical force R acting downward at A, as shown.

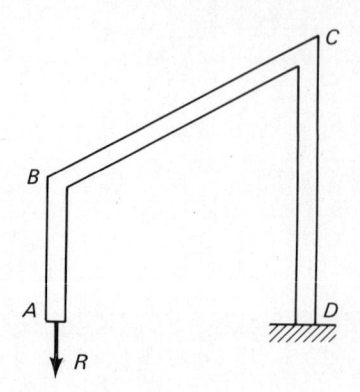

Figure SP4-D

(e) Use Castigliano's theorem to calculate the vertical displacement of point A caused by the loading in part (a). Use bending moment expressions developed in parts (a) and (d).
(f) Use the method of virtual work and the bending moment expressions developed in part (b) to calculate:
   (1) The horizontal displacement of point A caused by a horizontal unit force acting toward the right at A
   (2) The rotations of end A caused by a horizontal unit force acting to the right at A
   (3) The rotation of end A caused by a counterclockwise unit couple applied at A
   (4) The horizontal displacement of end A caused by a counterclockwise unit couple acting at A.
(g) Use Castigliano's theorem to determine the vertical displacement of point A caused by a horizontal unit force to the right acting at A. Use the bending moments found in parts (b) and (d).
(h) Use Castigliano's theorem to determine the vertical displacement of point A caused by a counterclockwise unit couple acting at A. Use the bending moments found in parts (b) and (d).
(i) Use Castigliano's theorem to determine the vertical displacement of point A caused by a unit force acting downward at A. Use the bending moment found in part (d).
(j) Action (1) is defined as a vertical force or displacement at point A. Action (2) is defined as a rotation or couple at point A. Action (3) is defined as a horizontal force or displacement at A. Based on the above definitions, calculate the following flexibility coefficients:

$$\delta_{11}, \delta_{12}, \delta_{13}, \delta_{21}, \delta_{22}, \delta_{23}, \delta_{31}, \delta_{32}, \text{ and } \delta_{33}$$

Hint: Study the definitions of these flexibility coefficients and compare them to the displacements and their causing forces in parts (f), (g), (h), and (i). Be sure to indicate the units of each flexibility coefficient.

(k)

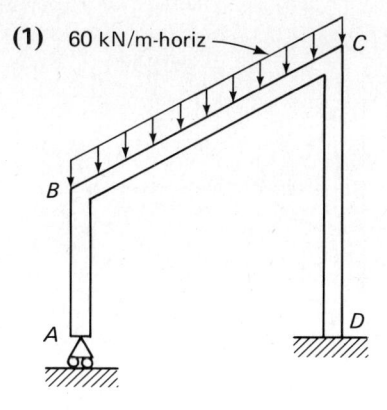

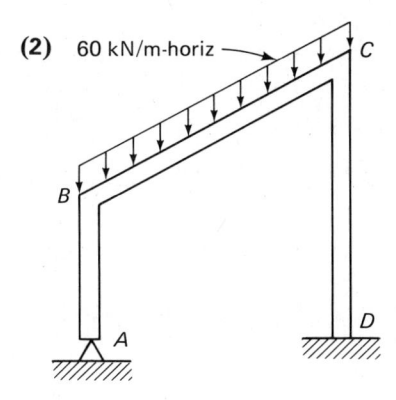

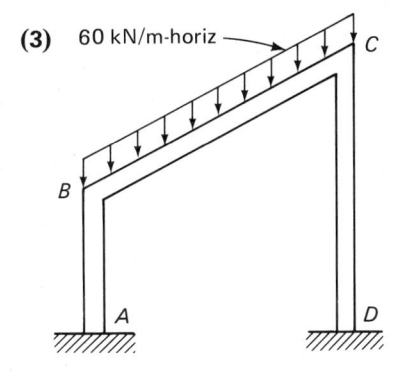

(1) 60 kN/m-horiz

(2) 60 kN/m-horiz

(3) 60 kN/m-horiz

Figure SP4-K

Use solutions for parts (c), (e), and (j) as real load displacements and flexibility coefficients for a primary structure with a free end at A. Calculate the support reactions of cases (1)–(3). Draw the shear and moment diagrams for each case.

(l) Use the method of least work to find the support reactions.

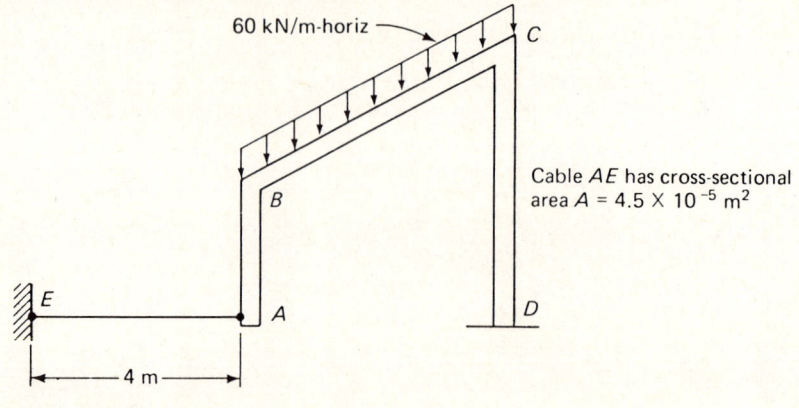

Figure SP4-L

Cable *AE* has cross-sectional area $A = 4.5 \times 10^{-5}$ m^2

60 kN/m-horiz

4 m

(m) Use the slope deflection method to find the support reactions.

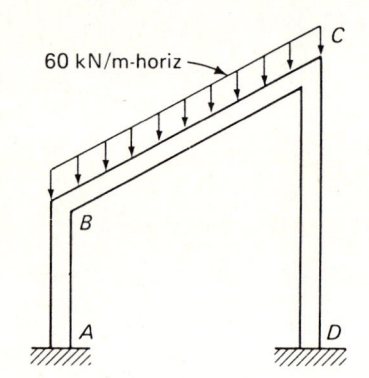

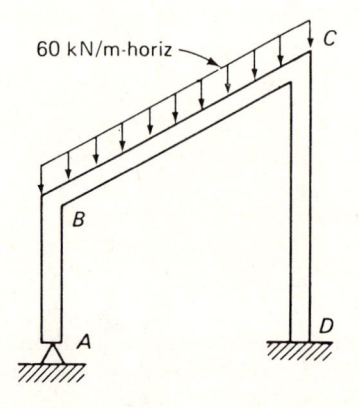

60 kN/m-horiz

Figure SP4-M

(n) Use the slope deflection method to find the support reactions.

60 kN/m-horiz

Figure SP4-N

(o) Use moment distribution to find the support reactions.

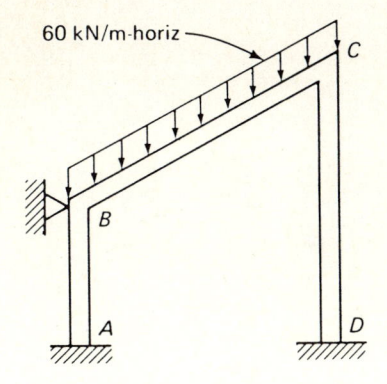

Figure SP4-O

(p) Use moment distribution to find the support reactions.

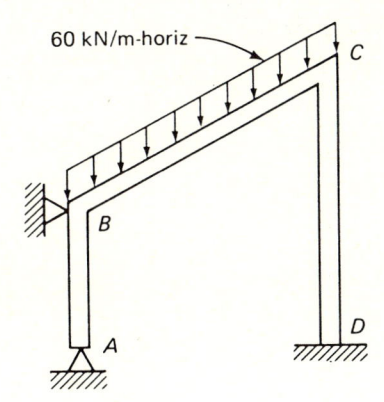

Figure SP4-P

(q) Use moment distribution to find the support reactions.

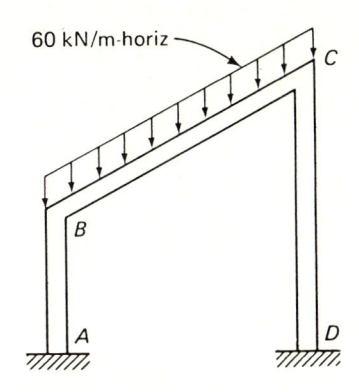

Figure SP4-Q

Hint: Use the solution of part (p) for the sidesway prevented part of the analysis for this case.

(r)

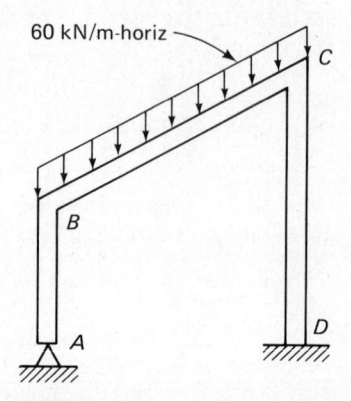

Figure SP4-R

Use moment distribution to determine the support reactions.
Hint: Use the solution of part (p) for the sidesway prevented part of the analysis for this case.

(s)

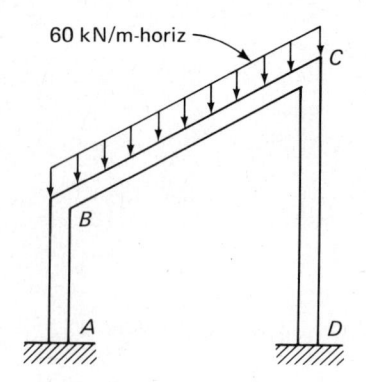

Figure SP4-S

Use slope deflection to determine the support reactions.

Sequenced Problem 5

(a) A special 24-ft-long beam is fabricated from a W21 × 127 standard steel beam by welding 1 in thick plates to its top and bottom flanges for a length of 8 ft at its two ends.

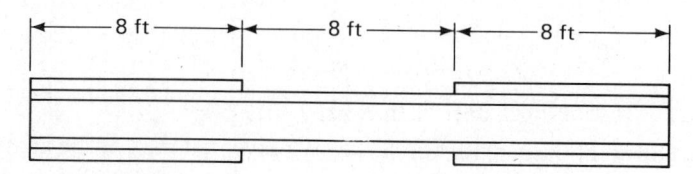

Figure SP5-A

Properties of the W21 × 127 beam are $I = 3020$ in⁴, depth $= 21.24$ in, and flange width $= 13.06$ in. The plates are each

$t = 1.00$ in (thickness)

$b = 13.06$ in (width)

$l = 8$ ft (length)

Determine the fixed end moments for a uniformly distributed force w for this beam. Take advantage of symmetry.
(b) Determine the stiffnesses and carry-over factors of the special beam of part (a). Take advantage of symmetry.
(c) Use moment distribution to determine the end moments for the structure. Sketch the moment diagram.

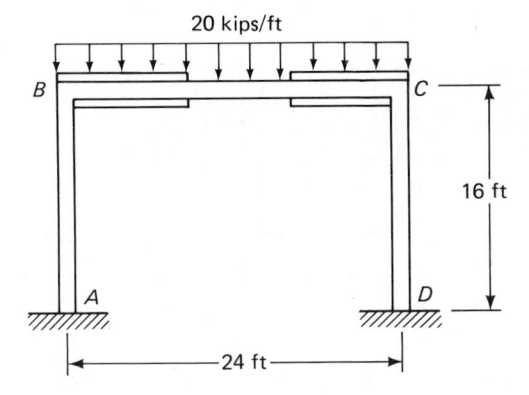

Figure SP5-C

Members AB and CD are W13 × 398 standard steel beam with $I = 6010$ in⁴, while member BC is the member described in part (a).
(d)

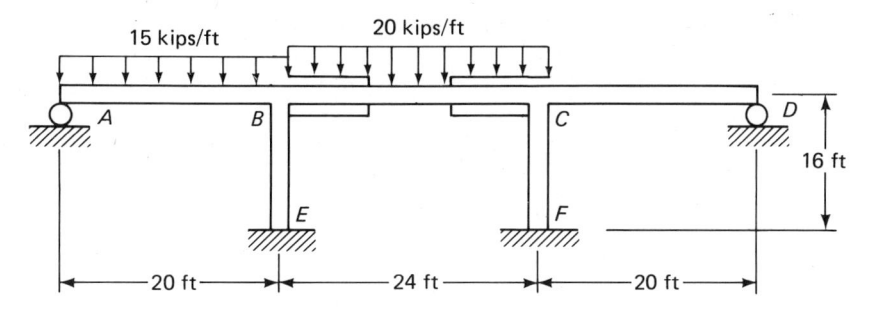

Figure SP5-D

Use moment distribution to determine the end moments for the structure loaded as shown. Sketch the moment diagram. Members are:

AB, CD: W21 × 96 standard steel beam with $I = 2100$ in⁴

BC: Special beam described in part (a)

BE, CF: W14 × 398 standard steel beam with $I = 6010$ in⁴.

388　SEQUENCED PROBLEMS

Notice that sidesway is possible, and that some advantage may be found in the use of modified stiffnesses.

Sequenced Problem 6

(a) Determine the end moments of the structure.

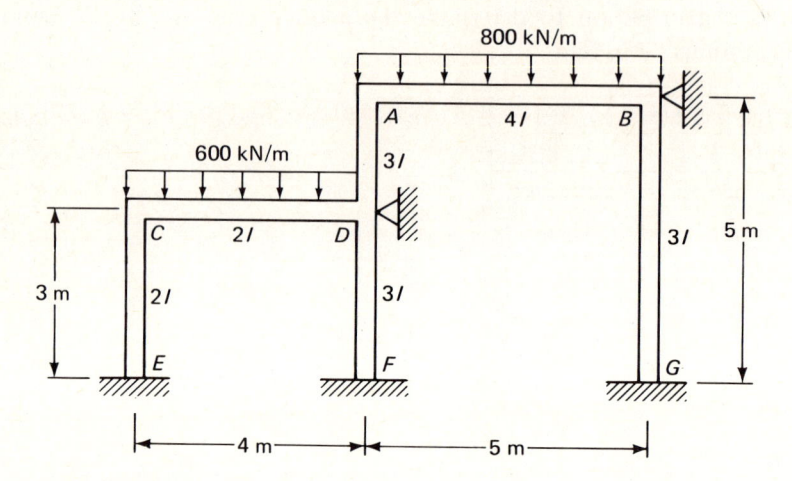

Figure SP6.A

(b) Determine the lateral forces applied by the supports at D and B for part (a).
(c) Determine the end moments of the structure. Members are the same as in part (a).

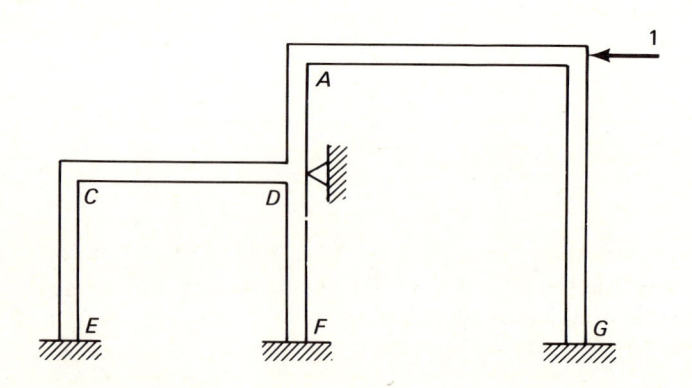

Figure SP6-C

(d) Determine the end moments of the structure.

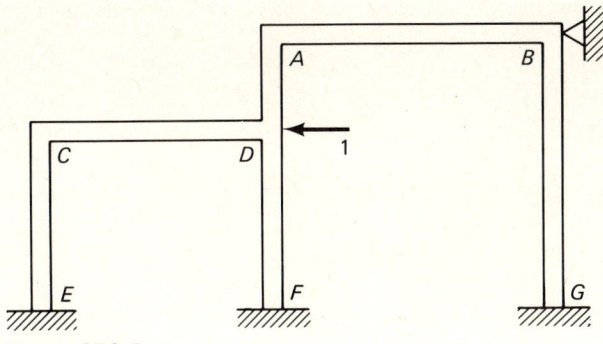

Figure SP6-D

(e) Combine the solutions for parts (a), (b), (c), and (d) to find the end moments for the structure.

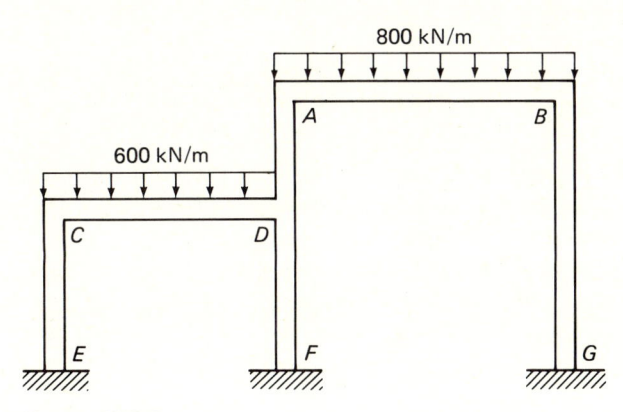

Figure SP6-E

Sequenced Problem 7

(a) Evaluate the rotational stiffnesses and carry-over factors for the 10-m-long haunched beam CD.

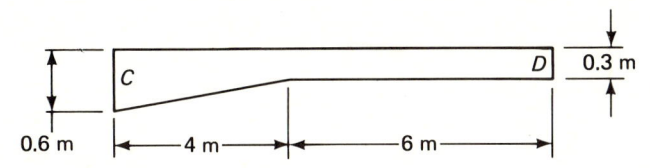

Figure SP7.A

It has rectangular cross sections throughout, with a uniform thickness of 0.15 m. Assume that it is steel with $E = 200 \times 10^6$ kN/m².

(b) Determine the fixed end moments for a concentrated downward force P applied 4 m from C on the beam described in part (a).

(c) Both members AB and CD are the same as the member described in part (a), while members BC, BE, and CF are prismatic, having rectangular cross sections, with 0.15-m thicknesses. The material is steel, $E = 200 \times 10^6$ kN/m². Determine the end moments and the horizontal force at A necessary to prevent sidesway for the loading shown.

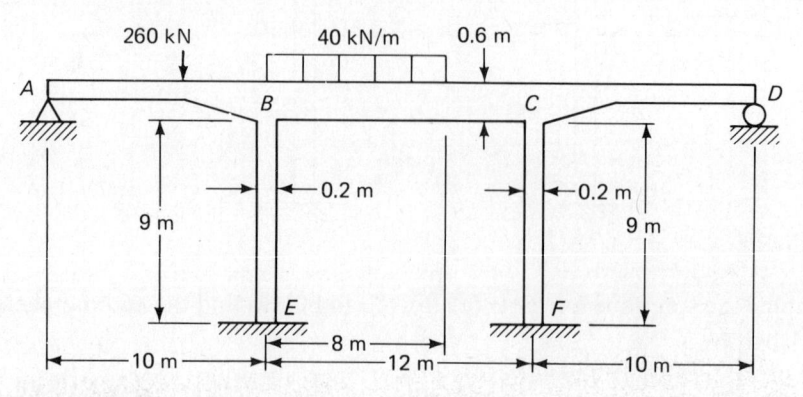

Figure SP7.C

(d) Assume that for the structure of part (c) the pin support at A has been replaced with a roller. Induce a sidesway and scale the end moments to a horizontal force at A that offsets the force found necessary to prevent sidesway in part (c). Superpose these end moments with those found in part (c).
(e) Draw the moment diagram of the structure with sidesway not prevented, and loaded as in part (c).

Sequenced Problem 8

(a)

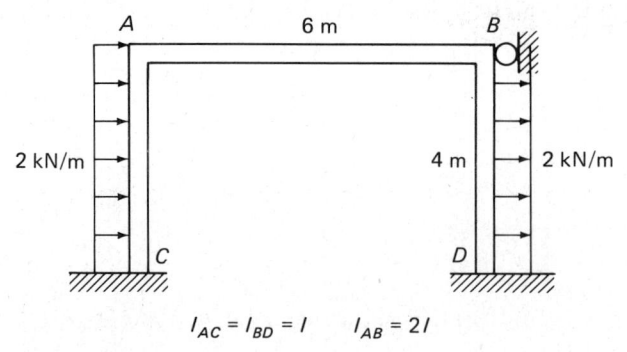

Figure SP8.A

$I_{AC} = I_{BD} = I \qquad I_{AB} = 2I$

Use slope deflection or moment distribution to determine the end moments and support reactions.

(b)

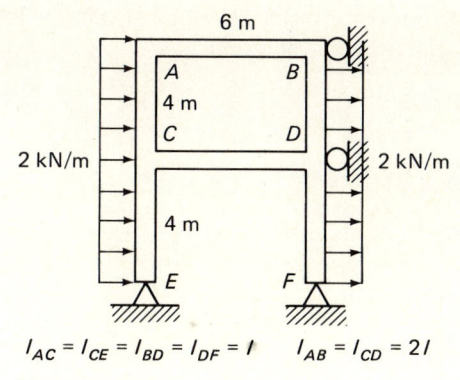

[Figure: Frame with horizontal beam AB at top (6 m), beam CD below at 4 m from top, legs extending 4 m down to pinned supports E and F. Distributed load 2 kN/m applied horizontally on both left and right sides.]

$I_{AC} = I_{CE} = I_{BD} = I_{DF} = I$ $I_{AB} = I_{CD} = 2I$

Figure SP8.B

Use slope deflection or moment distribution to determine the end moments and support reactions. Compare M_{CA} in this part to M_{CA} in part (a). Are they the same? Should they be?

(c) Find the end moments for the structure of part (b) with sidesway prevented at D, but induced to have sidesway at B by force P_B. Evaluate the force preventing sidesway at D.

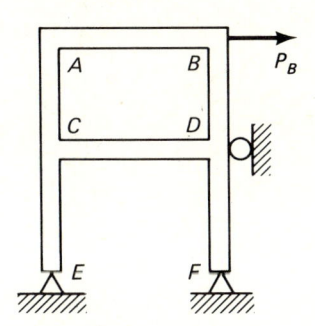

[Figure: Same frame with horizontal force P_B applied at B, roller preventing sidesway at D.]

Figure SP.C

(d) Find the end moments for the structure of part (b) with sidesway prevented at B, but induced at D by force P_D. Evaluate the force preventing sidesway at B.

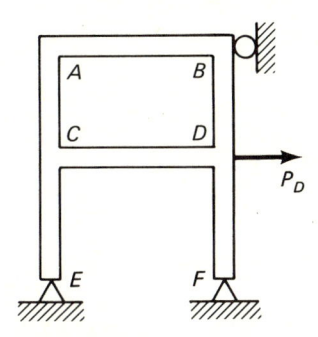

[Figure: Same frame with roller preventing sidesway at B, horizontal force P_D applied at D.]

Figure SP8.D

(e) Solve for values of P_B and P_D which, when applied to parts (c) and (d) and superposed with part (b), will give the end moments for the structure of part (b) loaded and supported as illustrated. Draw the moment diagram.

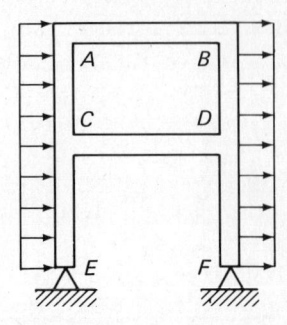

Figure SP8.E

Sequenced Problem 9

(a) Write a computer program to multiply two square or rectangular matrices up to 6×6. Use DO loops and subscripted variables. The program must include a feature to recognize whether the two matrices are conformable for multiplication.
(b) Write a computer program to produce the transpose of a square or rectangular matrix up to 10×10.
(c) Write a computer program to solve N simultaneous equations in N unknowns using Gauss elimination. It should be able to handle up to $N=30$ equations. It should test for and indicate whether the set of equations has no unique solution.
(d) Write a computer program to calculate the global element stiffness matrix for each member of a plane truss. Use the following sequence.
  (1) Read in necessary control nodal and member data:
    (i) Control data
      (α) Number of nodes
      (β) Number of members
    (ii) Nodal data
      (α) Node number
      (β) x coordinate
      (γ) y coordinate
      (δ) Global displacement subscripts
    (iii) Member data
      (α) Member number
      (β) Near end node
      (γ) Far end node
      (δ) Cross-sectional area
      (ε) Modulus of elasticity
  (2) Calculate for each member (element):
    (i) Direction cosines
    (ii) Coordinate transformation matrix $[T]$
    (iii) Local element stiffness matrix

(iv) Global element stiffness matrix (programs from parts (a) and (b) may be used as subroutines in generating the global element stiffness matrices)
- **(e)** Repeat part (d), but for members of space trusses rather than plane trusses.
- **(f)** Write a computer program to add to parts (d) and (e) to calculate the global structure stiffness matrix as global element stiffnesses are generated.
- **(g)** Write a computer program that takes nodal forces and displacements as inputs, assembles them with the structure global stiffness matrix to form stiffness equations, solves for unknown displacements using Gauss elimination [subroutine from part (c)], and develops support reactions and member end forces.
- **(h)** Write a computer program to calculate the global stiffness matrix for each member of a plane rigid frame. Use the same sequence as for part (d), except that member data must include the member moment of inertia.
- **(i)** Extend part (f) to include provision for bending members of a rigid frame.
- **(j)** Extend part (g) to provide for rigid frames for which member end forces include the end shear and end moments.

Appendix A: Matrix Algebra

A.1 INTRODUCTION

The bare essentials of matrix algebra are offered in this appendix. It would be better if the topic were studied in greater depth in more formal mathematics courses. With limited time, however, such courses rarely are part of engineering curricula.

A.2 DEFINITIONS

A *matrix* is an array of numbers, with m rows and n columns, which obey certain prescribed arithmetic operations. Notationally a matrix is enclosed in square brackets. Thus

$$[A] = \begin{bmatrix} A_{11} & A_{12} & \cdots & A_{1n} \\ A_{21} & A_{22} & \cdots & A_{2n} \\ \vdots & & & \vdots \\ A_{m1} & A_{m2} & \cdots & A_{mn} \end{bmatrix}$$

The numbers in the array are called *elements*. They may be referred to by subscripts, the first indicating the row and the second the column where the element is located.

The *order of a matrix* describes the size of the matrix. A matrix of order $m \times n$ has m rows and n columns.

A *row matrix* has one row and thus is of order $1 \times n$.

$$[R] = \begin{bmatrix} 1 & 6 & 5 & -2 \end{bmatrix}$$

A *column* matrix has one column and thus is of order $m \times 1$.

Occasionally column matrices are called *vectors*, but are not necessarily vectors in the sense of a magnitude and direction. They may be designated by special braces $\{\}$. Thus

$$\{C\} = [C] = \begin{Bmatrix} 1 \\ 7 \\ 4 \\ 6 \end{Bmatrix} = \begin{bmatrix} 1 \\ 7 \\ 4 \\ 6 \end{bmatrix}$$

A *square matrix* has the same number of rows as columns,

$$[S] = \begin{bmatrix} 8 & 7 & 1 & 9 \\ 7 & 6 & 5 & -1 \\ 1 & 5 & 12 & 0 \\ 9 & -1 & 0 & 14 \end{bmatrix}$$

The *main diagonal*, also referred to as *the diagonal*, is found in a square matrix and consists of those elements in a line from the upper left to lower right corner. These elements have matching row and column subscripts.

$$[D] = \begin{bmatrix} D_{11} & 5 & 6 \\ 5 & D_{22} & 7 \\ 6 & 7 & D_{33} \end{bmatrix}$$

In the illustration D_{11}, D_{22}, and D_{33} constitute the main diagonal.

A *diagonal matrix* is a square matrix that has zero-valued elements everywhere except on the main diagonal.

$$[D] = \begin{bmatrix} 6 & 0 & 0 & 0 \\ 0 & 7 & 0 & 0 \\ 0 & 0 & 8 & 0 \\ 0 & 0 & 0 & 5 \end{bmatrix}$$

A *symmetric matrix* is a square matrix with equal elements at image locations about the main diagonal. Thus in a symmetric matrix $H_{ij} = H_{ji}$.

$$[H] = \begin{bmatrix} 10 & -5 & 9.5 \\ -5 & 12 & 2.7 \\ 9.5 & 2.7 & 6 \end{bmatrix}$$

The *identity matrix* is a diagonal matrix in which each diagonal element is 1. The identity matrix is represented by $[I]$.

$$[I] = \begin{bmatrix} 1 & 0 & 0 \\ 0 & 1 & 0 \\ 0 & 0 & 1 \end{bmatrix}$$

The *null matrix* has every element equal to zero. In matrix algebra, the null matrix plays the role of zero.

$$[O] = \begin{bmatrix} 0 & 0 & 0 & 0 \\ 0 & 0 & 0 & 0 \\ 0 & 0 & 0 & 0 \\ 0 & 0 & 0 & 0 \end{bmatrix}$$

Two matrices are *equal* if they are identical, element by element. Equal matrices must be of the same order.

$$[A] = \begin{bmatrix} 5 & 6 & 7 \\ 9 & 1 & 3 \\ 5 & 0 & 6 \\ 10 & 11 & 0 \end{bmatrix}$$

$$[B] = \begin{bmatrix} 5 & 6 & 7 \\ 9 & 1 & 3 \\ 5 & 0 & 6 \\ 10 & 11 & 0 \end{bmatrix}$$

$$[A] = [B]$$

The *transpose* of a *matrix* is the matrix obtained when the rows and columns of the original matrix are interchanged. The transpose of a matrix is indicated notationally by a superscript T. Thus

$$[B] = \begin{bmatrix} 3 & 5 & 7 & 9 \\ 2 & 0 & 6 & 3 \end{bmatrix}$$

$$[B]^T = \begin{bmatrix} 3 & 2 \\ 5 & 0 \\ 7 & 6 \\ 9 & 3 \end{bmatrix}$$

A.3 ALGEBRAIC OPERATIONS

Two matrices may be *added* if they are of the same order. The process is carried out by adding the elements of the two matrices that are in matching locations.

$$\begin{bmatrix} 5 & 3 & 1 & 2 \\ 6 & 2 & 1 & 5 \\ 0 & 8 & -9 & -10 \end{bmatrix} + \begin{bmatrix} 2 & 1 & 5 & -9 \\ -8 & -6 & 5 & 10 \\ 7 & -1 & 12 & 5 \end{bmatrix} = \begin{bmatrix} 7 & 4 & 6 & -7 \\ -2 & -4 & 6 & 15 \\ 7 & 7 & 3 & -5 \end{bmatrix}$$

A matrix may be *subtracted* from another if they are of the same order. This is carried out by subtracting the elements in matching locations.

$$\begin{bmatrix} 5 & 3 & 1 & 2 \\ 6 & 2 & 1 & 5 \\ 0 & 8 & -9 & -10 \end{bmatrix} - \begin{bmatrix} 2 & 1 & 5 & -9 \\ -8 & -6 & 5 & 10 \\ 7 & -1 & 12 & 5 \end{bmatrix} = \begin{bmatrix} 3 & 2 & -4 & 11 \\ 14 & 8 & -4 & -5 \\ -7 & 9 & -21 & -15 \end{bmatrix}$$

Multiplication of a matrix by a scalar is accomplished by multiplying each element of the matrix by the scalar.

$$3 \begin{bmatrix} 1 & -2 & 3 \\ 2 & 5 & -9 \end{bmatrix} = \begin{bmatrix} 3 & -6 & 9 \\ 6 & 15 & -27 \end{bmatrix}$$

A *matrix* may *multiply* another *matrix* if the two matrices are *conformable*, that is, if the first matrix has the same number of columns as there are rows in the second matrix. The product is a matrix having the same number of rows as found in the first matrix and same number of columns as found in the second matrix.

$$[C] = \begin{bmatrix} A_{11} & A_{12} & A_{13} \\ A_{21} & A_{22} & A_{23} \end{bmatrix} \begin{bmatrix} B_{11} & B_{12} & B_{13} & B_{14} \\ B_{21} & B_{22} & B_{23} & B_{24} \\ B_{31} & B_{32} & B_{33} & B_{34} \end{bmatrix}$$

In developing $[C]$ we determine the elements C_{ij} by the following algorithm:

$$C_{ij} = \sum_{l=1}^{n} (A_{il} \times B_{lj})$$

that is, taken from left to right, the elements of row i of $[A]$ are multiplied successively by the elements of column j of $[B]$ starting at the top, after which these products are added.

$$\begin{bmatrix} 8 & 3 & 9 \\ -2 & 1 & 6 \\ 7 & -1 & 2 \end{bmatrix} \begin{bmatrix} 2 & 1 & -2 \\ 0 & 5 & 3 \\ -1 & -1 & 2 \end{bmatrix} = \begin{bmatrix} 7 & 14 & 11 \\ -10 & -3 & 19 \\ 12 & 0 & -13 \end{bmatrix}$$

One of the most common uses of matrix multiplication is to write sets of simultaneous equations in a compact form. For example,

$$a_{11}x_1 + a_{12}x_2 + a_{13}x_3 = b_1$$
$$a_{21}x_1 + a_{22}x_2 + a_{23}x_3 = b_2$$
$$a_{31}x_1 + a_{32}x_2 + a_{33}x_3 = b_3$$

may be written as

$$\begin{bmatrix} a_{11} & a_{12} & a_{13} \\ a_{21} & a_{22} & a_{23} \\ a_{31} & a_{32} & a_{33} \end{bmatrix} \begin{Bmatrix} x_1 \\ x_2 \\ x_3 \end{Bmatrix} = \begin{Bmatrix} b_1 \\ b_2 \\ b_3 \end{Bmatrix}$$

or

$$[a]\{x\} = \{b\}$$

Matrix multiplication is *associative* and *distributive*, but *not commutative*. Thus

$$[A][B][C] = [[A][B]][C] = [A][[B][C]]$$

and

$$[A][B+C] = [A][B] + [A][C]$$

while

$$[A][B] \neq [B][A]$$

There is *no defined matrix division* process. Thus in solving the equation
$$[a]\{x\} = \{b\}$$

the seemingly natural step

$$\{x\} = \frac{\{b\}}{[a]}$$

has no meaning. Instead, the equation may be solved by premultiplying by $[a]^{-1}$, the inverse of $[a]$, defined according to the relation

$$[a]^{-1}[a] = [I]$$

Thus

$$[a]^{-1}[a]\{x\} = [a]^{-1}\{b\}$$
$$[I]\{x\} = [a]^{-1}\{b\}$$
$$\{x\} = [a]^{-1}\{b\}$$

Direct calculation of $[a]^{-1}$ from $[a]$ is extremely laborious for cases that are of practical interest in structural analysis. For that reason, direct inversion will not be considered in this appendix.

A.4 MATRIX PARTITIONING

Occasionally it is convenient to divide matrices into parts called submatrices. We can treat the submatrices themselves as elements of a matrix and carry out various matrix operations. For example, if

$$[A] = \begin{bmatrix} 1 & 2 & \vdots & 0 & 0 \\ 4 & 5 & \vdots & 0 & 0 \\ \hdashline 0 & 0 & \vdots & 3 & 6 \end{bmatrix} \quad \text{and} \quad [B] = \begin{bmatrix} 5 & 4 & 3 \\ 2 & 1 & 6 \\ \hdashline 0 & 7 & 9 \\ 0 & 8 & 11 \end{bmatrix}$$

$[A][B]$ could be calculated as

$$\begin{bmatrix} A_{11} & A_{12} \\ A_{21} & A_{22} \end{bmatrix} \begin{bmatrix} B_1 \\ B_2 \end{bmatrix} = \begin{bmatrix} A_{11}B_1 + A_{12}B_2 \\ A_{21}B_1 + A_{22}B_2 \end{bmatrix}$$

The top element is

$$A_{11}B_1 + A_{12}B_2 = \begin{bmatrix} 1 & 2 \\ 4 & 5 \end{bmatrix} \begin{bmatrix} 5 & 4 & 3 \\ 2 & 1 & 6 \end{bmatrix} + \begin{bmatrix} 0 & 0 \\ 0 & 0 \end{bmatrix} \begin{bmatrix} 0 & 7 & 9 \\ 0 & 8 & 11 \end{bmatrix}$$

$$= \begin{bmatrix} 9 & 6 & 15 \\ 30 & 21 & 42 \end{bmatrix} + \begin{bmatrix} 0 & 0 & 0 \\ 0 & 0 & 0 \end{bmatrix}$$

$$= \begin{bmatrix} 9 & 6 & 15 \\ 30 & 21 & 42 \end{bmatrix}$$

The bottom element is

$$A_{21}B_1 + A_{22}B_2 = \begin{bmatrix} 0 & 0 \end{bmatrix} \begin{bmatrix} 5 & 4 & 3 \\ 2 & 1 & 6 \end{bmatrix} + \begin{bmatrix} 3 & 6 \end{bmatrix} \begin{bmatrix} 0 & 7 & 9 \\ 0 & 8 & 11 \end{bmatrix}$$

$$= \begin{bmatrix} 0 & 0 & 0 \end{bmatrix} + \begin{bmatrix} 0 & 69 & 93 \end{bmatrix}$$

$$= \begin{bmatrix} 0 & 69 & 93 \end{bmatrix}$$

Thus

$$[A][B] = \begin{bmatrix} 9 & 6 & 15 \\ 30 & 21 & 42 \\ 0 & 69 & 93 \end{bmatrix}$$

A.5 SOLUTION OF SIMULTANEOUS EQUATIONS BY GAUSS ELIMINATION

For a system of n equations

$$a_{11}x_1 + a_{12}x_2 + a_{13}x_3 + \cdots + a_{1n}x_n = b_1$$
$$a_{21}x_1 + a_{22}x_2 + a_{23}x_3 + \cdots + a_{2n}x_n = b_2$$
$$a_{31}x_1 + a_{32}x_3 + a_{33}x_3 + \cdots + a_{3n}x_n = b_3$$
$$\vdots$$
$$a_{n1}x_1 + a_{n2}x_2 + a_{n3}x_3 + \cdots + a_{nn}x_n = b_n$$

in matrix form

$$\begin{bmatrix} a_{11} & a_{12} & a_{13} & \cdots & a_{1n} \\ a_{21} & a_{22} & a_{23} & \cdots & a_{2n} \\ a_{31} & a_{32} & a_{33} & \cdots & a_{3n} \\ \vdots & & & & \\ a_{n1} & a_{n2} & a_{n3} & \cdots & a_{nn} \end{bmatrix} \begin{Bmatrix} x_1 \\ x_2 \\ x_3 \\ \vdots \\ x_n \end{Bmatrix} = \begin{Bmatrix} b_1 \\ b_2 \\ b_3 \\ \vdots \\ b_n \end{Bmatrix}$$

we eliminate, row by row, $a_{21}, a_{31}, \ldots, a_{n1}$ as follows. For row 2, multiply each member of row 1, including b_1, by a_{21}/a_{11} and, column by column, subtract them from the members in row 2. For row 3, multiply each member of row 1, including b_1 by a_{31}/a_{11} and, column by column, subtract them from the members in row 3. Continue this through row n. Then, in the same way, eliminate $a_{32}, a_{42}, \ldots, a_{n2}$. This is done by multiplying each member of the new row 2 by a_{32}/a_{22} and subtracting, column by column, from the new row 3.

Next multiply each member of the new row 2 by a_{42}/a_{22} and subtract, column by column, from the new row 4. This continues to row n. The process repeats, next eliminating a_{43}, a_{53}, and so on. After all rows have been revised, the matrices are of the form

$$\begin{bmatrix} A_{11} & A_{12} & A_{13} & \cdots & A_{1n} \\ 0 & A_{22} & A_{23} & \cdots & A_{2n} \\ 0 & 0 & A_{33} & \cdots & A_{3n} \\ 0 & 0 & 0 & \cdots & \\ & & & & A_{nn} \end{bmatrix} \begin{Bmatrix} x_1 \\ x_2 \\ x_3 \\ \vdots \\ x_n \end{Bmatrix} = \begin{bmatrix} B_1 \\ B_2 \\ B_3 \\ \vdots \\ B_n \end{bmatrix}$$

Now by back substitution we solve for x_n,

$$x_n = \frac{B_n}{A_{nn}}$$

Then for the $(n-1)^{th}$ equation,

$$A_{n-1,n-1} x_{n-1} + A_{n-1,n} x_n = B_{n-1}$$

and with x_n known from the preceding step, we solve for x_{n-1}. This back substitution continues until all x_i are found.

Answers to Selected Problems

2.1 a, b $R_A = 3$ kN↓, $R_B = 33$ kN↑
    c. No difference
2.2 a, b $V_D = -3$ kN, $M_D = -9$ kN·m
    c. No difference
2.4 a, b $R_A = 3.33$ kips↑, $M_A = 33.3$ kip-ft ↩
    $R_D = 2.67$ kips↑, $M_D = 21.3$ kip-ft ↪
    c. No difference
2.6 a, b $R_{A-y} = 20$ kips↑, $R_{A-x} = 0$, $M_A = 144$ kip-ft ↩
    c. No difference
2.8 $R_{Ay} = R_{Cy} = 17.5$ kips↑, $R_{Cx} = 8.33$ kips←, $R_{Ax} = 8.33$ kips→
2.9 A. $V = -3x$ kN $0 < x < 5$ m
        $M = -1.5x^2$ kN·m $0 < x < 5$ m
    B. $V = -x^2 + 12x - 36$ kN $0 < x < 6$ m
        $M = -\frac{1}{3}x^2 + 6x^2 - 36x + 72$ kN·m $0 < x < 6$ m
    C. $V = -6x + 54$ $0 < x < 30$ ft
        $M = -3x^2 + 54x$ $0 < x < 30$ ft
    D. For AB: $V = -12x + 26.8$ kN $0 < x < 5$ m
                      $M = -6x^2 + 26.8x$ kN·m $0 < x < 5$ m
        For BC: $V = -12x + 28$ kN $0 < x < 4$ m
                      $M = -6x^2 + 28x - 16$ kN·m $0 < x < 4$ m

3.1 (a) unstable
    (b) unstable
    (d) stable, statically determinate
3.2 (a) unstable
    (b) stable, statically determinate
    (e) stable, statically determinate

ANSWERS TO SELECTED PROBLEMS

     (g) stable, statically indeterminate to 6th degree
3.3 (b) stable, statically indeterminate to 1st degree
     (d) stable, statically indeterminate to 1st degree
     (e) stable, statically determinate
     (g) unstable
3.4 (a) stable, statically indeterminate to 1st degree
     (c) stable, statically indeterminate to 12th degree
     (e) stable, statically indeterminate to 13th degree
     (h) stable, statically indeterminate to 12th degree

4.1 (a) $BC = +31.6$ kips, $CH = -15.81$ kips, $HJ = +15$ kips
     (b) $AB = +63.2$ kips, $GH = -15$ kips, $FB = -31.6$ kips, $FH = -15.81$ kips
4.2 $AB = DE = -18.51$ kN, $AF = EG = -7.58$ kN
     $GD = BF = +14.10$ kN, $CG = CF = 7.70$ kN
     $CD = BC = -18.51$ kN
4.4 (b) For AB $V = -12$ kN $\big\} \begin{cases} 0 < x < 4 \text{ m} \\ x = 0 \text{ at } A \end{cases}$
                  $M = -12x - 36$ kN $\cdot$ m

        For BC $V = 6$ kN $\big\} \begin{cases} 0 < x < 6 \text{ m} \\ x = 0 \text{ at } B \end{cases}$
                   $M = 6x - 84$ kN $\cdot$ m

        For CD $V = 1.5x$ kN $\big\} \begin{cases} 0 < x < 8 \text{ m} \\ x = 0 \text{ at } D \end{cases}$
                   $M = 0.75x^2$ kN $\cdot$ m

        Alternate for CD: $V = -1.5x + 12$ kN $\big\} \begin{cases} 0 < x < 8 \text{ m} \\ x = 0 \text{ at } C \end{cases}$
                                    $M = -0.75x^2 + 12x - 48$ kN $\cdot$ m

5.1 Roof beam load, 2.23 kips downward, 5 ft apart
5.2 Roof beam load, 0.55 kips upward, 5 ft apart
5.3 Maximum upward force, supporting the column: 68.6 kips
5.6 $V_{A-\max} = 39.2$ kips
     $V_{C-\max} = -12.15$ kips
5.8 $R_{A-\max} = 750$ kN; $R_{D-\max} = 1925$ kN, $R_{G-\max} = 1292$ kN
     $R_{F=\max} = -1867$ kN; $M_{A-\max} = -1575$ kN $\cdot$ m
     $V_{\text{mid pt, } BC-\max} = \pm 75$ kN
     $M_{EF-\max} = 3200$ kN $\cdot$ m
     $V_{EF-\max} = -825$ kN

6.2 (a) $\theta_C = +\dfrac{54 \text{ kip-ft}^2}{EI}$, $y_C = -\dfrac{972 \text{ kip-ft}^3}{EI}$

     (b) $y_{\max} = -\dfrac{1000 \text{ kip-ft}^3}{EI}$ at $x = 7.94$ ft

     (c) $\theta_A = -\dfrac{189 \text{ kip-ft}^2}{EI}$, $y_{\max} = -\dfrac{1000 \text{ kip-ft}^3}{EI}$ at 7.94ft

     (d) $y_C = \dfrac{972 \text{ kip-ft}^3}{EI}$ downward, $\theta_C = +\dfrac{54 \text{ kip-ft}}{EI}$

ANSWERS TO SELECTED PROBLEMS   405

(e) $\theta_A = -\dfrac{189 \text{ kip-ft}^2}{EI}$, $\theta_B = +\dfrac{216 \text{ kip-ft}^2}{EI}$

6.4 (a), (b), (c) $\theta_A = -1.535 \times 10^{-3}$ radian, $y_A = +4.61 \times 10^{-3}$ m
6.6 (a), (b), (c) $y_A = +6.25 \times 10^{-3}$ m, $y_E = +8.75 \times 10^{-3}$ m
6.7 (a) $\Delta_{D-\text{vertical}} = 0.0299$ m ↓
    (b) $\Delta_{D-\text{horizontal}} = 0.0594$ m →
    (c) $\theta_D = 0.00835$ radian ↻
    (d) $\Delta_{B-\text{horizontal}} = 0.00455$ m →
        $\theta_B = 0.00262$ radian ↻
6.8 (a) 0.0113 ft ↓,   (b) 3.11×10^{-4} ft/kip
6.10 $\Delta_{C-\text{vertical}} = 0.0278$ m ↓
6.11 $\Delta_{C-\text{horizontal}} = 0.485$ ft →

7.2 $AB = -25.3$ kips, $AD = -52.6$ kips, $AC = -36.3$ kips,
    $BC = +20.2$ kips, $CD = +20.2$ kips, $CE = -36.3$ kips
7.3 $AB = +100.8$ kN, $BE = -60.8$ kN, $DE = -91.2$ kN,
    $BD = -121.2$ kN, $AE = +109.6$ kN
7.4 $AB = -516$ kN, $BE = -344$ kN, $DE = -516$ kN,
    $BD = +620$ kN, $AE = +620$ kN
7.6 $R_{A-x} = 0$ (assumes no tension initially)

$R_{A-y} = \dfrac{Pb^2}{L^3}(3a+b) \uparrow$

$M_A = \dfrac{Pab^2}{L^2}$ ↻

$R_{B-x} = 0$

$R_{B-y} = \dfrac{Pa^2}{L^3}(a+3b) \uparrow$

$M_B = \dfrac{Pa^2 b}{L^2}$ ↻

7.8 $R_{A-y} = 6.67$ kN ↓, $R_{A-x} = 12$ kN →, $M_A = 40$ kN·m ↻,
    $R_D = 12.67$ kN ↑

8.2 $T = 62.6$ kN
8.4 $R_{A-x} = 22.7$ kips →, $R_{A-y} = R_{C-y} = 17.5$ kips ↑, $R_{C-x} = 22.7$ kips ←
8.6 $T_{AD} = T_{DE} = 18.47$ kips

9.1 (a) $R_A = 0.519P \uparrow$, $R_{B-y} = 0.481P \uparrow$, $R_{B-x} = 0$, $M_B = 0.1482PL$ ↻
    (c) $R_{A-y} = \tfrac{49}{64}P \uparrow$, $R_{A-x} = 0$ (assumes no axial forces)
    $M_A = \tfrac{17}{64}PL$ ↻, $R_{C-y} = \tfrac{15}{64}P \uparrow$, $R_{C-x} = 0$, $M_C = \tfrac{15}{64}PL$ ↻
    (e) $R_{A-y} = R_{C-y} = \tfrac{3}{16}wL \uparrow$, $R_{B-y} = \tfrac{5}{8}wL \uparrow$, $R_{A-x} = 0$

9.2 (a) $ILm_A = \dfrac{3x^2}{4L} - \dfrac{x^3}{4L^2} - x$ for AB

$$ILm_A = -\frac{x^3}{4L^2} + \frac{3x}{4} - \frac{L}{2} \quad \text{for } BC, \quad 0<x<L$$

$$ILr_A = -\frac{3x^2}{4L^2} + \frac{x^3}{4L^3} + 1 \quad \text{for } AB$$

$$ILr_A = \frac{x^3}{4L^3} - \frac{3x}{4L} + \frac{1}{2} \quad \text{for } BC, \quad 0<x<L$$

$$ILv_D = -\frac{3x^2}{4L^2} + \frac{x^3}{4L^3} \quad \text{for } AD, \quad 0<x<\frac{L}{2}$$

$$ILv_D = -\frac{3x^2}{4L^2} + \frac{x^3}{4L^3} + 1 \quad \text{for } DB, \quad \frac{L}{2}<x<L$$

$$ILv_D = \frac{x^3}{4L^3} - \frac{3x}{4L} + \frac{1}{2} \quad \text{for } BC, \quad 0<x<L$$

(c) $$ILm_D = \frac{3x^2}{8L} - \frac{x^3}{8L^2} \quad \text{for } AD, \quad 0<x<\frac{L}{2}$$

$$ILm_D = \frac{3x^2}{8L} - \frac{x^3}{8L^2} - x + \frac{L}{2} \quad \text{for } DB, \quad \frac{L}{2}<x<L$$

$$ILm_D = -\frac{x^3}{8L^2} + \frac{3x}{8} - \frac{L}{4} \quad \text{for } BC, \quad 0<x<L$$

$$ILv_B = \frac{x^3}{4L^3} - \frac{3x^2}{4L^2} \quad \text{for } AB, \quad 0<x<L$$

$$ILv_B = \frac{x^3}{4L^3} - \frac{3x}{4L} + \frac{1}{2} \quad \text{for } BC, \quad 0<x<L$$

9.2 (f) $$ILm_D = \frac{x}{4} \quad \text{for } AB, \quad 0<x<2L$$

$$ILm_D = -\frac{x}{2} + \frac{L}{2} \quad \text{for } BC, \quad 0<x<L$$

$$ILm_D = \frac{x^3}{2L^2} - \frac{x}{2} \quad \text{for } CD, \quad 0<x<L$$

$$ILr_D = -\frac{3x}{4L} \quad \text{for } AB, \quad 0<x<2L$$

$$ILr_D = \frac{3x}{2L} - \frac{3}{2} \quad \text{for } BC, \quad 0<x<L$$

$$ILr_D = -\frac{x^3}{2L^3} + \frac{3x}{2L} \quad \text{for } CD, \quad 0<x<L$$

10.1 (a) $R_{A-y} = 35.4$ kN↑, $M_A = 67.9$ kN·m ↩
$R_{A-x} = 0$ (assumes no axial force)
$R_B = 10.3$ kN↑

ANSWERS TO SELECTED PROBLEMS **407**

$R_{C-y}=14.3$ kN↑, $M_C=5.9$ kN·m ↻
$R_{C-x}=0$
(c) $R_{A-y}=46.05$ kN↑, $R_{A-x}=0$, $R_B=53.95$ kN↑
$R_C=53.95$ kN↑, $R_D=46.05$ kN↑
(e) $R_{A-y}=R_{C-y}=6$ kN↑, $R_{A-x}=0$, $R_B=48$ kN↑
(g) $M_A=11$ kip-ft ↻, $R_{A-y}=3.3$ kips↓, $R_{A-x}=0$ (assumed)
$M_D=11$ kip-ft ↻, $R_{D-y}=3.3$ kips↑, $R_{D-x}=0$
$R_B=3.3$ kips↓, $R_C=3.3$ kips↑

10.2 $M_{AB}=-250.2$ kip-ft, $M_{BA}=-195.3$ kip-ft
$M_{BC}=+195.3$ kip-ft, $M_{CB}=160.2$ kip-ft
$M_A=250.2$ kip-ft ↻, $R_{A-y}=27.8$ kip↑
$R_{A-x}=0$ (assumed), $R_B=42.0$ kips↓
$R_{D-y}=14.19$ kip↑, $R_{D-x}=0$ (assumed)
$M_C=160.2$ kip-ft ↻

10.4 (a) $R_{A-y}=102.7$ kip↑, $R_{A-x}=0$ (assumed)
$M_A=628.9$ kip-ft ↻, $R_B=134.4$ kips↓
$R_C=75.9$ kips↑, $R_{D-y}=44.2$ kips↓
$R_{D-x}=0$ (assumed), $M_D=118$ kip-ft ↻
$M_{AB}=-628.9$ kip-ft, $M_{BA}=-398.4$ kip-ft
$M_{BC}=+398.4$ kip-ft, $M_{CB}=235.9$ kip-ft
$M_{CD}=-235.9$ kip-ft, $M_{DC}=-118.0$ kip-ft

10.5 $M_{AB}=11.55$ kN·m, $M_{BA}=5.67$ kN·m
$M_{BC}=-5.67$ kN·m, $M_{CB}=0.405$ kN·m
$M_{CD}=-0.405$ kN·m, $M_{DC}=13.975$ kN·m
$V_{AB}=4.305$ kN, $V_{DC}=7.695$ kN
$M_A=11.55$ kN·m ↻, $R_{A-y}=0.876$ kN↑
$R_{A-x}=4.305$ kN→, $R_{D-y}=5.124$ kN↑
$R_{D-x}=7.695$ kN→, $M_D=13.975$ kN·m ↻

10.8 $M_{AB}=-404$ kip-ft, $M_{BA}=-336.8$ kip-ft
$M_{BC}=+336.8$ kip-ft, $M_{CB}=273.2$ kip-ft
$M_{CD}=-273.2$ kip-ft, $M_{DC}=-125.4$ kip-ft
$M_{DE}=125.4$ kip-ft, $M_{ED}=15.45$ kip-ft

10.10 $M_{AB}=-15$ kip-ft, $M_{BA}=+15$ kip-ft, $M_{BC}=-15$ kip-ft
$M_{CB}=+15$ kip-ft $A_x=8.33$ kips→
$A_y=17.5$ kips↑, $C_x=8.33$ kips←
$C_y=17.5$ kips↑, $M_A=15$ kip-ft ↻
$M_C=15$ kip-ft ↻

11.1 (a), (c), (e), (g) see answers for 10.1 (a), (c), (e), (g)
11.2 $M_{AB}=0$, $M_{BA}=-M_{BC}=-190$ kip-ft, $M_{CB}=-M_{DC}=142$ kip-ft
11.4 $M_{AB}=M_{DC}=0$, $M_{BA}=-M_{BC}=+81.02$ kip-ft
$M_{CB}=-M_{CD}=+110.86$ kip-ft
11.7 $M_{DA}=+19.94$ kN·m, $M_{AD}=-M_{AB}=+39.88$ kN·m
$M_{BA}=+46.42$ kN·m, $M_{BE}=-16.03$ kN·m
$M_{BC}=-30.38$ kN·m, $M_{CB}=-1.68$ kN·m
$M_{EB}=-8.01$ kN·m
11.8 $M_{CA}=+43.7$ kip-ft$=-M_{DB}$
$M_{AC}=M_{AB}=M_{BA}=-M_{BD}=+87.4$ kN·m

ANSWERS TO SELECTED PROBLEMS

12.2 See answers for 10.2
12.4 $M_{EA}=+6.62$ kip-ft, $M_{AE}=-M_{AB}=+15.75$ kip-ft
$M_{BA}=+15.07$ kip-ft, $M_{BF}=-7.65$ kip-ft, $M_{BC}=-7.42$ kip-ft
$M_{CB}=-3.92$ kip-ft, $M_{CG}=+1.93$ kip-ft, $M_{CD}=+1.99$ kip-ft
$M_{DC}=+1.56$ kip-ft, $M_{DH}=-1.56$ kip-ft, $M_{HD}=-2.04$ kip-ft
$M_{FB}=-6.37$ kip-ft, $M_{GC}=-1.57$ kip-ft
12.6 $M_{AB}=-147.3$ kip-ft, $M_{BA}=-28.8$ kip-ft, $M_{BC}=+28.8$ kip-ft
$M_{CB}=+94.0$ kip-ft, $M_{CD}=-94.0$ kip-ft, $M_{DC}=-179.9$ kip-ft

13.1 $S_{AB}=29920$ kN·m, $S_{BA}=60580$ kN·m
$C_{AB}=0.7086$, $C_{BA}=0.350$
13.2 $FEM_{AB}=-13.84$ kN·m, $FEM_{BA}=+12.02$ kN·m
13.3 $S_{AB}=9152$ ft-kip, $S_{BA}=20070$ ft-kip
$C_{AB}=0.7353$, $C_{BA}=0.3352$
13.4 $FEM_{AB}=-25.6$ kip-ft, $FEM_{BA}=+74.0$ kip-ft
13.6 $M_{AB}=M_{CB}=0$, $M_{BA}=+74.2$ kip-ft, $M_{BC}=-18.6$ kip-ft
$M_{BD}=-55.6$ kip-ft, $M_{DB}=-27.8$ kip-ft
13.7 $M_{AB}=0$, $M_{BA}=58.97$ kip-ft, $M_{BC}=-33.80$ kip-ft
$M_{BD}=-25.17$ kip-ft, $M_{DB}=+25.33$ kip-ft
$M_{CB}=0$
13.9 $M_{AC}=26.4$ kip-ft, $M_{CA}=52.7$ kip-ft
$M_{CE}=13.1$ kip-ft, $M_{CD}=-65.8$ kip-ft
$M_{DC}=36.9$ kip-ft, $M_{DF}=-7.4$ kip-ft
$M_{DB}=-29.5$ kip-ft, $M_{BD}=-14.8$ kip-ft

14.2 $AD=+35.6$ kips, $BD=+93.1$ kips, $CD=+31.7$ kips
$A_x=-24.1$ kips, $A_y=+26.3$ kips, $B_x=0$, $B_y=+93.1$ kips
$C_x=+24.1$ kips, $C_y=+20.6$ kips
14.3 $AE=-3.23$ kips, $BE=-7.20$ kips, $CE=-3.72$ kips
$CF=-1.00$ kip, $DF=+12.77$ kips, $EF=-9.36$ kips
$A_x=2.18$ kip, $A_y=-2.38$ kip, $B_x=0$, $B_y=-7.20$ kips
$C_x=-2.18$ kip, $C_y=-3.19$ kip
14.4 $AE=DE=+6.36$ kips, $BE=CE=-22.7$ kips
$A_x=-1.609$ kips, $A_y=-6.034$ kips, $A_z=-1.207$ kips
$B_x=-5.742$ kips, $B_y=+21.53$ kips, $B_z=+4.306$ kips
$C_x=-5.742$ kips, $C_y=+21.53$ kips, $C_z=-4.306$ kips
$D_x=-1.609$ kips, $D_y=-6.034$ kips, $D_z=+1.207$ kips
14.6 $E_x=-8.95$ kips, $E_y=-3.42$ kips, $E_z=0$
$F_x=-7.80$ kips, $F_y=6.33$ kips, $F_z=0$
$G_x=0$, $G_y=-4.08$ kips, $G_z=-1.608$ kips
$H_x=1.754$ kips, $H_y=6.17$ kips, $H_z=-8.39$ kips
$AB=5.23$ kips, $BD=7.64$ kips, $CD=-2.57$ kips
$AC=-0.748$ kips, $AE=-4.04$ kips, $BF=0.1676$ kips
$DH=0$, $CG=5.54$ kips, $BC=1.107$ kips
$AD=3.49$ kips, $BH=-11.34$ kips, $AF=-10.15$ kips
$CH=2.28$ kips, $AG=-2.17$ kips, $BE=11.66$ kips
14.8 $D_x=163.6$ kips, $D_y=614$ kips, $D_z=-122.7$ kips
$A_x=-163.6$ kips, $A_y=-614$ kips, $A_z=-122.7$ kips

ANSWERS TO SELECTED PROBLEMS

$B_x = -163.6$ kips, $B_y = 614$ kips, $B_z = 122.7$ kips
$C_x = 163.6$ kips, $C_y = -614$ kips, $C_z = 122.7$ kips
$ED = EB = -646$ kips, $EA = EC = 646$ kips

14.9 $A_x = 96.79$ kN, $A_y = 49.31$ kN, $M_A = 343.96$ kN·m
$D_x = 61.21$ kN, $D_y = 49.31$ kN, $M_D = 206.97$ kN·m

14.11 $A_x = D_x = 79.22$ kN, $A_y = D_y = 23.98$ kN
$M_A = 279.74$ kN·m, $M_D = 654.00$ kN·m

14.14 $A_x = D_x = 22.24$ kips, $A_y = D_y = 40$ kips
$M_A = M_D = 72.69$ kip-ft

SP-1(a) $AB = BC = CD = DE = -500$ kN
$BF = DH = -200$ kN, $CG = 0$
$AF = EH = +583.1$ kN
$CF = CH = -194.4$ kN
$FG = GH = +666.7$ kN

SP-1(b) $AB = BC = +1.25$, $CD = DE = 0.4167$
$BF = +1$, $DH = CG = 0$
$AF = -1.458$, $EH = -0.486$
$CF = -0.486$, $CH = +0.486$
$FG = GH = -0.833$

SP-1(c) $AB = BC = -593.75$ kN, $CD = DE = -531.25$ kN
$BF = -275$ kN, $DH = -200$ kN, $CG = 0$
$AF = 692.6$ kN, $EH = 619.6$ kN, $CF = -158$ kN
$CH = -230.9$ kN, $FG = GH = 729.2$ kN

SP-1(d) $AB = BC = CD = DE = +1$
All other members $= 0$

SP-1(e) $AB = BC = CD = DE = +0.833$
$BF = DH = CG = 0$
$AF = EH = -0.9718$
$CF = CH = +0.9718$
$FG = GH = -1.667$

SP-1(f) $AB = AF = ED = CD = DH = CH = GH = EH = 0$
$BG = CF = +1$
$BC = FG = -0.8575$
$BF = CG = -0.5145$

SP-1(g) $\Delta_B = \Delta_D = -0.01678$ m (downward)
$\Delta_C = -0.0228$ m (downward)

SP-1(h) $\Delta L_{AB} = \Delta L_{DE} = +0.00646$ m
$\Delta L_{FG} = \Delta L_{GH} = -0.00361$ m

SP-1(i) 1.609×10^{-5} m/kN

SP-1(j) (1) 3.298×10^{-6} m/kN←, (2) 1.539×10^{-5} m/kN→
(3) -3.29×10^{-6} m/kN

SP-1(k) $\delta_{11} = 1.609 \times 10^{-5}$ m/kN
$\delta_{12} = \delta_{21} = -3.298 \times 10^{-6}$ m/kN
$\delta_{22} = 1.539 \times 10^{-5}$ m/kN

SP-1(l) 8.65×10^{-3} m ↓

SP-1(m) -1.118×10^{-3} m

SP-1(n) $AB = BC = -31.75$ kN, $CD = DE = +30.25$ kN

$AF = 692.5$ kN, $FG = GH = +729.2$ kN
$EH = +619.6$ kN, $BF = -275$ kN
$CF = -158$ kN, $CG = 0$
$CH = -230.9$ kN, $DH = -200$ kN

SP-1(o) $AB = -593.75$ kN, $BC = -653.3$ kN
$CD = DE = -531.25$ kN
$AF = 692.5$ kN, $FG = 669.6$ kN
$GH = 729.2$ kN, $EH = 619.6$ kN
$BF = -310.7$ kN, $CF = -88.5$ kN
$CG = -35.7$ kN, $CH = -230.9$ kN
$DH = -200$ kN, $BG = +69.5$ kN

SP-1(p) $AB = 9.65$ kN, $BC = -156$ kN
$CD = DE = +72.2$ kN
$AF = 692.5$ kN, $FG = 564$ kN
$GH = 729$ kN, $HE = 619.6$ kN
$BF = -374$ kN, $CF = +35.2$ kN
$CG = -99.4$ kN, $CH = -231$ kN
$DH = -200$ kN, $BG = +193.2$ kN

SP-2(a) $AB = 357.1$ kips
$BC = 261.6$ kips
$CD = 174.4$ kips
$AE = 200$ kips
$EB = -245.8$ kips
$BF = 50$ kips
$CF = -87.2$ kips
$CG = 100$ kips
$EF = -214.2$ kips
$FG = -142.8$ kips
$DG = -142.8$ kips

SP-2(b) $AB = -0.7143$
$AE = -1$
$EB = +1.229$
$BF = -1$
All others $= 0$

SP-2(c) $AB = 428.5$ kips
$BC = 261.6$ kips
$CD = 174.4$ kips
$AE = 300$ kips
$EB = -368.7$ kips
$BF = 150$ kips
$CF = -87.2$ kips
$CG = 100$ kips
$EF = -214.2$ kips
$FG = -142.8$ kips
$DG = -142.8$ kips

SP-2(d) $AB = -1.229$
$AE = -0.5736$

$EB = 0.7045$
$BH = 1$
All others $= 0$

SP-2(e) $BC = -0.7097$
$BF = -0.4069$
$CF = 0.7097$
$CG = -0.8137$
$FG = -0.5183$
$BG = 1$
All others $= 0$

SP-2(f) -0.01674 ft
SP-2(g) -0.00988 ft
SP-2(h)   0
SP-2(i) $\delta_{11} = 1.097 \times 10^{-4}$ ft/kip, $\delta_{12} = \delta_{21} = 0$
$\delta_{22} = 8.84 \times 10^{-5}$ ft/kip

SP-2(j) $AB = 124.3$ kips
$BC = 261.6$ kips
$CD = 174.4$ kips
$AE = 91.4$ kips
$EB = -112.4$ kips
$BF = 50$ kips
$CF = -87.2$ kips
$CG = 100$ kips
$EF = -214.2$ kips
$FG = -142.8$ kips
$GD = -142.8$ kips
$BH = 189.4$ kips

SP-2(k) $AB = 124.3$ kips
$BC = 197.7$ kips
$CD = 174.4$ kips
$AE = 91.4$ kips
$EB = -112.4$ kips
$BF = 13.4$ kips
$FC = -23.3$ kips
$CG = 26.7$ kips
$EF = -214.2$ kips
$FG = -195.2$ kips
$GD = -142.8$ kips
$BG = 90.1$ kips
$BH = 189.4$ kips

SP-3(a) AB: $M = 0$, $0 < x < 5$ ft
                $M = -80x + 400$ kip-ft,   5 ft $< x <$ 10 ft
        BC: $M = -2.5x^2 - 400$ kip-ft, $0 < x < 12$ ft

SP-3(b)   (1) AB: $m = -x$ ft, $0 < x < 10$ ft
               BC: $m = -10$ ft, $0 < x < 12$ ft
       (2) AB: $m = -1$, $0 < x < 10$ ft
               BC: $m = -1$, $0 < x < 12$ ft

412 ANSWERS TO SELECTED PROBLEMS

SP-3(c) $\Delta_A = 0.15314$ ft ↓, $\theta_A = 0.0159808$ radian ccw

SP-3(d) AB: $M = 0$, $0 < x < 10$ ft
            BC: $M = -H_x$, $0 < x < 12$ ft

SP-3(e) 8.01792×10^{-2} ft ←

SP-3(f)   (1)   0.00363733 ft/kip
            (2)   0.0004304 ft/kip-ft
            (3)   0.0004304 radian/kip
            (4)   0.00006304 radian/ft-kip

SP-3(g) 1.3824×10^{-3} ft/kip

SP-3(h) 1.3824×10^{-4} ft/kip-ft

SP-3(i) 1.10592×10^{-3} ft/kip

SP-3(j) $\delta_{11} = 3.63733 \times 10^{-3}$ ft/kip, $\delta_{12} = 4.304 \times 10^{-4}$ ft/kip-ft,
         $\delta_{13} = 1.3824 \times 10^{-3}$ ft/kip, $\delta_{21} = 4.304 \times 10^{-4}$ radian/kip,
         $\delta_{22} = 6.304 \times 10^{-5}$ radian/kip-ft, $\delta_{23} = 1.3824 \times 10^{-4}$ radian/kip,
         $\delta_{31} = 1.3824 \times 10^{-3}$ ft/kip, $\delta_{32} = 1.3824 \times 10^{-4}$ ft/kip-ft,
         $\delta_{33} = 1.10592 \times 10^{-3}$ ft/kip

SP-3(k)   (1) $A_y = 42.1$ kips ↑, $C_x = 60$ kips →
               $C_y = 37.9$ kips ↑, $M_C = 339$ kip-ft ↻
            (2) $A_x = 37.9$ kips →, $A_y = 27.7$ kips ↑
               $C_x = 22.1$ kips →, $C_y = 52.3$ kips ↑
               $M_C = 28.6$ kip-ft ↻
            (3) $A_x = 33.2$ kips →, $A_y = 42.2$ kips ↑
               $C_x = 26.8$ kips, $C_y = 37.8$ kips ↑
               $M_A = 107.3$ kip-ft ↺, $M_C = 47.4$ kip-ft ↻

SP-3(l) $D_x = 65.5$ kips →, $D_y = 0$
         $C_x = 5.5$ kips ←, $C_y = 80$ kips ↑
         $M_C = 26$ kip-ft ↻

SP-3(m) $A_x = 33.2$ kips →, $A_y = 42.2$ kips ↑, $M_A = 107.3$ kip-ft ↻
          $C_x = 26.8$ kips →, $C_y = 37.8$ kips ↑, $M_C = 47.3$ kip-ft ↻

SP-3(n) $A_x = 37.8$ kips →, $A_y = 27.7$ kips ↑,
         $C_x = 22.2$ kips →, $C_y = 52.3$ kips ↑, $M_C = 28.6$ kip-ft ↻

SP-3(o)   See answers to SP-3(m)

SP-3(p)   See answers to SP-3(n)

SP-3(q)   See answers to SP-3(k)(1)

SP-3(r)   See answers to SP-3(k)(1)

SP-4(a) AB: $M = 0$, $0 < x < 2.5$ m
            BC: $M = -23.34x^2$ kN · m, $0 < x < 5.1$ m
            CD: $M = -607.5$ kN · m, $0 < x < 4.9$ m

SP-4(b)   (1) AB: $m = -x$, $0 < x < 2.5$ m
                BC: $m = -0.4706x - 2.5$ kN · m, $0 < x < 5.1$ m
                CD: $m = x - 4.9$ kN · m, $0 < x < 4.9$ m
            (2) AB: $m = -1$, $0 < x < 2.5$ m
                BC: $m = -1$, $0 < x < 5.1$ m
                CD: $m = -1$, $0 < x < 4.9$ m

SP-4(c) $\Delta_{A-\text{horiz}} = 1.63823 \times 10^{-1}$ m →
         $\theta_A = 5.8427 \times 10^{-2}$ radian

SP-4(d) AB: $M = 0$, $0 < x < 2.5$ m

ANSWERS TO SELECTED PROBLEMS **413**

BC: $M = -0.882\,Rx$, $0 < x < 5.1$ m
CD: $M = -4.5\,R$, $0 < x < 4.9$ m

SP-4(e) $\Delta_{A-\text{vert}} = 0.250347$ m $\downarrow$

SP-4(f) (1) 1.60667×10^{-3} m/kN
(2) 5.1590×10^{-4} radian/kN
(3) 2.2987×10^{-4} radian/kN $\cdot$ m
(4) 5.1590×10^{-4} m/kN $\cdot$ m

SP-4(g) 1.36677×10^{-3} m/kN

SP-4(h) 4.74188×10^{-4} m/kN $\cdot$ m

SP-4(i) 1.947565×10^{-3} m/kN

SP-4(j) $\delta_{11} = 1.94757 \times 10^{-3}$ m/kN
$\delta_{12} = 4.74188 \times 10^{-4}$ m/kN $\cdot$ m
$\delta_{13} = 1.366772 \times 10^{-3}$ m/kN
$\delta_{21} = 4.74188 \times 10^{-4}$ radian/kN
$\delta_{22} = 2.2987 \times 10^{-4}$ radian/kN $\cdot$ m
$\delta_{23} = 5.1590 \times 10^{-4}$ radian/kN
$\delta_{31} = 1.366772 \times 10^{-3}$ m/kN
$\delta_{32} = 5.1590 \times 10^{-4}$ m/kN $\cdot$ m
$\delta_{33} = 1.60667 \times 10^{-3}$ m/kN

SP-4(k) (1) $A_y = 128.5$ kN, $D_x = 0$, $D_y = 141.5$ kN,
$M_D = 29.25$ kN $\cdot$ m
(2) $A_x = 18.33$ kN, $A_y = 141.4$ kN, $D_x = -18.33$ kN,
$D_y = 128.6$ kN, $M_D = 28.8$ kN $\cdot$ m
(3) $A_x = 22.1$ kN, $A_y = 140.8$ kN, $M_A = 13.25$ kN $\cdot$ m
$D_x = 22.1$ kN, $D_y = 129.2$ kN, $M_D = 39.35$ kN $\cdot$ m

SP-4(l) $E_x = 79.9$ kN, $D_x = 79.9$ kN, $D_y = 270$ kN,
$M_D = 607.5$ kN $\cdot$ m

SP-4(m) $A_x = 31.4$ kN, $A_y = 143$ kN, $M_A = 26.2$ kN $\cdot$ m,
$B_x = 12.7$ kN, $D_x = 18.7$ kN, $D_y = 127$ kN,
$M_D = 30.55$ kN $\cdot$ m, $M_A = 26.2$ kN $\cdot$ m

SP-4(n) $A_x = 18.3$ kN, $A_y = 141.4$ kN, $D_x = 18.3$ kN
$D_y = 128.6$ kN, $M_D = 28.8$ kN $\cdot$ m

SP-4(o) $A_x = 31.39$ kN, $A_y = 26.16$ kN, $M_A = 26.16$ kN $\cdot$ m
$B_x = 12.69$ kN, $D_x = 18.70$ kN, $D_y = 127$ kN
$M_D = 30.54$ kN $\cdot$ m

SP-4(p) $A_x = 17.44$ kN, $A_y = 141.0$ kN, $B_x = 1.89$ kN
$D_x = 19.33$ kN, $D_y = 129.0$ kN, $M_D = 31.58$ kN $\cdot$ m

SP-4(q) $A_x = 22$ kN, $A_y = 140.8$ kN, $M_A = 13.19$ kN $\cdot$ m
$D_x = 22$ kN, $D_y = 129.2$ kN, $M_D = 39.26$ kN $\cdot$ m

SP-4(r) $A_x = 18.29$ kN, $A_y = 141.4$ kN, $D_x = 18.29$ kN
$D_y = 128.6$ kN, $M_D = 28.74$ kN $\cdot$ m

SP-4(s) $A_x = 22.0$ kN, $A_y = 140.8$ kN, $M_A = 13.2$ kN $\cdot$ m
$D_x = 22.0$ kN, $D_y = 129.2$ kN, $M_D = 39.3$ kN $\cdot$ m

SP-5(a) $FEM_{AB} = -FEM_{BA} = -7.721 \times 10^3 w$ kip-inch
where w is kips/inch

SP-5(b) $S_{AB} = S_{BA} = 2.361 \times 10^6$ inch kip/radian
$C_{AB} = C_{BA} = 0.5932$

SP-5(c) $M_{AB} = -M_{DC} = +427$ kip-ft

414 ANSWERS TO SELECTED PROBLEMS

SP-5(d) $M_{BA} = -M_{BC} = M_{CB} = -M_{CD} = +854$ kip-ft
$M_{AB} = M_{DC} = 0$
$M_{BA} = +796$ kip-ft, $M_{BC} = -1206$ kip-ft
$M_{BE} = +410$ kip-ft, $M_{CB} = +692$ kip-ft
$M_{CD} = -153$ kip-ft, $M_{CF} = -539$ kip-ft
$M_{EB} = +302$ kip-ft, $M_{FC} = -173$ kip-ft

SP-6(a) $M_{FD} = -317$ kN·m, $M_{EC} = +274$ kN·m
$M_{CE} = -M_{CD} = 547$ kN·m, $M_{DC} = 690$ kN·m
$M_{DF} = -633$ kN·m, $M_{DA} = -57$ kN·m
$M_{AD} = -M_{AB} = 1322$ kN·m, $M_{GB} = -459$ kN·m
$M_{BA} = -M_{BG} = 918$ kN·m

SP-6(b) $B = 358$ kN, $D = 678$ kN

SP-6(c) $M_{FD} = -0.328$ m, $M_{EC} = +0.0458$ m
$M_{CE} = -M_{CD} = 0.0915$ m, $M_{DC} = -0.289$ m
$M_{DF} = -0.647$ m, $M_{DA} = 0.935$ m
$M_{AD} = -M_{AB} = +0.732$ m, $M_{GB} = 0.406$ m
$M_{BA} = -M_{BG} = -0.386$ m

SP-6(d) $M_{FD} = 0.574$ m, $M_{EC} = 0.236$ m
$M_{CE} = -M_{CD} = 0.130$ m, $M_{DC} = -0.020$ m,
$M_{DF} = 0.635$ m, $M_{DA} = -0.616$ m
$M_{AD} = -M_{AB} = -0.336$ m, $M_{GB} = -0.042$ m
$M_{BA} = -M_{BG} = 0.083$ m

SP-6(e) $M_{FD} = 58.3$ kN·m, $M_{EC} = 420.7$ kN·m, $M_{CE} = 624.9$ kN·m
$M_{CD} = -624.9$ kN·m, $M_{DC} = 689.6$ kN·m, $M_{DF} = -205.4$ kN·m
$M_{DA} = -484.1$ kN·m, $M_{AD} = -M_{AB} = 1079.2$ kN·m
$M_{BA} = -M_{BG} = 987$ kN·m, $M_{GB} = -502.9$ kN·m

SP-7(a) $S_{CD} = 69126$ kN·m/radian
$S_{DC} = 32899$ kN·m/radian
$C_{CD} = 0.4182$
$C_{DC} = 0.8787$

SP-7(b) $FEM_{CD} = -2.480P$ kN·m when $P = $ kN
$FEM_{DC} = +0.5234P$ kN·m when $P = $ kN

SP-7(c) $M_{AB} = 0$, $M_{BA} = 713.8$ kN·m, $M_{BE} = -10.3$ kN·m
$M_{BC} = -703.5$ kN·m, $M_{EB} = -5.2$ kN·m
$M_{CB} = +39.9$ kN·m, $M_{CD} = -33.2$ kN·m
$M_{CF} = -6.8$ kN·m, $M_{FC} = -3.4$ kN·m
$A_x = 2.85$ kN

SP-7(d) $M_{AB} = 0$, $M_{BA} = 714.7$ kN·m, $M_{BE} = -3.9$ kN·m
$M_{BC} = -709$ kN·m, $M_{EB} = +1.3$ kN·m
$M_{CB} = +34.4$ kN·m, $M_{CD} = -32.3$ kN·m
$M_{CF} = -0.4$ kN·m, $M_{FC} = +3.1$ kN·m

SP-8(a) $M_{CA} = M_{DB} = -3.111$ kN·m
$M_{AC} = M_{BD} = -M_{AB} = -M_{BA} = 1.778$ kN·m
$A_x = D_x = 4.333$ kN

ANSWERS TO SELECTED PROBLEMS

$A_y = -D_y = 0.593$ kN
$C_x = 7.33$ kN

SP-8(b) $M_{AC} = M_{BD} = -M_{AB} = -M_{BA} = +1.699$ kN·m
$M_{CA} = M_{DB} = -3.33$ kN·m
$M_{CD} = M_{DC} = -0.490$ kN·m
$M_{CE} = M_{DF} = +3.82$ kN·m
$M_{EC} = M_{FD} = 0$

SP-8(c) $M_{AC} = M_{BD} = -M_{AB} = -M_{BA} = -0.972 P_B$
$M_{CA} = M_{DB} = -1.030 P_B$
$M_{CD} = M_{DC} = 0.748 P_B$
$M_{CE} = M_{DF} = 0.282 P_B$
$M_{EC} = M_{FD} = 0$
$F_D = 1.141 P_B$

SP-8(d) $M_{AC} = M_{BD} = -M_{AB} = -M_{BA} = +0.651 P_D$
$M_{CA} = M_{DB} = 0.768 P_D$
$M_{CD} = M_{DC} = -0.188 P_D$
$M_{CE} = M_{DF} = -0.581 P_D$
$M_{EC} = M_{FD} = 0$
$F_B = 0.7095 P_D$

SP-8(e) $P_B = 107.48$ kN, $P_D = 141.36$ kN
$M_{AC} = M_{BD} = -M_{AB} = -M_{BA} = -10.7$ kN·m
$M_{CA} = M_{DB} = -5.5$ kN·m
$M_{CD} = M_{DC} = +53.3$ kN·m
$M_{CE} = M_{DF} = -48.0$ kN·m
$M_{EC} = M_{FD} = 0$

Index

AASHTO Specification, 73
Active force diagram, 17
AISC code, 76
ANSI code, 71
Antisymmetric loading, 221
Antisymmetry, 220, 311
AREA Specification, 73
Array, 396

Banded matrix, 334
Bandwidth, 334
Beam, 2
   Cantilever, 34
   classification, 34
   Compound, 35
   deflection methods, 106
   deflection, virtual work, 118
   determinacy, 35
   overhanging, 35
   sign conventions, 11
   simple, 34
   stability, 35
Bending moment, 10
Bridge impact loads, 73
Bridge loads, 72

Camber, 106
Cantilever Beam, 34
Carry over factor, 241
   non prismatic, 299
   numerical techniques, 301
Carry over moment, 241
Castigliano's Theorem, 132, 170
   for flexibility coefficients, 139
   for Rigid Frame deflection, 134
Cholesky's method, 333
Codes, 70
Column matrix, 396
Complex truss, 34
Composite structure deflections, 133
Compound beam, 35
Compound truss, 34
Compound truss analysis, 56
Computer considerations, 336
Concentrated forces, 70
Condition equations, 35, 39
Condition, Rigid Frame, 39
Conjugate Beam, 106
   Procedure, 110
   for Statically Indeterminate Beam, 184

Consistent deformations, equations for, 151
Consistent deformations, trusses, 160
Construction loads, 80
Cooper loads, 73
Cross, Professor Hardy, 202, 240

Dead load, 71
Deflections, composite structures, 130
   unit load method, 124
   virtual work, 124
Dependent joint translations, 228
Diagonal, 396
Diagonally dominant matrix, 157
Diagonal matrix, 397
Direction Cosines, 321, 336
Displacements, 320
Displacement matrix, 319
Distributed force, 98, 355
Distribution factor, 242
   clamped end, 247
   pin end, 250

Earth pressure loads, 80
Earthquake loads, 78
Elastic curve, 226
Element stiffness matrix, 319
Equilibrium, 8
Equilibrium equations, 10
Expansion joints, 80

First moment-area theorem, 114
Fixed end moments, 203, 244, 357
   non prismatic members, 300
   numerical techniques, 301
   sidesway induced, 262
   statically determinate members, 249
   table of, 203
Flexibility coefficients, 137, 150, 158
   Castigliano's Theorem, 139
   trusses, 140
   virtual work, 138
Flexibility matrix, 157
Force, active, 16
Force, inactive, 16
Free-body diagrams, 8
Freedoms, 328

Gauss elimination, 333, 400
Geometric instability, 39
Global coordinates, 320

418 INDEX

Global force/displacement relation, 323
Global to local transform, 321
Global stiffness, 318

Haunched members, 296
Highway bridge loads, 72
Horizontal force factor, 79

Ice loads, 75
Identity matrix, 397
Importance factor, 79
Independent joint translations, 228
Influence coefficients, 179
Influence lines, 81
   applications, 98
   defined, 85
   family of, 99
   girders, 91
   interpretation, 98
   the Müller–Breslau Principle, 86
   qualitative, 193
   sketching of, 193
   truss members, 91
   virtual work, 86
Internal forces and moments, 10
Internal hinge, 35
Internal instability, 30

Joint imbalance, 247
Joint translation, 262

Least work, 170
Line load defined, 70
Live load, 71
   reductions, 71
   reductions, bridges, 73
Loads, 70
   building use, 71
   classification, 70
   classification, live, 70
   classification, surface, 70
   combinations, 70, 80
   concentrated forces, 70
   determination, 70
   shear relations, 11
   earth pressure, 80
   ice, 75
   rain, 75
   snow, 74
   wind, 76
   misalignment, 88
   settlement, 80
   shrinkage, 80
Local coordinates, 319
Local stiffness, 318
Locking joints, 240

Maney, Professor G. A., 202
Matrix, 396
   Algebra, 398
   element, 396
   methods, rigid frames, 343
   inversion, 333
   order of, 396
   partitioning, 399
   transpose, 397
Matrix methods, 318
   methods, plane trusses, 323
   methods, space trusses, 336
Maxwell's relation, 188
Member global stiffness matrix, 322
Member loads between joints, 355
Method of consistent deformations, 148
Misalignment loads, 80
Modified stiffness, 251
   antisymmetric member, 253
   far end pinned, 312, 252
   non prismatic members, 310
   symmetric member, 253
Moment-area method, 114
Moment diagrams, 11
Moment distribution, 225, 240
   joint translation, 262
   non prismatic members, 305, 312
   procedure, 244
   sidesway, 262
   table, 246
Muller-Breslaü principle, 86, 186
   statically indeterminate beams, 186
Multiple degrees of sidesway, 282

Newmark, Professor N. M., 78
Nodes, 355
Non prismatic members, 296
   integral formulas, 296
   numerical procedures, 301
Null matrix, 397

Overhanging beam, 35

Partitioned matrix, 331
Period of vibration, 79
Pin joint, 38
Ponding, 75
Primary structure, 149, 160, 170, 171
   selection of, 149, 156

Railroad bridge loads, 72
Redundants, 149
Relative stiffness, 243
Resonance factor, 79
Rigid frame, 2
   analysis of, 56
   defined, 38

INDEX

deflection by unit load method, 124
deflection by virtual work, 124
   moment diagram, 58
   moment equations, 60
   shear diagram, 58
   stability, 39
   static determinacy, 39
Rigid joint, 38
Rotation, 123
Rotational stiffness, 240
   carry over, relation with, 307
   non prismatic members, 296
Row matrix, 396

Second moment-area theorem, 116
Seismic loads, 76
Sequenced problems, 369
Settlement loads, 80
Shear diagrams, 11
Shear force in beams, 10
Shear and moment diagram construction, 15
Shear and moment relations, 11
Shrinkage loads, 80
Sidesway, 240, 262
   antisymmetric structures, 270
   fixed end moment, far end pinned, 270
   fixed end moments, 309
   multiple degrees, 225, 282
   symmetric structures, 270
Simple beam, 34
Simple truss, 33
Simpson's rule, 301
Slope calculations, unit load method, 123
Slope deflection method, 202
   equations, 207
   fixed end moments, 203
   non prismatic members, 312
   notation, 202
   rigid frames, 214
   sign conventions, 202
Slope by virtual work, 123
Snow loads, 74
Sparse matrix, 334
Square matrix, 396
Stability, defined, 28
Stability, rigid body, 28
Statically indeterminate beams
   conjugate beam method, 184
   consistent deformations, 151
   least work, 171
   slope deflection, 208
Staticaily indeterminate composite structures
   least work, 176
   trusses, consistent deformations, 160
Statically indeterminate rigid frames
   least work, 173
   slope deflection, 214

Stepped members, 296
Stiffness, 240
   axial force members, 318
   non-prismatic members, 299
   numerical techniques, 301
   variable EI, 296
Strain energy, 171
Structural design, 2
Structural engineering, 2
Structure defined, 2
Successive integration, 106
Superposition, 106
   principle of, 3
   sidesway induced/prevented, 266
Support, 328
   clamped, 9
   displacements, 358
   knife edge, 8
   pin, 8
   roller, 9
   settlement, 211
Surface load defined, 70
Symmetry, 220, 311
Symmetric matrix, 397

Tapered members, 296
Thermal deflections truss, 129
Thermal loads, 80
T-matrix, 328
Torsion, 130
Truss, 2, 30
   analysis, 48
   classification, 33
   complex, 34
   compound, 34
   deflections due manufacturing errors, 129
   deflections due random causes, 129
   deflections, unit load method, 127
   deflections, virtual work, 127
   determinacy, 30
   methods of joints, 48
   method of sections, 52
   simple, 33
   stability, 30
Two-force member, 30

UBC code, 71
Unit displacement, 87
Unit force, 120
Unit load method, 118
   flexibility coefficients, 138
   truss deflections, 127

Virtual displacement, 16
Virtual work
   deflection calculations, 118

equations, 120
principle of, 15
truss deflections, 127

Wheel loads, 101
Wind loads, 76
Wind velocity pressure, 77